MECHANICS OF FLIGHT

TENTH EDITION

A. C. KERMODE CBE, MA, CEng, FRAeS

Revised and edited by
R. H. BARNARD PhD, CEng, FRAeS
and D. R. PHILPOTT PhD, CEng, MRAeS, MAIAA

Longman Group Limited
Longman House, Burnt Mill, Harlow
Essex CM20 2JE, England
and Associated Companies throughout the world

First published by Pitman Books Ltd
Ninth edition published by Longman Group UK Ltd 1987
Tenth edition 1996

British Library Cataloguing in Publication Data
A catalogue entry for this title is available from the British Library

ISBN 0-582-23740-8

Library of Congress Cataloging-in-Publication data

A catalog entry for this title is available from the Library of Congress.

Set by 3 in $10\frac{1}{2}$/12 Ehrhardt
Produced by Longman Singapore Publishers (Pte) Ltd
Printed in Singapore

CONTENTS

PREFACE TO TENTH EDITION

The first edition of this classic book was written more than sixty years ago. Since then, it has been updated and improved at regular intervals, seven times by the original author, and twice by the current editors. Only one of the original drawings has survived, yet like the axe that has had three new handles and four new heads, it still has the same good feel to it. The fact that a book originally aimed largely at biplane riggers is still popular, shows the soundness of its basic approach.

The late A. C. Kermode was a high-ranking Air Force officer, and his book attained the standing of a minor gospel in the RAF. Since ex-RAF personnel have frequently ended up as flying instructors or civilian pilots, the revered status has spread. To undertake a new revision is therefore something of a challenge.

It would be impossible to incorporate details of all of the developments in aviation that have taken place in recent years, so we have concentrated on what has always been the strength of the book, the descriptions of the basic principles and practice of flight, even though that now means flight at speeds that would not have been dreamed of at the time of the first edition.

In the area of dynamics, we have changed to the modern practice of using free-body diagrams rather than static equivalents, and despite a fear of possible charges of heresy, we have also dared to change a couple of items that we felt were not quite correct in earlier editions. The number of 'can you answer these' questions has been reduced, but answers are now provided.

R. H. Barnard
D. R. Philpott
1995

CHAPTER 1

MECHANICS

FLYING AND MECHANICS

The flight and manoeuvres of an aeroplane provide glorious examples of the principles of mechanics. However, this is not a book on mechanics. It is about flying, and is an attempt to explain the flight of an aeroplane in a simple and interesting way; the mechanics are only brought in incidentally – as an aid to understanding. In the opening chapter I shall try to sum up some of the principles with which we are most concerned in flying.

MASS AND INERTIA

There is a natural tendency for things to continue doing what they are already doing. A body that is at rest, tends to remain at rest. A body that is moving, tends to continue moving – at the same speed and in the same direction.

This is the first principle of mechanics – it is, in effect, what is sometimes called **Newton's First Law of Motion.**

If a book lies on the table; it is at rest, it is **in equilibrium**.

If a train is running along the level track at sixty kilometres an hour; it is not at rest, **but it is in equilibrium,** that is to say it is continuing to run at the same speed in the same direction, and in fact it will continue to do so, unless some change takes place.

Things at rest, and things moving steadily, are both in equilibrium, and they both have a tendency to continue in the same state of rest or motion; in short, they both have the property called **inertia**.

Inertia is a property of all bodies – **it is a quality**. We can only measure inertia in terms of **mass – which is a quantity**. The mass of a body is a measure of how difficult it is to start or stop. It is sometimes described as the 'quantity of matter in a body' (in fact Newton himself defined it rather like that!), but just what this is meant to convey it is hard to know – one might just as well simply say that mass is mass.

If mass is a quantity, **we must have a unit in which it can be measured**. Until we became metrically minded, the unit chiefly used in the English-speaking world was the mass of a lump of metal which was carefully preserved in London; now it is the mass of a different lump of metal equally carefully preserved in Paris, but the principle is just the same, i.e. all other masses are compared with this one.

MOMENTUM

There are **two** quantities that decide the difficulty of starting or stopping a body, its **mass** and its **velocity**. **The combined quantity, mass multiplied by velocity**, is called **momentum**.

A body of 10 kg of mass moving at 2 m/s of velocity has 20 kg m/s of momentum; so does a body with 5 kg of mass moving at 4 m/s of velocity. The first has the greater mass, the second the greater velocity; but both have the same momentum, both are equally difficult to stop. A train has large mass, and, compared with a bullet, low velocity; a bullet has a small mass and a high velocity. Both are difficult to stop, and both can do considerable damage to anything that tries to stop them quickly!

FORCE

We have said that there is a **tendency** for things to continue doing what they are doing. We all know that they don't always continue so to do. **What then compels them to change? What makes them start, or stop, go faster or slower, or even to go round a corner?** The answer is a **force**.

We are familiar not only with the word force, but with its physical effects. Every push or pull is a force. The motor exerts a force on a car, to accelerate it; the brakes exert a force when they are used to stop it. Forces try to alter things; to change the momentum of objects. As we shall see, however, they do not always succeed.

FORCES IN EQUILIBRIUM

If two tug-of-war teams pulling on a rope are well matched, there may for a while be no movement, just a lot of shouting and puffing! Both teams are exerting the same amount of force on the two ends of the rope.

The forces are therefore in equilibrium and there is no change of momentum. There are, however, other more common occurrences of forces in equilibrium. If you push down on an object at rest on a table, the table will resist the force with an equal and opposite force of reaction, so the forces are in equilibrium. Of course if you press too hard, the table might break, in which case the forces will no longer be in equilibrium, and a sudden and unwanted acceleration will occur. As another example, consider a glider being towed behind a small aircraft as in Fig. 1.1. If the aircraft and glider are flying straight and level at constant speed, then the pulling force exerted by the aircraft on the tow-rope must be exactly balanced by an equal and opposite aerodynamic resistance or drag force acting on the glider. The forces are in equilibrium.

Pull applied
by towing
aircraft
1000 N

Aerodynamic
resistance
force
1000 N

Fig. 1.1   Forces in equilibrium

Some people find it hard to believe that these forces really are exactly equal. Surely, they say, the aircraft must be pulling forward just a bit harder than the glider is pulling backwards; otherwise, what makes them go forward? Well, what makes them go forward is the fact that they are going forward, and the law says that they will continue to do so unless there is something to alter that state of affairs. If the forces are balanced then there is nothing to alter that state of equilibrium, and the aircraft and glider keep moving at constant speed.

FORCES NOT IN EQUILIBRIUM

In the case of the glider mentioned above, what would happen if the pilot of the towing aircraft suddenly opened the engine throttle? The pulling force on the tow-rope would increase, but at first the aerodynamic resistance on the glider would not change. The forces would therefore no longer be in equilibrium. The air resistance force is still there of course, so some of the pull on the tow-rope must go into overcoming it, but the remainder of the force will cause the glider to accelerate as shown in Fig. 1.2 which is called a free-body diagram.

This brings us to **Newton's Second Law**, which says in effect that if

Pull applied by towing aircraft 1200 N	Aerodynamic resistance force 1000 N

Fig. 1.2   **Forces not in equilibrium**

the forces are not in balance, then the acceleration will be proportional to force and inversely proportional to the mass of the object:

$$a = F/m$$

where a is the acceleration, m is the mass of the body, and F is the force. This relationship is more familiarly written as

$$F = m \times a$$

INERTIA FORCES

In the above example, of the accelerating glider, the force applied to one end of the rope by the aircraft is greater than the air resistance acting on the glider at the other end. As far as the rope is concerned, however, the force it must apply to the glider tow-hook must be equal to the air resistance force plus the force required to accelerate the glider. In other words, the forces on the two ends of the rope are in equilibrium (as long as we ignore the mass of the rope). The extra force that the rope has to apply to produce the acceleration is called an **inertia force**. An inertia force is the force that has to be applied in order to cause a mass to accelerate.

As far as the rope is concerned, it does not matter whether the force at its far end is caused by tying it to a wall to create a reaction, or attaching it to a glider which it is causing to accelerate, the effect is the same, it feels an equal and opposite pull at the two ends. From the point of view of the glider, however, the situation is very different; if there were a force equal and opposite to the pull from the rope, no acceleration would take place. The forces on the glider are not in equilibrium.

Great care has to be taken in applying the concept of an inertia force. When considering the stresses in the tow-rope it is acceptable to apply the pulling force at one end, and an equal and opposite force at the other end due to the air resistance plus the inertia of the object that it is causing to accelerate. When considering the motion of the aircraft and glider, however, no balancing inertia force should be included, or there would be no acceleration. A free-body diagram should be drawn as in Fig. 1.2.

This brings us to the much misunderstood **third law of Newton**: to every action there is an equal and opposite reaction. If a book rests on a table then the table produces a reaction force that is equal and opposite to the weight force. However, be careful; the force which is accelerating the glider produces a reaction, but the reaction is not a force, but an acceleration of the glider.

WEIGHT

There is one particular force that we are all familiar with; it is known as the force due to gravity. We all know that any object placed near the earth is attracted towards it. What is perhaps less well known is that this is a mutual attraction like magnetism. The earth is attracted towards the object with just as great a force as the object is attracted towards the earth.

All objects are mutually attracted towards each other. The force depends on the masses of the two bodies and the distance between them, and is given by the expression

Fig. 1A   Weight and thrust
The massive Antonov An-255 Mriya, the heaviest aircraft ever built, with a maximum take-off weight of 5886 kN (600 tonnes). The six Soloviev D-18T turbofans deliver a total maximum thrust of 1377 kN.

$$F = \frac{Gm_1m_2}{d^2}$$

where G is a constant which has the value $6.67 \times 10^{-11}\,\mathrm{N\,m^2/kg^2}$, m_1 and m_2 are the masses of the two objects, and d is the distance between them. Using the above formula you can easily calculate the force of attraction between two one kilogram masses placed one metre apart. You will see that it is very small. If one of the masses is the earth, however, the force of attraction becomes large, and it is this force that we call the force of gravity. In most practical problems in aeronautics, the objects that we consider will be on or relatively close to the surface of the earth, so the distance d is constant, and as the mass of the earth is also constant, we can reduce the formula above to a simpler one

$$F = m \times g$$

where m is the mass of the object and g is a constant called the gravity constant which takes account of the mass of the earth and its radius. It has the value $9.81\,\mathrm{m/s^2}$ in the SI system, or $32\,\mathrm{ft/s^2}$ in the Imperial or Federal systems.

The force in the above expression is what we know as **weight**. Weight is the force with which an object is attracted towards the centre of the earth. In fact g is not really a constant because the earth is not an exact sphere, and large chunks of very dense rock near the surface can cause the force of attraction to increase slightly locally. For most practical aeronautical calculations we can ignore such niceties. We cannot, however, use this simple formula once we start looking at spacecraft or high-altitude missiles.

Weight is an example of what is known as a **body force**. Body forces unlike mechanical forces have no visible direct means of application. Other examples of body forces are electrostatic and electromagnetic forces.

When an aircraft is in steady level flight, there are two vertical forces acting on it, as shown in Fig. 1.3. There is an externally applied force,

Fig. 1.3 **Aerodynamic and body forces**

the lift force provided by the air flowing over the wing, and a body force, the weight.

THE ACCELERATION DUE TO GRAVITY

All objects near the surface of the earth have the force of gravity acting on them. If there is no opposing force, then they will start to move, to accelerate. The rate at which they accelerate is independent of their mass.

The force due to gravity (weight) $F = m \times g$

but, from Newton's Second Law, $F = m \times$ acceleration

By equating the two expressions above, we can see that the acceleration due to gravity will be numerically equal to the gravity constant g, and will be independent of the mass. Not surprisingly, many people confuse the two terms 'gravity constant' and 'acceleration due to gravity', and think that they are the same thing. The numerical value is the same, but they are different things. If a book rests on a table, then the weight is given by the product of the gravity constant and the mass, but it is not accelerating. If it falls off the table, it will then accelerate at a rate equal to the value of the gravity constant.

This brings us to the old problem of the feather and the lump of lead; which will fall fastest? Well, the answer is that in the vacuum of space, they would both fall at the same rate. In the atmosphere, however, the feather would be subjected to a much larger aerodynamic resistance force in relation to the accelerating gravity force (the weight), and therefore, the feather would fall more slowly.

For all objects falling through the atmosphere, there is a speed at which the aerodynamic resistance is equal to the weight, so they will then cease to accelerate. This speed is called the **terminal velocity** and will depend on both the shape, the density, and the orientation of the object. A man will fall faster head first than if he can fall flat. Free–fall sky-divers use this latter effect to control their rate of descent in free fall.

MASS WEIGHT AND g

The mass of a body depends on the amount of matter in it, and it will not vary with its position on the earth, nor will it be any different if we place it on the moon. The weight (the force due to gravity) will change,

however, because the so-called gravity constant will be different on the
moon, due to the smaller mass of the moon, and will even vary slightly
between different points on the earth, as described previously. Also,
therefore, the rate at which a falling object accelerates will be different.
On the moon it will fall noticeably slower, as can be observed in the
apparently slow-motion moon-walking antics of the Apollo astronauts.

UNITS

The system of units that we use to measure quantities, feet, metres etc.,
can be a great source of confusion. In European educational
establishments and most of its industry, a special form of the metric
system known as the **Système International** or **SI** is now in general
use. The basic units of this system are the kilogram for mass (not
weight) (kg), the metre for distance (m), and the second for time (s).

Temperatures are in degrees Celsius (or Centigrade) (°C) when
measured relative to the freezing point of water, or in Kelvin (K) when
measured relative to absolute zero; 0°C is equivalent to 273 K. A
temperature change of one degree Centigrade is exactly the same as a
change of one degree Kelvin, it is just the starting or zero point that is
different. Note that the degree symbol ° is not used when temperatures
are written in degrees Kelvin, for example we write 273 K.

Forces and hence weights are in newtons (N) not kilograms. Beware
of weights quoted in kilograms; in the old (pre-SI) metric system still
commonly used in parts of Europe, the name kilogram was also used for
weight or force. To convert weights given in kilograms to newtons,
simply multiply by 9.81.

The SI system is known as a coherent system, which effectively
means that you can put the values into formulae without having to
worry about conversion factors. For example, in the expression relating
force to mass and acceleration: $F = m \times a$, we find that a force of 1
newton acting on a mass of 1 kilogram produces an acceleration of 1
m/s^2. Contrast this with a version of the old British 'Imperial' system
where a force of 1 pound acting on a mass of 1 pound produces an
acceleration of $32.18\,\text{ft/sec}^2$. You can imagine the problems that the
latter system produces. Notice how in this system, the same name, the
pound, is used for two different things, force and mass.

Because aviation is dominated by American influence, American
Federal units and the similar Imperial (British) units are still in
widespread use. Apart from the problem of having no internationally
agreed standard, the use of Federal or Imperial units can cause
confusion, because there are several alternative units within the system.

In particular, there are two alternative units for mass, the pound mass, and the slug (which is equivalent to 32.18 pounds mass). The slug may be unfamiliar to most readers, but it is commonly used in aeronautical engineering because, as with the SI units, it produces a coherent system. A force of 1 pound acting on a mass of one slug produces an acceleration of 1 ft/sec². The other two basic units in this system are, as you may have noticed, the foot and the second. Temperatures are measured in degrees Fahrenheit.

You may find all this rather confusing, but to make matters worse, in order to avoid dangerous mistakes, international navigation and aircraft operations conventions use the foot for altitude, and the knot for speed. The knot is a nautical mile per hour (0.5145 m/s). A nautical mile is longer than a land mile, being 6080 feet instead of 5280 feet. Just to add a final blow, baggage is normally weighed in kilograms (not even newtons)!

To help the reader, most of the problems and examples in this book are in SI units. If you are presented with unfamiliar units or mixtures of units, convert them to SI units first, and then work in SI units. One final tip is that when working out problems, it is always better to use basic units, so convert millimetres or kilometres to metres before applying any formulae. In the real world of aviation, you will have to get used to dealing with other units such as slugs and knots, but let us take one step at a time. Below, we give a simple example of a calculation using SI units.

EXAMPLE
The mass of an aeroplane is 2000 kg. What force, in addition to that required to overcome friction and air resistance, will be needed to give it an acceleration of 2 m/s² during take-off?

SOLUTION

$$\text{Force} = ma$$
$$= 2000 \times 2$$
$$= 4000 \text{ newtons}$$

This shows how easy is the solution of such problems if we use the **SI** units.

Many numerical examples on the relationship between forces and masses involve also the principles of simple kinematics, and the reader who is not familiar with these should read the next paragraph before he tackles the examples.

KINEMATICS

The previous section may seem a little hard going. In comparison, much of what follows is plain sailing. It will help us in working examples if we summarise the relations which apply in kinematics, that is, the study of the movement of bodies irrespective of the forces acting upon them.

We shall consider only the two simple cases, those of uniform velocity and uniform acceleration.

Symbols and units will be as follows –

$$\text{Time} = t \text{ (sec)}$$
$$\text{Distance} = s \text{ (metres)}$$
$$\text{Velocity (initial)} = u \text{ (metres per sec)}$$
$$\text{Velocity (final)} = v \text{ (metres per sec)}$$
$$\text{Acceleration} = a \text{ (metres per sec per sec)}$$

UNIFORM VELOCITY
If velocity is uniform at u metres per sec clearly

Distance travelled = Velocity × Time
$$\text{or } s = ut$$

UNIFORM ACCELERATION
Final velocity = Initial velocity + Increase of velocity.

$$\text{or } v = u + at$$

Distance travelled = Initial velocity × Time
$$+ \tfrac{1}{2} \text{ Acceleration × Time squared}$$

i.e. $s = ut + \tfrac{1}{2}at^2$

Final velocity squared = Initial velocity squared
$$+ 2 \times \text{Acceleration × Distance}$$

or $v^2 = u^2 + 2as$

With the aid of these simple formulae – all of which are founded on first principles – it is easy to work out problems of uniform velocity or uniform acceleration. For instance –

EXAMPLE 1
If, during a take-off run an aeroplane starting from rest attains a velocity of 90 km/h in 10 seconds, what is the average acceleration?

ANSWER

Initial velocity $u = 0$
Final velocity $v = 90 \, \text{km/h} = 25 \, \text{m/s}$
Time $t = 10 \, \text{sec}$
$a = ?$

Since we are concerned with u, v, t and a, we use the formula

$v = u + at$
$25 = 0 + 10a$
$a = 25/10 = 2.5 \, \text{m/s}^2$

EXAMPLE 2
How far will the aeroplane of the previous example have travelled during the take-off run?

$u = 0$, $v = 25 \, \text{m/s}$, $t = 10 \, \text{sec}$, $a = 2.5 \, \text{m/s}^2$

To find s, we can either use the formula

Final velocity squared = Initial velocity squared
$\qquad\qquad\qquad\quad + \, 2 \times \text{Acceleration} \times \text{Distance}$

$s = ut + \tfrac{1}{2}at^2$
$\quad = 0 + \tfrac{1}{2} \times 2.5 \times 10^2$
$\quad = \mathbf{125 \, m}$

or $v^2 = u^2 + 2as$
$25 \times 25 = 0 + 2 \times 2.5 \times s$
$\therefore s = (25 \times 25)/(2 \times 2.5)$
$\quad = \mathbf{125 \, m}$

EXAMPLE 3
A bomb is dropped from an aeroplane which is in level flight at 200 knots at a height of 3500 m. Neglecting the effect of air resistance, how long will it be before the bomb strikes the ground, and how far horizontally before the target must the bomb be released?

ANSWER
To find the time of fall we are concerned only with the vertical velocity, which was zero at release.

$\therefore u = 0$
$a = \text{acceleration of gravity} = 9.81 \, \text{m/s}^2$
$s = \text{vertical distance from aeroplane to ground} = 3500 \, \text{m}$
$t = ?$

We need the formula connecting u, a, s and t, i.e.

$$s = ut + \tfrac{1}{2}at^2$$
$$\therefore 3500 = 0 + \tfrac{1}{2} \times 9.81 \times t^2$$
$$\therefore t^2 = (3500/9.81) \times 2 = 713$$
$$\therefore t = 27 \text{ sec (approx)}$$

Since we are neglecting the effect of air resistance the horizontal velocity of the bomb will, throughout the fall, remain the same as it was at the moment of release, i.e. the same as the velocity of the aeroplane, namely 200 knots or, converting into metres per second, $(200 \times 1852)/3600 = \mathbf{103\,m/s}$ (approx).

Therefore the distance that the bomb will travel forward during the falling time of $27\,s$ will be $103 \times 27 = \mathbf{2781\,m}$.

This, of course, is the distance before the target that the bomb must be released.

Note that in this example we have neglected air resistance. Since we are interested in flying this may seem rather a silly thing to do, because we are only able to fly by making use of the same principles that are responsible for air resistance. In fact, too, the effects of air resistance on bombs are of vital importance and are always taken into account when bombing. But it is better to learn things in their most simple form first, then gradually to add the complications. As these complications are added we get nearer and nearer to the truth, but if we are faced with them all at once the picture becomes blurred and the fundamental principles involved fail to stand out clearly.

Other examples on kinematics will be found in Appendix 3, and the reader who is not familiar with examples of this type is advised to work through them.

MOTION ON CURVED PATHS

It has already been emphasised that bodies tend to continue in the same state of motion, and that this involves direction as well as speed. It is clear, therefore, that **if we wish to make a body change its motion by turning a corner or travelling on a curved path, we must apply a force to it in order to make it do so**, and that this will apply even if the speed of the body does not change. This is a force exactly similar to the one that is required to accelerate the train out of the station, that is to say **the force must be proportional to the mass of the body and to the acceleration which it is desired to produce**. But what is the acceleration of a body that is going round a corner? Is there, in fact, any acceleration at all if the speed remains constant? And another rather important question – in what direction is the acceleration?

Let us deal with the last question first. There is another part of Newton's Second Law which has not so far been mentioned, namely that **the rate of change of momentum of the body will be in the direction of the applied force.** If the mass of the body does not change as it goes round the corner – and this applies to solid bodies such as trains, cars or aeroplanes – **the acceleration must be in the direction of the force.** But is there any acceleration if the speed does not change? Yes – because velocity is what we call a vector quantity, that is to say, it has both magnitude and direction, while speed has only magnitude. Thus if the direction of motion changes, the velocity changes even though the speed remains unaltered. But at what rate does the velocity change? – in other words, what is the acceleration? and in what direction is it?

CENTRIPETAL FORCE AND CENTRIPETAL ACCELERATION

Here we are going to cheat. The reader who is not familiar with the answer, and who doesn't like cheating, must consult books on mechanics. To save space and time let us take a short cut by saying that we all know the direction of the force as a result of practical experience. Swing a stone round on the end of a piece of string. **In what direction does the string pull on the stone to keep it on its circular path? Why, towards the centre of the circle, of course.** Yes, and since force and acceleration are in the same direction, the **acceleration must also be towards the centre.**

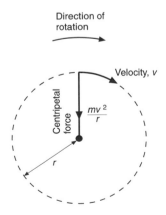

Fig. 1.4 Centripetal force

We know too that **the greater the velocity of the stone, and the smaller the radius of the circle on which it travels, the greater is the pull in the string, and therefore the greater the acceleration.** It is true that this does not tell us the exact value of the acceleration, but if we are slightly more sensitive to forces we may have realised that the velocity is more important than the radius or, if we have learnt text-book mechanics, that the **acceleration is actually given by the simple formula** v^2/r, where v is the velocity of the body and r the radius of the circle.

The force towards the centre is called centripetal force (centre-seeking force), and will be equal to the mass of the body $\times$ the centripetal acceleration, i.e. to $m \times v^2/r$ **(Fig. 1.4)**.

We have made no attempt to prove that the acceleration is v^2/r – the proof will be found in any textbook on mechanics – but since it is not easy to conceive of an acceleration towards the centre as so many metres per second per second when the body never gets any nearer to the centre, it may help if we translate the algebraic expression into some actual figures. Taking the simple example of a stone on the end of a piece of string, if the stone is whirled round so as to make one revolution per second, and the length of the string is 1 metre, the distance travelled by the stone per second will be $2\pi r$, i.e. $2\pi \times 1$ or 6.28 m. Therefore

$$v = 6.28 \, \text{m/s}, r = 1 \, \text{m}$$

$$\therefore \text{acceleration towards centre} = v^2/r$$
$$= (6.28 \times 6.28)/1$$
$$= \textbf{39.5 m/s}^2 \text{ (approx)}$$

Notice that this is nearly four times the acceleration of gravity, or nearly $4g$. Since we are only using this example as an illustration of principles, let us simplify matters by assuming that the answer is $4g$, i.e. $39.24 \, \text{m/s}^2$.

This means that the velocity of the stone towards the centre is changing at a rate 4 times as great as that of a falling body. Yet it never gets any nearer to the centre! No, but what would have happened to the stone if it had not been attached to the string? It would have obeyed the tendency to go straight on, and in so doing would have departed farther and farther from the centre. The acceleration of $4g$ may, in a sense, be taken as the rate at which it is being prevented from doing this.

What centripetal force will be required to produce this acceleration of $4g$? The mass of the stone $\times 4g$.

So, if the mass is $1/2$ kg, the centripetal force will be $1/2 \times 4g = 2 \times 9.81 = 19.62$, say 20 newtons.

Therefore the pull in the string is 20 N in order to give the mass of 1/2 kg an acceleration of $4g$.

Notice that the **force is 20 newtons**, the **acceleration is $4g$**. There is a horrible tendency to talk about 'g' as if it were a force; it is not, it is an acceleration.

Now this is all very easy provided the centripetal force is the only force acting upon the mass of the stone – but is it? Unfortunately, no. There must, in the first place, be a force of gravity acting upon it.

If the stone is rotating in a **horizontal circle** its weight will act at right angles to the pull in the string, and so will not affect the centripetal force. But of course a stone cannot rotate in a horizontal circle, with the string also horizontal, unless there is something to support it. So let us imagine the mass to be on a table – but it will have to be a smooth, frictionless table or we shall introduce yet more forces. We now have, at least in imagination, the simple state of affairs illustrated in Fig. 1.5.

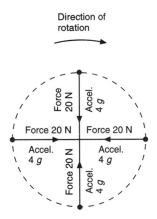

Fig. 1.5   Stone rotating in a horizontal circle, supported on a table

Now suppose that we rotate the stone in a **vertical** circle, like an aeroplane looping the loop, the situation is rather different (Fig. 1.6). Even if the stone were not rotating, but just hanging on the end of the string, there would be a tension in the string, due to its weight, and this as near as matters would be very roughly 5 newtons, for a mass of $\frac{1}{2}$ kg. If it must rotate with an acceleration of $4g$ the string must also provide a centripetal force of 20 newtons. So when the stone is at the bottom of the circle, D, the total pull in the string will be 25 N. When the stone is in the top position, C, its own weight will act towards the centre and this will provide 5 N, so the string need only pull with an additional 15 N to produce the total of 20 N for the acceleration of $4g$. At the side positions,

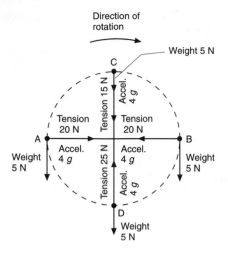

Fig. 1.6   Stone rotating in a vertical circle

A and B, the weight of the stone acts at right angles to the string and the pull in the string will be 20 N.

To sum up: **the pull in the string varies between 15 N and 25 N, but the acceleration is all the time 4***g* **and, of course, the centripetal force is all the time −20 N.** From the practical point of view what matters most is the pull in the string, which is obviously most likely to break when the stone is in position D and the tension is at the maximum value of 25 N.

To complicate the issue somewhat, suppose the stone rotates in a horizontal circle, but relies on the pull of the string to hold it up (Fig. 1.7), and that the string has been lengthened so that the radius on which the stone is rotating is still 1 metre. The string cannot of course be horizontal since the pull in it must do two things – **support the weight of the stone** and **provide the centripetal force**.

Here we must introduce a new principle.

A force of 5 N, vertically, is required to support the weight.

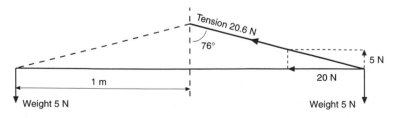

Fig. 1.7   Stone rotating in a horizontal circle, with string support

A force of 20 N, horizontally, is required to provide the centripetal force.

Now five plus twenty does not always make twenty-five! It does not in this example, and for the simple reason that they are not pulling in the same direction. We must therefore represent them by vectors (Fig. 1.7), and the diagonal will represent the total force which, by Pythagoras' Theorem, will be

$$\sqrt{(20^2 + 5^2)} = \sqrt{425} = 20.6\,\text{N}$$

The tangent of the angle of the string to the vertical will be $20/5 = 4.0$. So the angle will be approx $76°$. Expressing the angle, θ, in symbols –

$$\tan \theta = \frac{\text{Centripetal force}}{\text{Weight}} = (m \times v^2/r)/W$$

$$= (m \times v^2/r)/mg$$

(*mg* being the weight expressed in newtons)

$$= v^2/rg$$

This angle θ represents the correct angle of bank for any vehicle, whether it be bicycle, car or aeroplane, to turn a corner of radius r metres, at velocity v metres per second, if there is to be no tendency to slip inwards or to skid outwards.

CENTRIFUGAL FORCE

Whether by accident or design, I am not quite sure, we have managed to arrive so far without mentioning the term **centrifugal force**. This is rather curious because centrifugal force is a term in everyday use, while centripetal force is hardly known except to the student of mechanics.

Consider again the stone rotating, on a table, in a horizontal circle. We have established the fact that there is an **inward** force on the stone, **exerted by the string**, for the set purpose of providing the acceleration **towards the centre** – yes, centripetal force, however unknown it may be, is a real, practical, physical force. But is there also an outward force?

The situation is similar to that of the accelerating aircraft towing a glider that we described earlier. There is an outward reaction force on the outer end of the **string** caused by the fact that it is accelerating the stone inwards: an inertia force, and we could call this a centrifugal reaction force. This keeps the string in tension in a state of equilibrium just as if it were tied to a wall and pulled. Note, however, that **there is no outward force on the stone**, only an inward one applied by the string to produce the centripetal acceleration. As with the accelerating glider described previously and shown in Fig. 1.2, the forces on the two ends of the string are in balance, but the forces on the object, the stone or the glider are not, and hence acceleration occurs.

The concept of inertia forces is a difficult one. In a free-body diagram of the horizontally whirling stone, the only externally applied horizontal force is the inward force applied by the string. This force provides the necessary acceleration. There may be outward forces on the internal components of the system like the string, but not on the overall system. Note that if you let go of the string, the stone will not fly outwards, it will fly off at a tangent.

To sum up motion on curved paths. **There is an acceleration** (v^2/r) **towards the centre, necessitating a centripetal force of** mv^2/r.

At this stage, the reader is advised to try some numerical questions on motion on curved paths in Appendix 3.

THE MECHANICS OF FLIGHT

A knowledge of the principles of mechanics – and particularly of the significance and meaning of the force of gravity, of accelerated motion, of centripetal and centrifugal force and motion on curved paths – will help us to understand the movements and manoeuvres of an aeroplane. But even more will this knowledge help us to understand the movement of satellites and space craft.

This sounds like astronautics rather than the mechanics of flight, but in recent years the two subjects have merged one with the other to such an extent that it is impossible to say where one begins and the other ends.

WORK, POWER, AND ENERGY

These three terms are used frequently in mechanics, so we must understand their meaning. This is especially important because they are common words too in ordinary conversation, but with rather different shades of meaning.

A force is said to do work on a body when it moves the body in the direction in which it is acting, and the amount of work done is measured by the product of the force and the distance moved in the direction of the force. Thus if a force of 10 newtons moves a body 2 m (along its line of action), it does 20 newton metres (Nm) of work. **A newton metre, the unit of work, is called a joule (J).** Notice that, according to mechanics, you do no work at all if you push something without succeeding in moving it – no matter how hard you push or for how long you push. Notice that you do no work if the body moves in the opposite direction, or even at right angles to the direction in which you push. Someone else must be doing some pushing – and some work! Notice that brain-work does not count – unless perhaps some pen-pushing is required!

Power is simply the rate of doing work. If the force of 10 N moves the body 2 m in 5 seconds, then the power is 20 Nm (20 joules) in 5 seconds, or 4 joules per second. **A joule per second (J/s) is called a watt (W), the unit of power.** So the power used in this example is 4 watts. If the reader has studied electricity he will already be familiar with the watt as a unit of electrical power, just one example of the general trend towards the realisation that all branches of science are inter-related. Note the importance of the time, i.e. of the rate at which the work is done; the word power, or powerful, is apt to give an impression of size and brute force, but the word as used in mechanics throws light on the great power of a small aero-engine as compared with its bigger, but slower running, brother of say the Diesel type. The unit of 1 watt is small for practical use, and kilowatts are more often used. The old unit of a horse-power was never very satisfactory but, as a matter of interest, it was the equivalent of 745.7 watts (Fig. 1B).

A body is said to have **energy** if it has the ability to do work, and the **amount of energy reckoned by the amount of work that it can do. The units of energy will therefore be the same as those of work.** The amount of work a body can do **should** be the same as the amount that has been done on it, that is to say, it should be able to give back what it has been given. Unfortunately it usually cannot do so, but that is beside the point at the moment. We know that petrol can do work by driving a car or an aeroplane, a man can do work by propelling a bicycle or even by walking, a chemical battery can drive an electric motor which can do work on a train, an explosive can drive a shell at high speed from the muzzle of a gun. All this means that **energy can exist in many forms, heat, light, sound, electrical, chemical, magnetic, atomic – and, most useful of all, mechanical**. A little thought will convince us how much of our time and energy is spent in converting, or trying to convert, other forms of energy into mechanical energy, the eventual form which enables us to get somewhere. The human body is simply a form of engine – not a simple form of engine – in which the energy contained in food is converted into useful, or useless, work. Unfortunately there is a tendency for energy to slip back again, we might almost say deteriorate, into other forms, and our efforts to produce mechanical energy are not always very efficient.

Even mechanical energy can exist in more than one form; a weight that is high up can do work in descending, and it is said to possess **potential energy** or **energy of position**; a mass that is moving rapidly

can do work in coming to rest, and it is therefore said to have **kinetic energy** or **energy of motion**; a spring that is wound up, a gas that is compressed, even an elastic material that is stretched, all can do work in regaining their original state, and all possess energy which is in a sense potential but which is given various names according to its application.

In figures, a weight of 50 newtons raised to a height of 2 metres above its base has 100 joules of potential energy or, to be more correct, it has 100 joules **more potential energy than it had when at its base**. This was the work done to raise it to the new position, and it is the work that it should be able to do in returning to its base.

In symbols, W **newtons at height** h **metres has –**
Wh **joules of energy.**

What is the kinetic energy of mass of m kg moving at v m/s?

We don't know, of course, how it got its kinetic energy, but as this should not matter very much, let us suppose that it was accelerated uniformly at a metres per second per second from zero velocity to v metres per second by being pushed by a constant force of F newtons.

If the distance travelled during the acceleration was s metres, then the work done, i.e. its kinetic energy, will be Fs joules.

But $v^2 = u^2 + 2as$ (and $u = 0$)
$$\therefore v^2 = 2as$$
$$\therefore s = v^2/2a$$
But $F = ma$
So K.E. $= Fs = ma \times v^2/2a = \frac{1}{2}mv^2$ **joules.**

Thus the kinetic energy of 2 kg moving at 10 m/s

$$= \frac{1}{2}mv^2$$
$$= \frac{1}{2} \times 2 \times 10^2$$
$$= 100 \text{ joules}$$

ENERGY AND MOMENTUM

Let us be sure that we understand the differences between energy and momentum, because we shall be concerned with this later on.

Fig. 1B Power (opposite)
(By courtesy of the Lockheed Aircraft Corporation, USA)
The Lockheed C-5 Galaxy with four turbofan engines, each of 183 kN thrust represents a total power of 183 000 kW at the maximum level speed of about 900 km/h.

Energy is $1/2\, mv^2$. Momentum is mv.

So the mass of 2 kg, moving at 10 m/s, has 100 units of energy (joules), but 2×10, i.e. 20 units, of momentum (kg × m/s).

Yes, but there is more to it than that.

Consider two bodies colliding, e.g. billiard balls.

The total momentum after the collision is the same as the total momentum before; the momentum lost by one ball is exactly the same as the momentum gained by the other. This is the principle of the **conservation of momentum**. (In considering this it must be remembered that momentum has **direction**, because velocity has direction.) The law will apply whether the balls rebound, or whether they stick together, or whatever they do.

But the total mechanical energy after the collision will not be the same as before; energy will be dissipated, it will go into the air in the form of heat, sound, etc.; the total energy of the universe will not be changed by the collision – but that of the balls will be.

So momentum is a more permanent property than energy, the latter is often wasted and we shall sometimes find it unfortunate that in order to give a body momentum we must also give it energy.

PRESSURE

The words pressure and force are often confused, which is unfortunate, for **pressure means force per unit area**. If the force is evenly distributed we need only divide the force by the area over which it acts to get the pressure. Thus a force of 10 newtons distributed over an area of 2 square metres, gives a pressure of 5 newtons per square metre.

FLUID PRESSURE

In the mechanics of flight we shall be chiefly concerned with fluid pressure, that is the pressure exerted by a liquid or gas. The reason why a fluid exerts pressure is because its molecules are in rapid motion and bombard any surface that is placed in the fluid; each molecule exerts a tiny force on the surface but the combined effect of the bombardment of millions upon millions of molecules results in an evenly distributed pressure on the surface.

Pressure is a scalar quantity. That means that it has a magnitude, but unlike a vector, there is no direction involved. When a fluid at rest is in contact with a surface, the pressure produces a force which acts at right angles to the surface (Fig. 1.8). Note that the force *does* have a direction

Fig. 1.8   Force due to pressure acts at right angles to the surface

whereas the pressure that causes it does not. It is not surprising that people often confuse cause and effect when talking about pressure. You will find many old books that say that pressure acts equally in all directions. It does not. Pressure does not act in any direction; it is the force **due to pressure** that acts in a direction. The direction of the force is always at right angles to the surface that the pressure is exposed to. We measure pressure in terms of the force that it will produce on an area, so pressure has the units of newtons per square metre (N/m²).

The pressure in a fluid increases with the depth below the surface. If the thickness, or density, of the fluid is constant – as is very nearly true in a **liquid** because it is practically incompressible – the increase of pressure will be proportional to the depth (Fig. 1.9). In a gas there is an increase of density with depth because the lower layers are compressed by the weight of the gas above them.

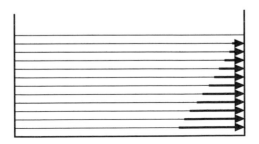

Fig. 1.9   Pressure increases with depth below the surface

DENSITY

Density is defined as the mass per unit volume of a substance (kilograms per cubic metre). Notice that **it is mass and not weight**.

Newton actually used density as the fundamental conception and defined the mass of a body as the quantity of matter in it obtained by multiplying the density by the volume. Be that as it may, the forces experienced by bodies moving through the air are largely reactions caused by the mass of air being forced to change its motion, and it is therefore essential that we think of density in terms of mass. It must be admitted, however, that there are times when it is convenient to use the conception of **weight divided by volume** and this is sometimes called **specific weight**. If we do this, we must realise that we are thinking of a fundamentally different property of the fluid, measured in different units, i.e. newtons per cubic metre, instead of kilograms-mass per cubic metre.

PRESSURE HEAD

The fact that the pressure of a liquid increases almost uniformly with depth is often made use of in talking of **pressure in terms of the height or head of a liquid** (Fig. 1.10). Thus we may speak of 760 millimetres of mercury, or 10 metres of water, instead of so many newtons per square metre. How many? Well, let us see.

Fig. 1.10   **Pressure head**

First, we must realise that the pressure, i.e. the force per unit area, will be the same whatever the area of the column that we consider so, for convenience, let us take a column of 1 m² area. Then the volume of a mercury column, 760 mm high and 1 m² in area will be 0.76 m³.

The density of mercury is 13.6 × that of water, i.e. about

$13.6 \times 1000 = 13\,600 \text{ N/m}^2$

So the total weight on 1 m² at the bottom of the column is

$13\,600 \times 9.81 \times 0.76 = \textbf{101\,000 N approx}$

Therefore the pressure is 101 kN/m², which is well-known as being approximately the normal atmospheric pressure at sea-level. The pressure of $100\,kN/m^2$ is known as a **bar**, which is sometimes used as a unit of pressure, or more often in the form of **millibars**, a millibar being one thousandth of a bar or $100\,N/m^2$.

We really should not use the bar though, as this is not a basic SI unit. The proper unit to use for pressure is N/m^2. This is sometimes called the pascal (Pa).

ARCHIMEDES' PRINCIPLE

When a body is immersed in a fluid it experiences an upthrust, or apparent loss of weight, equal to the weight of the fluid displaced by the body (Figs 1.11 and 1.12).

This important principle is only an extension of the idea of the

Fig. 1.11 Archimedes' Principle
Pressure forces on a body in a fluid.

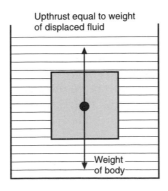

Fig. 1.12 Archimedes' Principle
Effect of pressures on a body in a fluid.

increase in the pressure of a fluid with depth, which means that there is a greater pressure pushing up on the body from underneath than there is pushing down on it from on top.

This principle is the clue to the floating of balloons and airships and of ships and submarines, but we are not much concerned with it in the mechanics of flight of an aeroplane.

THE BEHAVIOUR OF GASES

In the study of the flight of aircraft, we are really only interested in the behaviour of one particular gas, air. The most important relationship that we need to know is called the gas law which can be written as

$$\frac{p}{\rho} = RT$$

where p is the pressure, ρ is the density, T is the temperature measured relative to absolute zero (i.e. in degrees Kelvin in the SI system), and R is a constant called **the gas constant**.

If the gas is compressed so that its density increases, then either or both the other quantities, temperature or pressure must change. The way that they change depends on how the compression takes place. If the compression is very slow, and the gas is contained in a poorly insulated vessel so that heat is transferred out of the system, then the temperature will stay constant, and the pressure change will be directly proportional to the density change. This is called an isothermal process, and it involves a heat transfer from the gas to its surroundings. In this case the relationship between pressure and density is given by

$$\frac{p}{\rho} = \text{a constant}$$

If the compression takes place with no transfer of heat, a situation that commonly occurs when the compression is very rapid, then the change is said to be adiabatic. If furthermore, the change also takes place without any increase in turbulence, so that there is no increase in the disorder (entropy) of the system, then the process is called isentropic, and the relationship between pressure and density is given by

$$\frac{p}{\rho^{\gamma}} = \text{a constant}$$

where γ is the ratio of the specific heat at constant pressure to the specific heat at constant volume, and has a value of approximately 1.4 for air.

We cannot go much further down this path without becoming embroiled in the complexities of thermodynamics, however, and as the relationships above are the only ones that are relevant to the understanding of the contents of this book, we will not pursue the subject any further.

STATICS

It may seem rather inconsistent that, whereas in this chapter we have dealt with fluids at rest and omitted fluids in motion, in the mechanics of solid bodies our emphasis has been on moving bodies rather than on bodies in equilibrium. There is, however, method in our madness for the object has been to prepare the ground for what is to follow. Experience shows that the main difficulty of students in understanding the flight of an aeroplane is caused by confused ideas about the **dynamics** of solid bodies, in particular the inter-action of forces, masses and accelerations, and for this reason we have concentrated on that part of the subject. The majority of students, even if they have not learnt mechanics at school, find little or no difficulty in understanding **statics**. Much of the subject is, after all, common sense. The main principles of statics that we shall require later are summed up in the following paragraphs and if any student is not familiar with them he is advised to work through the examples in Appendix 3.

COMPOSITION AND RESOLUTION OF FORCES, VELOCITIES, ETC.

A force is a vector quantity – that is to say, it has magnitude and direction, and can be represented by a straight line, passing through the point at which the force is applied, its length representing the magnitude of the force, and its direction corresponding to that in which the force is acting.

As vector quantities, **forces can be added, or subtracted, to form a resultant force, or they can be resolved,** that is to say, split up into two or more component parts, by the simple process of drawing the vectors to represent them (Fig. 1.13).

Velocity and momentum are also vector quantities and can be represented in the same way by straight lines. Mass, on the other hand, is not; a mass has no direction, and this is yet another distinction between a force and a mass.

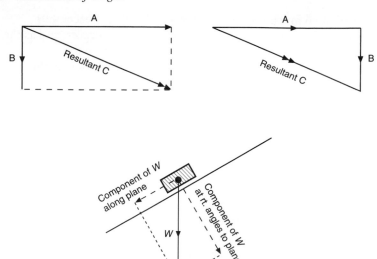

Fig. 1.13   Composition and resolution of vector quantities

THE TRIANGLE, PARALLELOGRAM, AND POLYGON OF FORCES

If three forces which act at a point are in equilibrium, they can be represented by the sides of a triangle taken in order (Fig. 1.14). This is called the principle of the **triangle of forces**, and the so called **parallelogram of forces** is really the same thing, two sides and the diagonal of the parallelogram corresponding to the triangle.

If there are more than three forces, the principle of the **polygon of forces** is used – **when any number of forces acting at a point are in equilibrium, the polygon formed by the vectors representing the forces and taken in order will form a closed figure, or,**

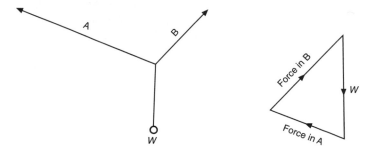

Fig. 1.14   Triangle of forces

conversely, **if the polygon is a closed figure the forces are in equilibrium**.

MOMENTS, COUPLES AND THE PRINCIPLES OF MOMENTS

The **moment** of a force about any point is the **product of the force and the perpendicular distance from the point to the line of action of the force**.

Thus the moment of a force of 10 N about a point whose shortest distance from the line of action is 3 m (Fig. 1.15) is $10 \times 3 = 30$ N-m.

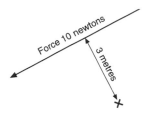

Fig. 1.15   Moment of a force
(Anti-clockwise)

Notice that, though both are measured by force $\times$ distance, there is a subtle but important distinction between a moment (unit N-m) and the work done by a force (unit Nm, or joules). The distance in the moment is merely a leverage and no movement is involved; **moments cannot be measured in joules**.

A moment about a point may act in a clockwise, or in an anti-clockwise direction.

If a body is in equilibrium under the influence of several forces in the same plane, the sum of the clockwise moments about any point is equal to the sum of the anti-clockwise moments about that point, or, what amounts to the same thing and is much shorter to express, **the total moment is zero**. This is called the **principle of moments,** and applies whether the forces are parallel or not.

When considering the forces acting on a body, the weight of the body itself is often one of the most important forces to be considered. The weight may be taken as acting through the **centre of gravity** which is defined as the point through which the resultant weight acts whatever position the body may be in.

Two equal and opposite parallel forces are called a **couple**. The moment of a couple is one of the forces multiplied by the distance

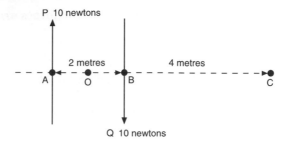

Fig. 1.16   A couple

between the two, i.e. by the arm of the couple. Notice that the moment is the same about any point (Fig. 1.16), and a couple has no resultant.

Moments about O. P $10 \times 1 = 10$ clock, Q $10 \times 1 = 10$ clock,
Total 20 clock.
Moments about A. P zero, Q $10 \times 2 = 20$ clock,
Total 20 clock.
Moments about B. P $10 \times 2 = 20$ clock, Q zero,
Total 20 clock.
Moments about C. P $10 \times 6 = 60$ clock, Q $10 \times 4 = 40$ anti,
Total 20 clock.

MECHANICS OF FLIGHT

We do not pretend to have covered all the principles of mechanics, nor even to have explained fully those that have been covered. All we have done has been to select some aspects of the subject which seem to form the chief stumbling blocks in the understanding of how an aeroplane flies; we have attempted to remove them as stumbling blocks, and perhaps even so to arrange them that, instead, they become stepping stones to the remainder of the subject. In the next chapter we will turn to our real subject – the *Mechanics of Flight*.

Before continuing, try to answer some of the questions below, and the numerical questions in Appendix 3.

CAN YOU ANSWER THESE?

These questions are tests not so much of mechanical knowledge as of mechanical sense. Try to puzzle them out. Some of them are easy, some difficult; the answers are given in Appendix 5.

1. A lift is descending, and is stopping at the ground floor. In what direction is the acceleration?
2. What is the difference between –
   (*a*) Pressure and Force?
   (*b*) Moment and Momentum?
   (*c*) Energy and Work?
3. Why does it require less force to pull a body up an inclined plane rather than lift it vertically? Is the same work done in each case?
4. Distinguish between the mass and weight of a body.
5. If the drag of an aeroplane is equal to the thrust of the propeller in straight and level flight, what makes the aeroplane go forward?
6. Is the thrust greater than the drag during take-off?
7. Can the centre of gravity of a body be outside the body itself?
8. Is an aeroplane in a state of equilibrium during –
   (*a*) A steady climb?
   (*b*) Take-off?
9. Are the following the same, or less, or more, on the surface of the moon as on the surface of the earth –
   (*a*) The weight of a given body as measured on a spring balance?
   (*b*) The apparent weight of a given body as measured on a weigh-bridge (using standard set of weights)?
   (*c*) The time of fall of a body from 100 m?
   (*d*) The time of swing of the same pendulum?
   (*e*) The thrust given by a rocket?
10. In a tug-o'-war does the winning team exert more force on the rope than the losing team?
11. Are the following in equilibrium –
    (*a*) A book resting on a table?
    (*b*) A train ascending an incline at a steady speed?
12. A flag is flying from a vertical flag pole mounted on the top of a large balloon. If the balloon is flying in a strong but steady east wind, in what direction will the flag point?

For numerical examples on mechanics see Appendix 3.

CHAPTER 2

AIR AND AIRFLOW – SUBSONIC SPEEDS

INTRODUCTION – SIGNIFICANCE OF THE SPEED OF SOUND

As was explained in Chapter 1, the remainder of the book will be concerned almost entirely with fluids in motion or, what comes to much the same thing, with motion through fluids. But it would be misleading even to start explaining the subject without a mention of the significance of the speed of sound.

The simple fact is that fluids behave quite differently when they move, or when bodies move through them, at speeds below and at speeds above the speed at which sound travels in that fluid. This virtually means that in order to understand modern flight we have to study **two** subjects – flight at speeds below that at which sound travels in air, and flight at speeds above that speed – in other words, flight at **subsonic** and flight at **supersonic** speeds.

To complicate things still further, the airflow at speeds near the speed of sound, **transonic speeds**, is complex enough to be yet a third subject in its own right. We shall cover these subjects as fully as we can, but we must not let our natural interest in supersonic flight tempt us to try to run before we can walk, and in the early chapters the emphasis will be on flight at subsonic speeds, though we shall point out from time to time where we may expect to find differences at supersonic speeds.

But first let us have a closer look at the fluid, air, with which we are most concerned.

INVISIBILITY OF THE ATMOSPHERE

Air is invisible, and this fact in itself makes flight difficult to understand. When a ship passes through water we can see the 'bow wave', the 'wash' astern, and all the turbulence which is caused; when an aeroplane makes

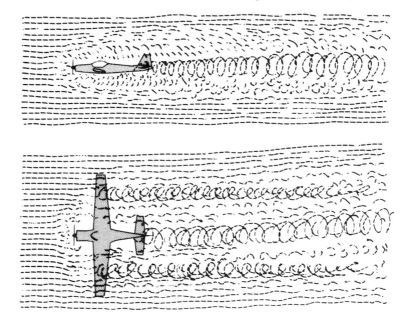

Fig. 2.1 'Seeing the air'

its way through air nothing appears to happen – yet in reality there has been even more commotion (Fig. 2.1).

If only we could see this commotion, many of the phenomena of flight would need much less explanation, and certainly if the turbulence formed in the atmosphere were visible no one could have doubted the improvement to be gained by such inventions as streamlining. After some experience it is possible to cultivate the habit of 'seeing the air' as it flows past bodies of different shapes, and the ability to do this is made easier by introducing smoke into the air or by watching the flow of water, which exhibits many characteristics similar to those of air.

DENSITY OF THE AIR

Another property of air which is apt to give us misleading ideas when we first begin to study flight is its low density. The air feels thin, it is difficult for us to obtain any grip upon it, and if it has any mass at all we usually consider it as negligible for all practical purposes. Ask anyone who has not studied the question what is the mass of air in any ordinary room – you will probably receive answers varying from 'almost nothing' up to 'about 5 kilograms.' Yet the real answer will be nearer 150 kilograms, and in a large hall may be over a metric tonne! Again, most of us who have tried to dive have experienced the sensation of coming down 'flat' onto the surface of the water; since then we have treated water with respect, realising that it has substance, that it can exert forces which have to be reckoned with. We have probably had no such experience with air, yet if we ever try we shall find that the opening of a parachute after a long drop will cause just such a jerk as when we encountered the surface of the water. It is, of course, true that the density of air – i.e. **the mass per unit volume** – is low compared with water (the mass of a cubic metre of air at ground level is roughly 1.226 kg – whereas the mass of a cubic metre of water is a metric tonne, 1000 kg, nearly 800 times as much); **yet it is this very property of air – its density – which makes all flight possible**, or perhaps we should say **airborne** flight possible, because this does not apply to rockets. The balloon, the kite, the parachute, and the aeroplane – all of them are supported in the air by forces which are entirely dependent on its density; the less the density, the more difficult does flight become; and for all of them flight becomes impossible in a vacuum. So let us realise the fact that, however thin the air may seem to be, it possesses the property of density.

INERTIA OF THE AIR

It will now be easy to understand that **air must also possess,** in common with other substances, **the property of inertia and the tendency to obey the laws of mechanics**. Thus air which is still will tend to remain still, while air which is moving will tend to remain moving and will resist any change of speed or direction (First Law); secondly, if we wish to alter the state of rest or uniform motion of air, or to change the direction of the airflow, we must apply a force to the air, and the more sudden the change of speed or direction and the greater the mass of air affected, the greater must be the force applied (Second Law); and, thirdly, the application of such a force upon the air will cause an equal and opposite reaction upon the surface which produces the force (Third Law).

PRESSURE OF THE ATMOSPHERE

As explained in Chapter 1, the weight of air above any surface produces a pressure at that surface – i.e. a force of so many newtons per square metre of surface. **The average pressure at sea-level due to the weight of the atmosphere is about 101 kN/m^2**, a pressure which causes the mercury in a barometer to rise about 760 mm. This pressure is sometimes referred to as 'one atmosphere', and high pressures are then spoken of in terms of 'atmospheres'. The higher we ascend in the atmosphere, the less will be the weight of air above us, and so the less will be the pressure.

DECREASE OF PRESSURE AND DENSITY WITH ALTITUDE

The rate at which the pressure decreases is much greater near the earth's surface than at altitude. This is easily seen by reference to Fig. 2.2; between sea-level and 10 000 ft (3480 m) the pressure has been reduced from 1013 mb to 697 mb, a drop of 316 mb; whereas for the corresponding increase of 10 000 ft between 20 000 ft (6096 m) and 30 000 ft (9144 m), the decrease of pressure is from 466 mb to 301 mb, a drop of only 165 mb; and between 70 000 ft (21 336 m) and 80 000 ft (24 384 m) the drop is only 17 mb.

This is because air is compressible; the air near the earth's surface is compressed by the air above it, and as we go higher the pressure becomes less, the air becomes less dense, so that if we could see a cross-section of the atmosphere it would not appear homogeneous – i.e. of uniform density – but it would become thinner from the earth's surface upwards, the final change from atmosphere to space being so gradual as to be indistinguishable. In this respect air differs from liquids such as water; in liquids there is a definite dividing line or surface at the top; and beneath the surface of a liquid the pressure increases in direct proportion to the depth because the liquid, being practically incompressible, remains of the same density at all depths.

TEMPERATURE CHANGES IN THE ATMOSPHERE

Another change which takes place as we travel upwards through the lower layers of the atmosphere is the gradual drop in temperature, a fact which unhappily disposes of one of the oldest legends about flying – that of Daedalus and his son Icarus, whose wings were attached by wax which melted because he flew too near the sun. In most parts of the world, the atmospheric temperature falls off at a steady rate called the **lapse rate** of about $-6.5°C$ for every 1000 metres increase in height up to about 11 000 metres. Above 11 000 metres, the temperature remains nearly constant until the outer regions of the atmosphere are reached. The portion of the atmosphere below the height at which the change occurs is called the **troposphere**, and the portion above, the **stratosphere**. The interface between the two is called the **tropopause**. The lapse rate and the height of the tropopause vary with latitude. In

Speed of sound m/s	Density kg/m³	Press. kN/m²	Temp. °K	Height above sea-level km	Thousands of feet	Temp. °C	Press. mb	Relative density ρ/ρ₀	Relative pressure p/p₀
299	0.040	2.5	222	25					
298	0.047	3.0	221	24	80	−52	28	0.036	0.028
298	0.055	3.5	220	23	75	−54	35	0.046	0.035
297	0.065	4.0	219	22					
296	0.076	4.7	218	21	70	−55	45	0.059	0.044
295	0.089	5.5	217	20	65	−56	57	0.075	0.056
295	0.104	6.5	217	19					
295	0.122	7.6	217	18	60	−56	72	0.095	0.071
295	0.142	8.8	217	17	55	−56	92	0.121	0.091
295	0.166	10.4	217	16					
295	0.195	12.1	217	15	50	−56	117	0.153	0.115
295	0.228	14.2	217	14	45	−56	148	0.194	0.145
295	0.267	16.6	217	13					
295	0.312	19.4	217	12	40	−56	188	0.247	0.186
295	0.365	22.7	217	11	35	−54	239	0.311	0.236
299	0.414	26.5	223	10	Cirrus clouds				
304	0.467	30.8	230	9	30	−44	301	0.375	0.298
308	0.526	35.7	236	8	Mt. Everest 25	−34	377	0.449	0.372
313	0.590	41.1	243	7					
317	0.660	47.2	249	6	20	−25	466	0.533	0.460
321	0.736	54.0	256	5	High Cumulus clouds 15	−15	572	0.629	0.565
325	0.819	61.7	262	4	Mt. Blanc				
329	0.909	70.1	269	3	10	−5	697	0.739	0.688
333	1.007	79.5	275	2	Stratus and Nimbus clouds				
337	1.112	89.9	282	1	5	5	843	0.862	0.832
					Snowdon				
340	1.225	101.3	288	Sea level		15	1013	1.000	1.000

Fig. 2.2   The international standard atmosphere
Based on the US Standard Atmosphere, 1962, which was prepared under the
sponsorship of NASA, the USAF and the US Weather Bureau.

Arctic regions, the rate of temperature change is lower, and the stratosphere does not start until around 15 500 m. The temperature in the stratosphere varies between about $-30°C$ at the equator to $-95°C$ in the Arctic. In temperate regions such as Europe the temperature in the stratosphere is around $-56.5°C$.

For aircraft performance calculations, it is normal practice to use a standard set of conditions called the **International Standard Atmosphere (ISA)**. This defines precise values of lapse rate, height of the tropopause, and sea-level values of temperature, pressure and density. For temperate regions the ISA value of the lapse rate is $-6.5°C$ per 1000 m, the tropopause is at 11 km, and the sea-level values of pressure and temperature are 101.325 kN/m^2, and $15°C$ respectively.

Modern long- and medium-range airliners cruise in or very close to the stratosphere, and supersonic airliners such as Concorde fly in the stratosphere well above the tropopause. When piston-engined aircraft first started to fly in the stratosphere, conditions were very uncomfortable for the crew. The low density and pressure meant that oxygen masks had to be worn, and at temperatures of $-56°C$, even the heavy fur-lined clothing was barely adequate. Nowadays, the cabins of high-flying airliners are pressurised, and the air is heated, so that the passengers are unaware of the external conditions. Nevertheless, above every seat there is an emergency oxygen mask to be used in the event of a sudden failure of the pressurisation system.

Despite the low external air temperature in the stratosphere, supersonic aircraft have the problem that surface friction heats the aircraft up during flight, so means have to be provided to keep the cabin cool enough.

EFFECT OF TEMPERATURE AND PRESSURE ON DENSITY

Although air is not quite a 'perfect gas' it does obey the gas law within reasonable limits, so we can say that

$$\frac{p}{\rho} = RT$$

It is often convenient to express the density in terms of the ratio of the density at some height ρ to the value at standard sea-level conditions ρ_0. This ratio ρ/ρ_0 is usually denoted by the Greek letter σ, and is called the **relative density**.

EXAMPLE
If the temperature at sea-level in the temperate ISA is $15°C$, and the lapse rate is $6.5°C$ per km, find the temperature and density at 6 km

altitude where the pressure is 47 200 N/m². The gas constant $R = 287$ J/kg K.

First we must convert all temperatures to absolute (Kelvin). Sea-level temperature is

$$273 + 15 = 288\,\text{K}.$$

Temperature at 6000 m is

$$288 + 6.5 \times 6 = 249\,\text{K}$$

The gas law states that

$$\frac{p}{\rho} = RT \quad \text{or} \quad \rho = \frac{p}{RT}$$

so the density at 6 km is

$$\frac{47\,200}{287 \times 249} = 0.66\,\text{kg/m}^3$$

VISCOSITY

An important property of air in so far as it affects flight is its **viscosity**. This is a measure of the resistance of one layer of air to movement over the neighbouring layer; it is rather similar to the property of friction between solids. It is owing to viscosity that eddies are formed when the air is disturbed by a body passing through it, and these eddies are responsible for many of the phenomena of flight. Viscosity is possessed to a large degree by fluids such as treacle and certain oils, and although the property is much less noticeable in air, it is none the less of considerable importance.

WINDS AND UP-AND-DOWN CURRENTS

Air flows from regions of high pressure to regions where the pressure is lower, and this is the cause of wind, or bodily movement of large portions of the atmosphere. Winds vary from the extensive trade winds caused by belts of high and low pressure surrounding the earth's surface to the purely local gusts and 'bumps', caused by local differences of temperature and pressure. On the earth's surface we are usually only concerned with the horizontal velocity of winds, but when flying the rising convection currents and the corresponding downward movements of the air are also important. The study of winds, of up-and-down convection currents, of cyclones and anti-cyclones, and the weather changes produced by them – all these form the fascinating science of *meteorology*, and the reader who is interested is referred to books on that subject.

In the lower regions of the atmosphere conditions are apt to be erratic; this is especially so within the first few hundred feet. It often happens that as we begin to climb the temperature rises instead of falling – called an **inversion** of temperature. This in itself upsets the stability of the air, and further disturbances may be caused by the sun heating some parts of the earth's surface more than others, causing thermal up-currents, and by the wind blowing over uneven ground, hangars, hills, and so on. On the windward side of a large building, or of a hill, the wind is deflected upwards, and on the leeward side it is apt to leave the contour altogether, forming large eddies which may result in a flow of air near the ground back towards the building or up the far side of the hill, that is to say in the opposite direction to that of the main wind. Even when the surface of the ground is comparatively flat, as on the average airfield, the wind is retarded near the ground by the roughness of the surface, and successive layers are held back by the layers below them – due to viscosity – and so the wind velocity gradually increases from the ground upwards. This phenomenon is called **wind gradient**. When the wind velocity is high it is very appreciable, and since most of the effect takes place within a few metres of the ground it has to be reckoned with when landing.

Quite apart from this wind gradient very close to the ground, there is often also a wind gradient on a larger scale. Generally, it can be said that on the average day the wind velocity increases with height for many thousands of feet, and it also tends to veer, i.e. to change in a clockwise direction (from north towards east, etc.); at the same time it becomes more steady and there are fewer bumps.

AIR SPEED AND GROUND SPEED

But our chief concern with the wind at the present moment is that we must understand that when we speak of the speed of an aeroplane we mean its speed relative to the air, or **air speed** as it is usually termed. Now the existence of a wind simply means that portions of the air are in motion relative to the earth, and although the wind will affect the speed of the aeroplane **relative to the earth** – i.e. its **ground speed** – it will not affect its speed relative to the air.

For instance, suppose that an aeroplane is flying from A to B (60 km apart), and that the normal speed of the aeroplane (i.e. its air speed) is 100 km/h (see Fig. 2.3). If there is a wind of 40 km/h blowing from B

Fig. 2.3

Fig. 2.4   Outward flight   **Fig. 2.5   Return flight**

towards A, the ground speed of the aeroplane as it travels from A to B will be 60 km/h, and it will take one hour to reach B, but the air speed will be 100 km/h (Fig. 2.4). If, when the aeroplane reaches B, it turns and flies back to A, the ground speed on the return journey will be 140 km/h (Fig. 2.5); the time to regain A will be less than half an hour, but the air speed will still remain 100 km/h – that is, the wind will strike the aeroplane at the same speed as on the outward journey. Similarly, if the wind had been blowing across the path, the pilot would have inclined his aeroplane several degrees towards the wind on both journeys so that it would have travelled crabwise, but again, on both outward and homeward journeys the air speed would have been 100 km/h and the wind would have been a headwind straight from the front as far as the aeroplane was concerned.

An aeroplane which encounters a head wind equal to its own air speed will appear to an observer on the ground to stay still, yet its air speed will be high. A free balloon flying in a wind travels over the ground, yet it has no air speed – a flag on the balloon will hang vertically downwards.

All this may appear simple, and it is in fact simple, but it is surprising how long it sometimes takes a student of flight to grasp **the full significance of air speed and all that it means**. There are still pilots who say that their engine is overheating because they are flying 'down wind'! It is not only a question of speed, but of direction also; a glider may not lose height in a rising current of air (it may, in fact, gain height), yet it is all the time descending **relative to the air**. In short, the only true way to watch the motion of an aeroplane is to imagine that one is in a balloon floating freely with the wind and to make all observations **relative to the balloon**.

Ground speed is, of course, important when the aeroplane is changing from one medium to another, such as in taking-off and landing, and also in the time taken and the course to be steered when flying cross-country – this is the science of *navigation*, and once again the student who is interested must consult books on that subject.

The reader may have noticed that we have not been altogether consistent, nor true to the SI system, in the units that we have used for speed; these already include m/s, km/h and knots. There are good reasons for this inconsistency, the main one being that for a long time to come it is likely to be standard practice to use knots for navigational purposes both by sea and by air, km/h for speeds on land, e.g. of cars, while m/s is not only the proper SI unit but it must be used in certain formulae. We shall continue to use these different units throughout the book as and when each is most appropriate, and the important thing to remember is that it is only a matter of simple conversion from one to the other –

1 knot $= 0.514\,\text{m/s} = 1.85\,\text{km/h}$

CHEMICAL COMPOSITION OF THE ATMOSPHERE

We have, up to the present, only considered the physical properties of the atmosphere, and, in fact, we are hardly concerned with its chemical or other properties. Air, however, is a **mixture of gases, chiefly nitrogen and oxygen**, in the proportion of approximately four-fifths nitrogen to one-fifth oxygen. Of the two main gases, nitrogen is an inert gas, but oxygen is necessary for human life and also for the proper combustion of the fuel used in the engine, therefore when at great heights the air becomes thin it is necessary to provide more oxygen. In the case of the pilot, this was formerly done by supplying him with pure oxygen from a cylinder, but in modern high-flying aircraft, the whole cabin is pressurised so that pilot, crew, and passengers can still breathe air of similar density, pressure, temperature, and composition as that to which they are accustomed at ground level, or at some reasonable height. As for the engines, it has always been preferable to provide extra air rather than oxygen, because although the oxygen is needed for combustion the nitrogen provides, as it were, the larger part of the working substance which actually drives the engine. In piston engines the extra air is provided by a process known as **supercharging**, which means blowing in extra air by means of a fan or fans; in jet aircraft the principle is fundamentally the same, though simpler because the engine is in itself a supercharger and this, combined with the 'ram effect' at the higher true speeds achieved at altitude, keeps up the supply of air as necessary.

THE INTERNATIONAL STANDARD ATMOSPHERE

The reader will have realised that there is liable to be considerable variation in those properties of the atmosphere with which we are concerned – namely, temperature, pressure, and density. Since the performance of engine, aeroplane, and propeller is dependent on these three factors, it will be obvious that the **actual** performance of an aeroplane does not give a true basis of comparison with other aeroplanes, and for this reason the **International Standard Atmosphere** has been adopted. The properties assumed for this standard atmosphere in temperate regions are those given in Fig. 2.2. If, now, the actual performance of an aeroplane is measured under certain conditions of temperature, pressure, and density, it is possible to deduce **what would have been** the performance under the conditions of the Standard Atmosphere, and thus it can be compared with the performance of some other aeroplane which has been similarly reduced to standard conditions.

THE ALTIMETER

The instrument normally used for measuring height is the **altimeter**, which was traditionally merely an aneroid barometer graduated in feet or metres instead of millimetres of mercury or millibars. As a barometer it will record the **pressure** of the air, and since the pressure is dependent on the **temperature** as well as the height, it is only possible to graduate the altimeter to read the height if we assume certain definite conditions of temperature. If these conditions are not fulfilled in practice, then the altimeter cannot read the correct height. Altimeters were at one time graduated on the assumption that the temperature remained the same at all heights. We have already seen that such an assumption is very far from the truth, and the resulting error may be as much as 900 m at 9000 m, the altimeter reading too high owing to the drop in temperature. Altimeters are now calibrated on the assumption that the temperature drops in accordance with the International Standard Atmosphere; this method reduces the error considerably, although the reading will still be incorrect where standard conditions do not obtain.

As a barometer the altimeter will be affected by changes in the pressure of the atmosphere, and therefore an adjustment is provided, so that the scale can be sèt (either to zero or to the height of the airfield above sea-level) before the commencement of a flight, but in spite of this precaution, atmospheric conditions may change **during** the flight, and it is quite possible that on landing on the same airfield the altimeter will read too high if the pressure has dropped in the meantime, or too low if the pressure has risen.

Although it is convenient to use SI units for calculations and numerical examples, it should be remembered that knots and feet are still the internationally approved units for speed and altitude respectively when dealing with aircraft operations. If you ever sit at the controls of an aircraft you will almost certainly find that the altimeter is calibrated in feet. We shall therefore use feet (ft) when referring to this instrument.

Modern altimeters are much more sensitive than the old types; some, instead of having one pointer, may have as many as three, and these are geared together like the hands of a clock so that the longest pointer makes one revolution in 1000 ft, the next one in 10 000 ft, and the smallest one in 100 000 ft. Unfortunately this has sometimes proved 'too much of a good thing,' and accidents have been caused by pilots mistaking one hand for another, and the more modern tendency is to show the height level in actual figures in addition to one or more pointers. But, like a sensitive watch, the altimeter is of little use unless it can be made free from error, and can be read correctly. In the modern types, if the pilot sets the altimeter to read zero height, which he can do simply by turning a knob, a small opening on the face of the instrument discloses the pressure of the air at that height – in other words, the reading of the barometer. Conversely – and this is the important point – if, while in the air, he finds out by radio the barometric pressure at the airfield at which he wishes to land, he can adjust the instrument so that this pressure shows in the small opening, and he can then be sure that his altimeter is reading the correct height above that aerodrome, and that when he 'touches down' it will read zero. The altimeter may be used in this way for instrument flying and night flying, that is to say when the height above the ground in the vicinity of the aerodrome is of vital importance, but for ordinary cross-country flying during the day it may be preferable to set the **sea-level atmospheric pressure** in the opening. Then the pilot will always know his height above sea-level and can compare this with the height, as shown on the map, of the ground over which he is flying. If this method is used, instead of the altimeter reading zero on landing, it will give the height of the aerodrome above sea-level. There is, however, a snag in this method in that the sea-level atmospheric pressure varies from place to place and so different pilots may set their altimeters differently, thereby increasing the risk of collision; for this reason modern practice for flying above 3000 ft is to set the altimeter to **standard sea-level pressure of 1013.2 mb,** which means, in effect, that all the altimeters may be reading the incorrect height, but that only aircraft flying at the **same** height can have the **same** altimeter readings. Above 3000 ft heights are referred to in terms of **flight levels** (or hundreds of feet), e.g. FL 35 is 3500 ft, FL 40 is 4000 ft, then FL 45, 50, 55 and so on. Increases in flight levels are in fives because of the **quadrantal system** which determines the height at which the pilot must fly for specific compass headings.

The question of altimeter setting has long been a matter for controversy among pilots – and even among nations.

In recent years there have been radical changes in aircraft instruments and displays. Instead of individual instruments there may now be a computer-screen type of display, but the altimeter display still looks quite similar to the traditional instrument. The information on which the display is based may also still be the external pressure, but there are now alternative, more accurate, height-reading devices such as radio or radar altimeters.

The reader who is particularly interested in altimeters and other instruments is referred to *Aircraft Instruments* by E. H. J. Pallett, a companion volume in the Introduction to Aeronautical Engineering Series.

AIR RESISTANCE OR DRAG

Whenever a body is moved through air, or other viscous fluid, there is produced a definite resistance to its motion. In aeronautical work this resistance is usually referred to as **drag.**

Drag is the enemy of flight, and efforts must be made to reduce the resistance of every part of an aeroplane to a minimum, provided strength and other essential factors can be maintained. For this reason many thousands of experiments have been carried out to investigate the problems of air resistance; in fact, in this, as in almost every branch of the subject, our knowledge is founded mainly on the mean result of accumulated experimental data. Nowadays, however, there are increasingly more accurate theoretical methods of estimating drag.

In experimental work it is usual to allow the fluid to flow past the body rather than to move the body through the fluid. The former method has the great advantage that the body is at rest, and consequently the measurement of any forces upon it is comparatively simple. Furthermore, since we are only concerned with the **relative** motion of the body and the fluid, the true facts of the case are fully reproduced provided we can obtain a flow of the fluid which would be as steady as the corresponding motion of the body through the fluid.

WIND TUNNELS

Many experiments are carried out on models in wind tunnels. There are several types of tunnel, but probably the most commonly used is the closed working-section, closed-return type shown in Figs 2.6 and 2B. The model is placed in the narrow working section and air enters through a contraction. The contraction makes the airflow speed in the working section more uniform, and also higher than in the rest of the circuit. Having a high flow speed in the whole circuit would increase the energy losses due to friction. The term closed-return refers to the fact that the air flows round in a complete circuit.

As an alternative, open circuit tunnels are sometimes used. In these, only the working section contraction and fan sections are required as illustrated in Fig. 2.7. The air is simply sucked in from atmosphere through the contraction to the working section, and then exhausted back into the atmosphere, rather like a large vacuum cleaner. The advantage

Fig. 2A   A small open jet wind tunnel
This type of tunnel is useful for teaching purposes, as the model is readily accessible. Though not often used for aeronautical applications these days, the open jet tunnel has found some favour for road vehicle aerodynamic testing.

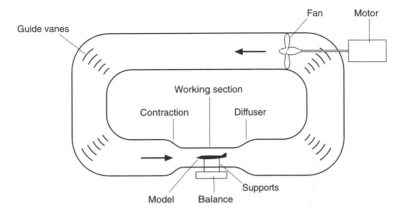

Fig. 2.6   A closed working-section closed return wind tunnel

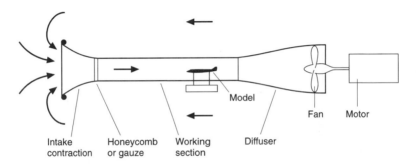

Fig. 2.7   A simple open return tunnel

of the open-return type of tunnel is that it takes up much less space, and costs less than a closed-return type. The principal disadvantages are that dust is drawn in, and the flow may be sensitive to external disturbances. Also, the pressure in the working section must be lower than atmospheric, since the air is drawn in from atmosphere and speeded up. This means that any small leaks around the working section will pull in a jet of air. This type of tunnel is frequently used in college and university laboratories.

Another type of tunnel that is sometimes employed is the open jet type illustrated in Figs 2A and 2.8. In this type of tunnel the working

---

Fig. 2B   Wind tunnel with return circuit (opposite)
(By courtesy of the Lockheed Aircraft Corporation, USA)
The Lockheed-Georgia subsonic closed-throat wind tunnel. Length of centre line 238 m. Mechanical balances measure lift, drag, side force and pitching, rolling and yawing moments. 6710 kW electric motor drives fan of 12 m diameter.

Fig. 2.8   An open jet tunnel

section is not enclosed, which gives it its main advantage, the fact that the flow in the working section is less constrained by the presence of walls. For teaching purposes, the open working section is useful, and this type of tunnel is also popular for automotive aerodynamic studies where the effects of wall constraint are less predictable than for aircraft. There are other types of tunnel such as the slotted wall, but let us not confuse ourselves with such subtleties at this stage.

Other common types of tunnel are of course the supersonic and transonic, but these are described later in the book, after compressible flow has been explained.

WIND-TUNNEL BALANCES

To measure the forces exerted by the airflow, the model is normally mounted on to a balance which may be a mechanical or an electrical type. The older mechanical force balance is rather like a simple weighing machine of the type used to weigh babies or patients in a hospital. As illustrated in Fig. 2.9, the force to be measured (such as the lift on the model) is applied via a series of levers and pivots or flexures to an arm upon which is placed a jockey weight. The moment of the applied force is balanced by moving the jockey weight along the arm. The position of the jockey weight therefore indicates the magnitude of the force. The further along the arm the jockey weight has to be moved, the greater is the force being balanced. The same principle is used to measure horizontal drag or thrust forces, as shown in Fig. 2.9. One arm has to be used for each of the three forces and three moments that make up the six 'components' illustrated in Fig. 2.10. In small college tunnels it is normal to have only three arms which measure the three most important components, lift, drag, and pitching moment. The design of the levers and pivots is very complicated because it is important that a

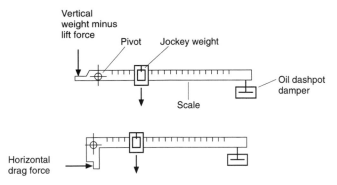

Fig. 2.9   A simple balance

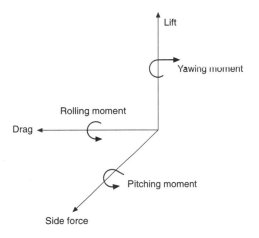

Fig. 2.10   The six 'components'

change in the vertical lift force does not affect the arm that is supposed to measure only the horizontal drag force. The balance unit which is large is situated outside the tunnel, and the model is attached to it by means of a number of rods or wires (see Fig. 2A). Originally the jockey weights were moved by hand, but nowadays they can either be moved automatically by a servo electric motor, or are left fixed, with the force of the arm being measured by an electrical force transducer. A force transducer is an instrument that produces an electrical output that is proportional to the applied force. This brings us to the second form of balance, the electronic type. The electronic force balance consists of a carefully machined block of metal that is attached to the model at one end and to a supporting structure at the other; this often takes the form of a single rod or 'sting' protruding from the rear of the model. Electrical resistance strain gauges attached to the block produce output voltages proportional to the applied forces. Up to six components can be measured. This type of balance is very small and compact and is normally contained within the model. One potential disadvantage is that there is usually a certain amount of unwanted interference between the different components; changes in lift affect the drag reading, etc. However, the computer-based data acquisition systems to which such balances are invariably attached are able to make corrections and allowances automatically.

SOURCES OF ERROR

Wind tunnel experiments on models, even at subsonic speeds, are liable to three main sources of error when used to forecast full-scale results. These are –

1. **Scale Effect**. It has been found that laws of resistance can be framed which apply well to bodies whose sizes are not very different, but that **these laws become less accurate when there is a great difference in size between the model and the full scale**. A similar effect is noticed **when the velocity of the model test differs appreciably from the full-scale velocity**.

Corrections can be applied which allow for this 'scale effect' and enable more accurate forecasts to be made. Readers who are interested will find an explanation of scale effect, and of the advantages of the compressed air tunnel, if they refer to Appendix 2. They will also be introduced to the important term Reynolds Number.

2. **Interference from Wind Tunnel Walls** (Fig. 2.11). The second error is due to the fact that in the wind tunnel the air stream **is confined to the limits of the tunnel**, whereas in free flight the air

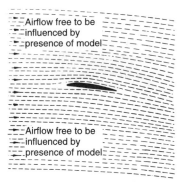

Fig. 2.11   **Interference from wind tunnel walls**

round the aeroplane is, for all practical purposes, unlimited in extent. In this case too corrections can be applied which considerably reduce the error.

3. **Errors in Model.** The smaller the scale of the model, the more difficult does it become to reproduce every detail of the full-scale body, and since very slight changes of contour may considerably affect the airflow, there will always be errors due to the discrepancies between the model and the full-scale body.

OTHER EXPERIMENTAL METHODS

Provided these three limitations are fully realised, and due allowances made for them, wind tunnel results can provide us with some very useful experimental data (Fig. 2C).

In addition to wind tunnel tests, experiments have been performed in the following ways –

1. **On whirling arms,** the model being attached to one end of a long arm, which is rotated about the other end.

Fig. 2C   Model in wind tunnel
(By courtesy of the Lockheed Aircraft Corporation, USA)
Model being prepared for flutter tests in wind tunnel.

2. **By experiments in water instead of air**.
3. **By experiments in actual flight**.

STREAMLINES AND FORM DRAG

Lines which show the direction of the flow of the fluid at any particular moment are called **streamlines**. A body so shaped as to produce the least possible resistance is said to be of **streamline shape**.

We may divide the resistance of a body passing through a fluid into two parts –

1. **Form Drag** or **Pressure Drag**
2. **Skin Friction** or **Surface Friction**

These two between them form a large part of the total drag of an aeroplane – in the high subsonic range, the major part. The sum of the two is sometimes called **profile drag** but this term will be avoided since it is apt to give an impression of being another name for form drag, whereas it really includes skin friction.

The total drag of an aeroplane is sometimes divided in another way in which the drag of the wings or lifting surfaces, **wing drag**, is separated from the drag of those parts which do not contribute towards the lift, the drag of the latter being called **parasite drag**. Figure 2D illustrates an old type of aeroplane in which parasite drag formed a large part of the total. Figure 2E tells another story.

1. **Form Drag**. This is the portion of the resistance which is due to the fact that when a viscous fluid flows past a body, the pressure on the forward-facing part is on average higher than that on the rearward-facing portion. The extreme example of this type of resistance is a flat plate placed at right angles to the wind. The resistance is very large and almost entirely due to the pressure difference between the front and rear faces, the skin friction being negligible in comparison (Fig. 2.12).

Experiments show that not only is the pressure in front of the plate greater than the atmospheric pressure, but that the pressure behind is less than that of the atmosphere, causing a kind of 'sucking' effect on the plate.

It is essential that form drag should be reduced to a minimum in all those parts of the aeroplane which are exposed to the air. This can be done by so shaping them that the flow of air past them is as smooth as possible, and much experimental work has been carried out with this in view. **The results show the enormous advantage to be gained by the streamlining of all exposed parts**; in fact, the figures obtained are so remarkable that they are difficult to believe without a practical

Fig. 2D Parasite drag
A flying replica of the Vickers Vimy bomber which was the first aeroplane to make a
non-stop flight across the Atlantic (in 1919). The extensive use of bracing wires and
struts creates a great deal of parasite drag.

demonstration. At a conservative estimate it can be said that a round
tube has not much more than half the resistance of a flat plate, while if
the tube is converted into the best possible streamline shape the
resistance will be only one-tenth that of the round tube or one-twentieth
that of the flat plate (Fig. 2.13).

The streamline shapes which have given the least resistance at
subsonic speeds have had a **fineness ratio** – i.e. a/b – of between 3 and
4 (see Fig. 2.14), and the maximum value of b should be about one-third

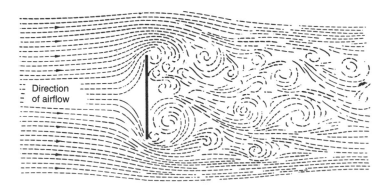

Direction
of airflow

Fig. 2.12   Form drag

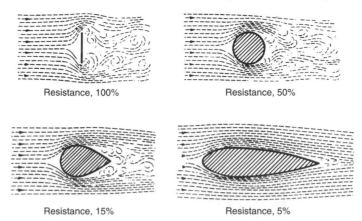

Resistance, 100%   Resistance, 50%

Resistance, 15%   Resistance, 5%

Fig. 2.13   Effect of streamlining

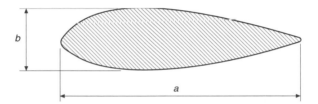

Fig. 2.14   Fineness ratio

of the way back from the nose. These dimensions, however, may vary considerably without increasing the resistance to any great extent.

It should be mentioned that although we now have a fair idea of the ideal shape for any **separate** body, it by no means follows that two bodies of this shape – e.g. a fuselage and a wing – will give the least resistance when **joined together**.

SKIN FRICTION AND BOUNDARY LAYER

Another consideration is that as we decrease the form drag the skin friction becomes of comparatively greater importance.

2. **Skin Friction.** Air is slowed up, and brought to a standstill, very close to a surface. If there is dust on an aeroplane wing before flight, it is

Fig. 2E   Streamlining
(By courtesy of the former British Aircraft Corporation)
Concorde 002 in contrast to the Vickers Vimy of Fig. 2D.

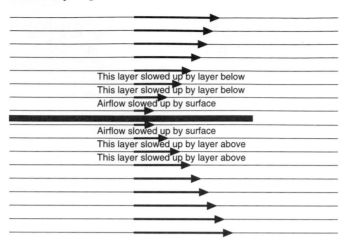

This layer slowed up by layer below
This layer slowed up by layer below
Airflow slowed up by surface

Airflow slowed up by surface
This layer slowed up by layer above
This layer slowed up by layer above

Fig. 2.15   Skin friction

usually still there after flight. **The layers of air near the surface retard the layers farther away** – owing to the friction between them, i.e. the **viscosity** – and so **there is a gradual increase in velocity as the distance from the surface increases.** The distance above the plate in which the velocity regains a value close to that of the free stream may be no more than a few millimetres over a wing.

The layer or layers of air in which the shearing action takes place, that is to say between the surface and the full velocity of the airflow, is called the **boundary layer.** Owing to the great importance of skin friction, and necessity of keeping it within reasonable limits, particularly at high speed, much patient research work has been devoted to the study of the boundary layer.

Now the boundary layer, like the main airflow, may be either **laminar** or **turbulent** (Figs 2.16 and 2F), and the difference that these two types of flow make to the total skin friction is of the same order as the effect of streamlining the main flow. It has been stated that if we could ensure a laminar boundary layer over the whole surface of a wing the skin friction would be reduced to about one-tenth of its value.

The turbulent layer is characterised by high frequency eddies superimposed on the average velocity at each distance from the surface, while in the laminar case the 'layers' of air flow smoothly over each other. The turbulent layer, other factors being equal, has a much higher degree of shear at the surface, and it is this which causes the skin friction to be much higher than it is for the laminar boundary layer. A smooth surface encourages a laminar layer, although other factors such as viscosity and the flow speed are also important. A smooth surface is also

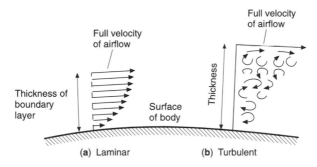

Fig. 2.16   Laminar and turbulent boundary layer
Thickness of layer greatly exaggerated.

important when the boundary layer is turbulent as in this case the skin friction is reduced by a high degree of surface finish, although it remains considerably above the laminar value.

The usual tendency is for the boundary layer to start by being laminar near the leading edge of a body, but there comes a point, called the **transition point, when the layer tends to become turbulent and thicker (Fig. 2.17). As the speed increases the transition point**

Fig. 2F   Investigating the boundary layer
(By courtesy of the former British Aircraft Corporation, Preston)
Wool tufts on model of a Lightning fuselage.

Fig. 2.17   **Transition point**

tends to move further forward, so more of the boundary layer is turbulent and the skin friction greater.

If this much is understood it will be obvious that the purpose of much research work has been to discover how the transition point moves forward, and how its movement can be controlled so as to maintain laminar flow over as much of the surface as possible.

But a further complication is that the behaviour of the boundary layer is very dependent on scale effect; this affects its thickness, whether it is laminar or turbulent, and how soon it separates from the surface at a stall; both the ordinary stall and the shock stall. This will be discussed later.

RESISTANCE FORMULA

Experiments show that, within certain limitations, it is true to say that the **total resistance** of a body passing through the air is dependent on the following factors –

(*a*) **The shape of the body.**
(*b*) **The frontal area of the body.**
(*c*) **The square of the velocity.**
(*d*) **The density of the air.**

Of these the velocity squared law is probably not strictly true at any speeds and is definitely untrue at very low and very high speeds: when the speeds are low it is truer to say that the resistance is proportional to the velocity, while the problem is complicated at extremely high speeds by the fact that the air may be compressed.

It is sometimes thought that the air is compressed in front of a body which is moving quite slowly through the air. We know that air is compressible, and is also to a large extent elastic, which means that after being compressed it will tend to return to its former position. But these properties do not come into play at speeds well below the speed of sound; at such speeds air behaves very like an almost incompressible liquid such as water. The passage of sound is, of course, caused by compression in the air, and it is only when speeds are reached in the neighbourhood of the speed of sound, about 340 m/s (661 knots), that appreciable compression of the air begins to take place. High-speed aircraft now fly in this region, and beyond it. It is also interesting to note that the velocity at which sound travels depends on the temperature of the air and becomes appreciably less at high altitudes, and thus the problem of reaching this critical velocity has become an important consideration in high-altitude flying. In such conditions there may be considerable departure from the velocity squared law, but for the low subsonic speeds – say, from 15 to 150 m/s – this law can be taken as accurate enough for practical purposes, so that double the speed means four times the resistance.

As regards frontal area, when we are considering bodies of very different dimensions we must remember the scale effect to which we have already referred; we should also notice that, if we have a **one-fifth scale model** of a body, the frontal area of the full-sized body will be **twenty-five times** that of the model.

The term 'frontal area' means the maximum projected area when viewed in the direction of normal motion, so the frontal area of an aircraft is the maximum cross-sectional area when viewed from the front. In some instances the surface area would be a more sensible area to take – no general rule can be laid down, and the student should remember that the chief object of experiments on resistance is to compare the resistance of bodies of a similar kind. We are not very much concerned with how the resistance of a wing compares with the resistance of a wheel, but we do wish to compare the resistance of wings of different sizes and shapes, and also the resistances of different types of wheels. Therefore, if we choose one method to measure the area of wings and another to measure the area of wheels (as indeed we do), it does not matter very much. It is all a question of **convenience**.

The law of the variation of the resistance with the density of the air is found to be very nearly correct at ordinary densities, and on first thoughts points to the advantages of flight at high altitudes.

Assuming (*a*), (*b*), (*c*) and (*d*) to be true, we can express the result by the following formula for bodies of the same shape –

$$R \propto \rho S V^2$$

or the general formula –

$$R = K \rho S V^2$$

where K is a coefficient depending on the shape of the body and found by experiment; ρ represents the density of the air; S the frontal area of the body; and V the velocity.

The units in this formula will correspond to those adopted in Chapter

1, i.e. the **resistance** (R) will be in **newtons**, the **density** (ρ) in **kilograms per cubic metre**, the **area** (S) in **square metres**, **velocity** (V) in **metres per second** and K, a **coefficient, merely a number**.

We shall therefore write the formula in its simplest form –

Resistance $= K\rho SV^2$

and the student must always remember that the density of the air, ρ, must be measured in kilograms per cubic metre, and that the answer will be in newtons.

THEORETICAL VALUE OF K

When this book was first written, there were no reliable theoretical models for determining the value of K, and all values were based on experimental data. Nowadays there are various analytical or computational methods, although experimentation is still probably more reliable, particularly for complex shapes. With the ever-increasing power of computers, and improvements in numerical techniques, the time will no doubt arrive when the above statement is no longer true.

BERNOULLI'S EQUATION

This equation can be written in many forms, but originally it was given by

$$\frac{p}{\rho} + \frac{V^2}{2} + gz = \text{a constant}$$

where z is the height, p is the pressure, ρ is the density, V is the flow speed, and g is the gravity constant.

The equation is sometimes called Bernoulli's integral, because it is obtained by integrating the Euler momentum equation for the case of a fluid with constant density. Since the equation involves a constant density, it should really only be applied to incompressible fluids. Completely incompressible fluids do not actually exist, although liquids are very difficult to compress. Air is definitely compressible, but nevertheless, airflow calculations using Bernoulli's equation give good answers unless the speed of the flow starts going above about half the speed of sound.

The terms in this equation all have the units of energy per unit mass, and the equation looks temptingly similar to the steady flow energy equation that you will meet if you ever study thermodynamics. The second and third terms do in fact represent kinetic energy per unit mass and potential energy per unit mass respectively. The true energy equation is, however, significantly different, and contains an important extra term, internal energy, which cannot be neglected in compressible airflows. However, whatever Bernoulli's equation is or is not, it remains a useful and simple means for getting approximately correct answers for low-speed flows. Aerodynamicists usually prefer it in the form below:

$$p + \tfrac{1}{2}\rho V^2 + gz = \text{a constant}$$

This is obtained by multiplying the original equation by the density, which makes all of the terms come out in the units of pressure (N/m^2). The last term is usually ignored because changes in height are small in most of the calculations that we perform for airflows around an aircraft, so we write

$$p + \tfrac{1}{2}\rho V^2 = \text{a constant}$$

The first term represents the local pressure of the air, and is called the **static pressure**. The second term $\tfrac{1}{2}\rho V^2$ is associated with the flow speed and is called **dynamic pressure**. For convenience it is sometimes represented by the letter q, but in this book we will use the full expression.

If we ignore the third term, as above, Bernoulli's equation says that adding the first two terms, the static and the dynamic pressure, produces a constant result. Therefore, if the flow is slowed down so that the dynamic pressure decreases, then to keep the equation in balance the static pressure must increase. If we bring the flow to rest at some point, then the pressure must reach its highest possible value, because the dynamic pressure becomes zero. This maximum value is called the **stagnation pressure** because it occurs at a point where the air has stopped or become stagnant. Using the version of Bernoulli's equation above, we can write

$$p + \tfrac{1}{2}\rho V^2 = \text{a constant} = p\,(\text{stagnation}) + 0$$

This gives the important result that

Static pressure + Dynamic pressure = Stagnation pressure

FORCE COEFFICIENTS

So important is the conception of dynamic pressure that instead of writing the resistance formula in the form

$$R = K\rho SV^2$$

we separate out the unit of dynamic pressure $\frac{1}{2}\rho V^2$, which would leave it as

$$R = 2K \times \tfrac{1}{2}\rho V^2 \times S$$

which can be simplified if we use a new coefficient, C, the value of which will be twice the value of K. Thus we arrive at the form

Resistance $= C \cdot \frac{1}{2}\rho V^2 \cdot S \cdot$ (or $C \cdot q \cdot S$.)

Although we have so far applied this formula to resistance only, it is really of far wider application; it can, in fact, be used to represent any force produced by the flow of air, and the reader will be well advised to be sure that he understands just what it means. Therefore, let us sum up the position by saying that **the aerodynamic force** experienced by any body depends on the **shape of the body** (represented by the coefficient C in the formula), **the pressure needed to bring air to rest when it is of the given density and flowing at the given velocity** (represented by $\frac{1}{2}\rho V^2$), and the **frontal area** (represented by S).

Because of the wider applications (to lift, and so on), the coefficient used for drag is usually distinguished by the suffix D and written C_D, thus

$$D = C_D \cdot \tfrac{1}{2}\rho V^2 \cdot S$$

From this formula we can estimate the resistance of bodies moving through the air, provided we know the value of C_D for the particular shape concerned. This is usually found by experiment, but in the absence of more accurate information, the following values may be used –

for a flat plate $\qquad C_D = 1.2$
for a circular tube $\quad\; C_D = 0.6$
for a streamline strut $C_D = 0.06$

EXAMPLE
Find the resistance of a flat plate, 15 cm by 10 cm, placed at right angles to an airflow of velocity 90 km/h. (Assume sea-level air density of 1.225 kg/m³.)

Data: $C_D = 1.2$
$\qquad \rho = 1.225\,\text{kg/m}^3$
$\qquad V = 90\,\text{km/h} = 25\,\text{m/s}$
$\qquad S = 15 \times 10 = 150\,\text{cm}^2 = 0.015\,\text{m}^2$

$\therefore$ Resistance $= C_D \cdot \frac{1}{2}\rho V^2;\, S = 1.2 \times 0.5 \times 1.225 \times 25 \times 25 \times 0.015$

$\qquad\qquad\qquad = \mathbf{6.89\,N}$

AIR SPEED MEASUREMENT

Bernoulli's equation gives rise to a simple method of measuring air speed. You can see that the dynamic pressure is related to the density and the speed, (dynamic pressure $= \frac{1}{2}\rho V^2$), so if we could measure the dynamic pressure and the density we could determine the speed. Fortunately there is a very simple way of measuring the dynamic pressure at least.

If we point a tube directly into the flow of air, and connect the other end to a pressure-measuring device, then that device will read the stagnation pressure. The reason for this is that the tube is full of air and its exit is blocked, so no air can flow down the tube; the oncoming air therefore is brought to rest as it meets the open end of the tube. This type of tube is called a **pitot tube** and provides a means of measuring stagnation pressure. A different result is obtained if we make a hole in the side of a wind tunnel or in the fuselage of an aircraft, and connect this via a tube to a pressure-measuring device. The hole will not impede the flow of air, so the pressure measured will be the local static pressure. A hole used for this purpose is called a **static vent** or tapping. Since static pressure plus dynamic pressure equals stagnation pressure as shown above, it follows that

Stagnation pressure − Static pressure = Dynamic pressure

If instead of connecting the static pressure vent and the pitot tube to two separate pressure-measuring devices, we connect them across one device which measures the **difference** in pressure then, from the above expression, we can see that we will obtain a measurement of the dynamic pressure. Thus we have a simple means of measuring the dynamic pressure, and if only we could find a way of measuring or assessing the density, we could determine the flow speed, but more of this later; let us first concentrate on the measurement of dynamic pressure, which we will show is actually just as important to the pilot as the speed.

The pressure difference measuring device used on aircraft consists of either a diaphragm or a capsule (similar to the type used in an aneroid barometer). The stagnation pressure is applied to one side, and the static pressure is applied to the other. The resulting deflection of the diaphragm can then either be amplified through a series of levers to cause a dial pointer to move, as on older mechanical devices, or can be used to produce a proportional electrical output to be fed into an appropriate electronic circuit. This instrument thus gives a reading that is proportional to the dynamic pressure, but as we shall see, it forms the basis of the **air speed indicator**.

Fig. 2.18   Concentric pitot-static tube

The pitot tube and static pressure hole are located at a suitable convenient position of the aircraft. The location of the static tapping is very important because it is essential to choose a position where the local static pressure is about the same as that in the free stream away from the aircraft. We need, therefore, to find a place where the flow speed is about the same as that in the free stream, and also is not too sensitive to change in the direction that the aircraft is pointing. The pitot and static holes are normally heated to avoid icing at low temperatures.

As an alternative to using separate pitot and static tubes, it is possible to use a combined device called the pitot-static tube which is illustrated in Fig. 2.18. The pitot-static tube consists of two concentric tubes. The inner one is simply a pitot tube, but the outer one is sealed at the front and has small holes in the side to sense the static pressure. The pitot-static tube is a very convenient device, and by mounting it on the wingtips or the nose it can be arranged so that it is well clear of interference from the flow around the aircraft. Pitot-static tubes are always used for accurate speed measurement on prototypes, but for civil and private aircraft separate pitot and static tappings are normally used. Pitot-static tubes are frequently used in wind tunnels.

THE AIR SPEED INDICATOR

As stated above, the pitot and static tube combination provides a means of measuring the dynamic pressure. It does not tell us the speed directly, but we can work out the speed if we know the density. Until recently, there was no simple method of measuring density, so all that could be done was to use the dynamic pressure-measuring device described above, and mark on the dial the speed that this dynamic pressure would correspond to at standard sea-level air density. This instrument is called the **Air Speed Indicator (ASI)**. You will see that since the instrument is calibrated assuming the standard sea-level value of air density, it does not give the true speed, unless the aircraft is flying at a height where the density just happens to be equal to the standard sea-level value.

Nowadays, there are devices which can measure the true air speed, but the air speed indicator described above is still an important item on any instrument panel. This is because the lift and other forces on the aircraft are dependent on dynamic pressure, and the air speed indicator gives a reading which is directly related to dynamic pressure. For example, if the dynamic pressure is too small, the wings will not be able to generate enough lift to keep the aircraft in steady level flight. The value of the dynamic pressure and hence the indicated speed at which this occurs will always be the same whatever the height. The pilot just has to remember to keep above this minimum indicated speed. If the pilot had only a true speed indicator, he would have to know what the minimum speed was at every height.

AIR SPEED INDICATOR CORRECTIONS

The speed indicated by the air speed indicator is called the **Instrument Indicated Air Speed (IIAS)**. There are several sources of error in this reading. Firstly, the instrument itself may not have been calibrated correctly, or may be suffering from some wear. This error is called instrument error. By recalibrating the instrument it is possible to determine what the correction should be at every indicated speed. The speed corrected for instrument error is called the **Indicated Air Speed (IAS)**. There will also be errors due to the positioning of the pitot and static tubes on the aircraft. It is virtually impossible to find a position where the static pressure is always exactly the same as the pressure in the free airstream away from the aircraft. To determine the correction for such position errors, the aircraft can be flown in formation with another aircraft with specially calibrated instruments. Once the position error correction has been applied, the speed is known as the **Calibrated Air Speed (CAS)** or the **Rectified Air Speed (RAS)**. Finally, for any aircraft that can fly faster than about 200 mph, it is necessary to apply a correction for compressibility, since Bernoulli's equation only applies to low-speed effectively incompressible flows.

After all the corrections have been applied, the resulting speed is called the **Equivalent Air Speed (EAS)**. Once the equivalent air speed has been obtained it is quite easy to estimate the **True Air Speed (TAS)** which is required for navigation purposes. For a light piston-engined aircraft, the corrections will be relatively small, and for simple navigational estimates, the pilot can assume that the speed read from his instrument the **IIAS** is roughly the same as the equivalent air speed **EAS**. The procedure for calculating the true air speed is as follows.

Suppose that at 6000 m, the air speed indicator reads 204 knots, i.e. 105 m/s. Ignoring any instrument or position errors, this means that the pressure on the pitot tube is the same as would be produced by a speed of 105 m/s at standard sea-level density of 1.225 kg/m³; but this pressure is $1/2\rho V^2$, i.e. $1/2 \times 1.225 \times 105 \times 105$.

Now according to the International Standard Atmosphere the air density at 6000 m is 0.66 kg/m³, and if the true air speed is V m/s, then the pressure on the pitot tube will be $1/2 \times 0.66 \times V^2$, which must be the same as

$$1/2 \times 1.226 \times 105 \times 105$$

so

$$0.66 V^2 = 1.226 \times 105^2$$

or

$$V = \sqrt{1.226/0.66} \times 105 = 1.36 \times 105$$
$$= 142.8 \, \text{m/s}.$$

Thus the indicated air speed is 105 m/s, and the true air speed approximately 143 m/s.

Note that, expressed in symbols, the true air speed (TAS) is the equivalent air speed (EAS) divided by the square root of the relative density (σ), or

$$TAS = EAS \sqrt{\rho}$$

A similar calculation for 12 200 m will reveal the interesting result that at that height the true air speed is slightly more than double the indicated air speed!

We have said that the instrument **indicated** air speed **may sometimes be** more useful to the pilot that the true air speed. For purposes of navigation, however, he must estimate his speed over the ground, and with traditional navigation methods, he must first determine the **true** air speed, then make corrections to allow for the speed of the atmospheric wind relative to the ground. The true air speed can be determined using the procedure above, but to do this we need to know the air density or the relative density. This can be obtained by using the altimeter reading, and tables for the variation of relative density in the ISA. In practice, as an alternative to calculations, the pilot can use tables showing the relationship between true air speed and indicated air speed at different heights in ISA conditions.

For high-speed aircraft, the indicated air speed has to be corrected for compressibility, and to do this we need a further instrument, one that indicates the speed relative to the local speed of sound. This instrument is called a **Mach meter**. However, we are again trying to run before we can walk, and we will leave the treatment of compressible flow to later chapters.

Nowadays navigation has been revolutionised by the introduction of ground-based radio, and satellite navigation systems which can give very accurate indications of position and speed relative to the ground. Despite these advances, however, aspiring pilots still have to learn the traditional methods of navigation in order to qualify for their licence. Old instruments like the air speed indicator and the altimeter are simple and reliable, and will not break down in the event of an electrical failure or a violent thunderstorm. Even on the most advanced modern airliners, an old mechanical air speed indicator and pressure altimeter are fitted, and will continue to work even if all the electrical systems have failed.

THE VENTURI TUBE

One of the most interesting examples of Bernoulli's Theorem is provided by the **venturi tube** (Fig. 2.19). This simple but effective instrument is nothing but a tube which gradually narrows to a throat, and then expands even more gradually to the exit. Its effectiveness as a means of causing a decrease of pressure below that of the atmosphere depends very much on the exact shape. If a photograph is taken, or a diagram made, of the flow of air or water through a venturi tube, it will be observed that the streamlines are closest together at the throat, and this gives an unfortunate impression that the fluid has been compressed at this point. Such an impression is the last thing we want to convey, and what we would like to be able to do is to show a cine film, or, better

Fig. 2.19   Flow through venturi tube

still, an actual experiment, which would make it quite clear that **while it is true that the streamlines are closer together at the throat, the velocity of flow is also higher**. This is the important point: **the dynamic pressure has gone up** and therefore, in accordance with Bernoulli's principle, **the static pressure has gone down**. If a tube is taken from the throat and connected to a U-tube containing water, the suction will be clearly shown.

An interesting experiment with a venturi tube is to place an ordinary pitot tube (**without a static**) facing the airflow at various positions in the tube. Connect the pitot tube to a U-tube, and leave the other side of the U-tube open to the atmospheric pressure outside the air stream. The pitot tube will record $p + \frac{1}{2}\rho V^2$, i.e. the static pressure in the stream plus the dynamic pressure, and the U-tube will therefore show the difference between this and the atmospheric pressure outside the stream. It will be found that $p + \frac{1}{2}\rho V^2$ **is very nearly constant, whether the pitot tube is placed in the free air stream in front of the venturi, or in the mouth, or the throat, or near the exit.** This is a convincing proof of Bernoulli's Theorem. The air speed increases from mouth to throat and then decreases again to the exit. The air speed increases very nearly in the same proportion as the area of cross-section of the venturi decreases, and this suggests **that there is little or no change in the density of the air.** Even more convincing evidence that the density does not change is provided by the flow of water through a venturi tube; the pattern of flow and the results obtained are very similar to those in air, and we know that water is for all practical purposes incompressible.

There are many practical examples of the venturi tube in everyday life, but there is no need to quote them, because we have sufficient examples in flying to illustrate this important principle. The choke tube in a carburettor is one; a wind tunnel is another, the experiments usually being done in the high-speed flow at the throat, and the air speed at this point is often measured by a single static hole in the side of the tunnel. A small venturi may be fitted inside a larger one, and the suction at the throat of the small venturi is then sufficient to drive gyroscopic instruments. But best example of all is the suction on top of an ordinary aerofoil section as explained in the next chapter.

Before reading this see if you can answer the following questions. If you can do so you have probably understood most of this chapter and you may proceed with confidence, but you should also try the numerical examples in Appendix 3.

CAN YOU ANSWER THESE?

1. Why is it difficult to find out the exact height of the atmosphere using a pressure altimeter?
2. What is meant by the density of air?
3. Which falls off more rapidly with height, the density or the pressure?
4. Why does an altimeter have an adjustment so that it can be set before each flight?
5. What are the chief differences between the atmosphere at sea-level and at 30 000 ft, or 10 000 m?
6. What do you understand by the term 'streamline shape'?
7. Distinguish between 'indicated' air speed and 'true' air speed.
8. What is the significance of $\frac{1}{2}\rho V^2$?
9. What is 'position error'?
10. What is the difference between the 'troposphere' and the 'stratosphere'?
11. What is the meaning of (*a*) subsonic, and (*b*) supersonic speeds?
12. What does the symbol '*q*' stand for?

For solutions see Appendix 5.
For numerical examples on air and airflow see Appendix 3.

CHAPTER 3

AEROFOILS – SUBSONIC SPEEDS

So far we have only considered the resistance, or drag, of bodies passing through the air. In the design of aeroplanes it is our aim to reduce such resistance to a minimum. We now come to the equally important problem of how to generate a force to lift or support the weight of an aircraft.

In the conventional aeroplane this is provided by **wings, or aerofoils, which are inclined at a small angle to the direction of motion, the necessary forward motion being provided by the thrust of a rotating airscrew, or by some type of jet or rocket propulsion.** These aerofoils are usually slightly curved, but in the original attempts to obtain flight on this system flat surfaces were sometimes used.

LIFTING SURFACES

If air flows past an aerofoil, a flat plate or indeed almost any shape that is inclined to the direction of flow, we find that the pressure of air on the top surface is reduced while that underneath is increased. This difference in pressure results in a net force on the plate trying to push it both upwards and backwards. In the case of a simple flat plate, you might imagine that the net force would act at right angles to the plate. This is not so, because there is also a tangential force caused by the different pressures that act on the small leading and trailing edge face areas. This tangential force though small, is by no means negligible. Rather surprisingly, the pressure at the leading edge is normally very low, and at small angles of inclination, the tangential force will act in the direction shown in Fig. 3.2. The reasons for the low pressure at the leading edge will be shown later. Note, that although the tangential force may be directed towards the front of the plate, the resultant of the

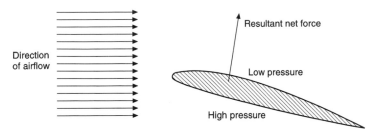

Fig. 3.1   Resultant force on an aerofoil due to pressure difference

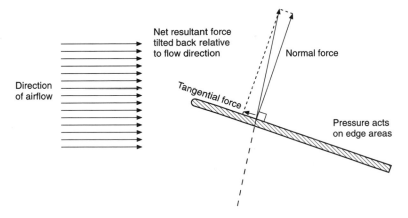

Fig. 3.2   Forces due to pressure differences in a flat plate

tangential and normal forces must always be tilted back relative to the local flow direction.

LIFT AND DRAG

The resultant or net force on the lifting surface may be conveniently split into two components relative to the airflow direction as follows –

1. The component at right angles to the direction of the airflow, called **LIFT** (Fig. 3.3).

2. The component parallel to the direction of the airflow, called **DRAG** (Fig. 3.3)

The use of the term 'lift' is apt to be misleading, for under certain conditions of flight, such as a vertical nose dive, it may act horizontally, and cases may even arise where it acts vertically downwards.

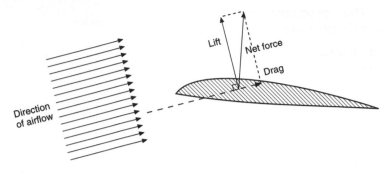

Fig. 3.3   Lift and drag shown for the case of a descending aircraft

AIRFLOW AND PRESSURE OVER AEROFOIL

It was soon discovered that a much greater lift, especially when compared with the drag, could be produced by using a curved surface instead of a flat one, and thus the modern **aerofoil** was evolved. The curved surface had the additional advantage that it provided a certain amount of thickness which was necessary for structural strength.

Experiments have shown that the air flows over an aerofoil (Fig. 3.4) much more smoothly than over a flat plate.

In Fig. 3.4, which shows the flow of air over a typical aerofoil, the following results should be noticed –

1. **There is a slight upflow before reaching the aerofoil.**
2. **There is a downflow after passing the aerofoil.** This downflow should not be confused with the downwash produced by the trailing vortices as described later.
3. **The air does not strike the aerofoil cleanly on the nose, but actually divides at a point just behind it on the underside.**

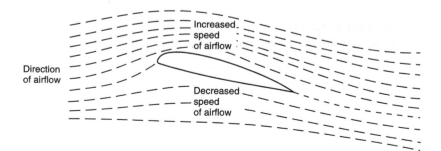

Fig. 3.4   Airflow over an aerofoil inclined at a small angle

4. The streamlines are closer together above the aerofoil where the pressure is decreased.

This last fact is at first puzzling, because, as in the venturi tube, it may lead us to think that the air above the aerofoil is compressed, and that therefore we should expect an increased pressure. The explanation is that the air over the top surface acts as though it were passing through a kind of bottleneck, similar to a venturi tube, and that therefore its velocity must increase at the narrower portions, i.e. at the highest points of the curved aerofoil.

The increase in kinetic energy due to the increase in velocity is accompanied by a corresponding decrease in static pressure. This is, in fact, an excellent example of Bernoulli's Theorem.

Another way of looking at it is to consider the curvature of the streamlines. In order that any particular particle of air may be deflected on this curved path, a force must act upon it towards the centre of the curve, so that it follows that the pressure on the outside of the particle must be greater than that on the inside; in other words, the pressure decreases as we move down towards the top surface of the aerofoil. This point of view is interesting because it emphasises the importance of curving the streamlines.

CHORD LINE AND ANGLE OF ATTACK

It has already been mentioned that the angle of inclination to the airflow is of great importance. On a curved aerofoil it is not particularly easy to define this angle, since we must first decide on some straight line in the aerofoil section from which we can ensure the angle to the direction of the airflow. Unfortunately, owing to the large variety of shapes used as aerofoil sections it is not easy to define this **chord line** to suit all aerofoils. Nearly all modern aerofoils have a convex under-surface; and the chord must be specially defined, although it is usually taken as **the line joining the leading edge to the trailing edge. This is the centre in the particular case of symmetrical aerofoils.**

We call the angle between the chord of the aerofoil and the direction of the airflow the angle of attack (Fig. 3.5).

This angle is often known as **the angle of incidence**; that term was avoided in early editions of this book because it was apt to be confused with the riggers' angle of incidence, i.e. the angle between the chord of the aerofoil and some fixed datum line in the aeroplane. Now that aircraft are no longer 'rigged' (in the old sense) there is no objection to the term angle of incidence; but by the same token there is no objection either to angle of attack, many pilots and others have become

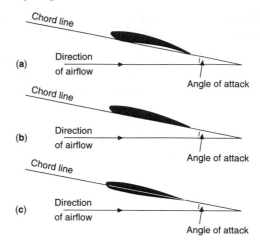

Fig. 3.5   Chord line and angle of attack
(*a*) Aerofoil with concave undersurface.
(*b*) Aerofoil with flat undersurface.
(*c*) Aerofoil with convex undersurface.

accustomed to it; it is almost universally used in America, and so we shall continue to use it in this edition.

[*Note.* If we wish to be precise we must be careful in the definition of the term 'angle of attack', because, as has already been noticed, the direction of the airflow is changed by the presence of the aerofoil itself, so that the direction of the airflow which actually passes over the surface of the aerofoil is not the same as that of the airflow at a considerable distance from the aerofoil. We shall consider the direction of the airflow to be that of the air stream at such a distance that it is undisturbed by the presence of the aerofoil.]

LINE OF ZERO LIFT

Now an aerofoil may provide lift **even when it is inclined at a slightly negative angle to the airflow**. And one may well ask how an aerofoil which is inclined at a negative angle can produce lift? The idea seems absurd, but the explanation of the riddle is simply that the **aerofoil is not really inclined at a negative angle**. Our curious chord may be at a negative angle, but the curved surfaces of the aerofoil are inclined at various angles, positive and negative, the net effect being that of a slightly **positive angle**, which produces lift.

If we tilt the nose of the aerofoil downwards until it produces **no lift**, it will be in an exactly similar position to that of a flat plate placed

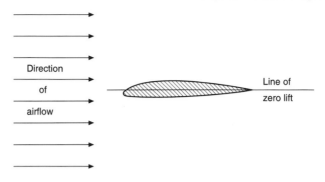

Fig. 3.6   **Line of zero lift**

edgewise to the airflow and producing no lift, and if we now draw a straight line through the aerofoil **parallel to the airflow** (Fig. 3.6) it will be the inclination of **this line** which settles whether the aerofoil provides lift or not.

Such a line is called the **line of zero lift** or **neutral lift line**, and would in some senses be a better definition of the chord line, but it can only be found by wind tunnel experiments for each aerofoil, and, even when it has been found, it is awkward from the point of view of practical measurements.

Nor is it of much significance in practical flight, except perhaps in a dive when the angle of attack may approach the no lift condition.

Note that for an aerofoil of symmetrical shape zero lift corresponds to zero angle of attack.

PRESSURE PLOTTING

As the angle of attack is altered the lift and drag change very rapidly, and experiments show that this is due to changes in the distribution of pressure over the aerofoil. These experiments are carried out by the method known as 'pressure plotting' (Fig. 3.7), in which small holes in the aerofoil surface (*a, b, c, d,* etc.) are connected to glass manometer tubes (*a, b, c, d,* etc.) containing water or other liquid; where there is a suction on the aerofoil the liquid in the corresponding tubes is sucked up, where there is an increased pressure the liquid is depressed. Such experiments have been made both on models in wind tunnels and on aeroplanes in flight, and the results are most interesting and instructive.

The reader is advised to work through Example No. 100 in Appendix 3. In this example the results of an actual experiment are given, together with a full explanation of how to interpret the results. In order to follow

Part plan view of aerofoil

Fig. 3.7   Pressure plotting

through to the end of this example it is necessary to have a knowledge of the lift formula given later in this chapter, but the actual 'pressure plotting' can be done without this. Nowadays, simple glass manometers are seldom used for this purpose except in elementary teaching laboratories. For more serious research work, pressure transducers are employed. These are devices that produce an electrical output that is proportional to the applied pressure. The output from such transducers may then be fed through an interface to a computer. The tedious process of reading the pressures and plotting the distribution can then be left to the computer which may also be used to calculate the resulting lift and pitching moment.

PRESSURE DISTRIBUTION

Figure 3.8 shows the pressure distribution, obtained in this manner, over an aerofoil at an angle of attack of 4°. Two points are particularly noticeable, namely –

1. **The decrease in pressure on the upper surface is greater than the increase on the lower surface.**
2. **The pressure is not evenly distributed,** both the decreased pressure on the upper surface and the increased pressure on the lower surface being most marked over the front portion of the aerofoil.

Both these discoveries are of extreme importance.

The first shows that, although both surfaces contribute, it is the **upper surface,** by means of its decreased pressure, which provides the greater part of the lift; at some angles as much as four-fifths.

The student is at first startled by this fact, as he feels that it is contrary to his ideas of common sense; but, as so often happens, once he

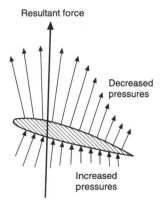

Fig. 3.8 Pressure distribution over an aerofoil

has learnt the truth, he is inclined to exaggerate it, and to refer to the area above the acrofoil as a 'partial vacuum' or even a 'vacuum'. Although, by a slight stretch of imagination, we might allow the term 'partial vacuum', the term 'vacuum' is hopelessly misleading. We find that the greatest height to which water in a manometer is sucked up when air flows over an ordinary aerofoil at the ordinary speeds of flight is about 120 to 150 mm; now, if there were a 'vacuum' over the top surface, the water would be sucked up about 10 m, i.e. 10 000 mm. Or, looking at it another way, suppose that there were a 'vacuum' over the top surface of an aerofoil and that the pressure underneath was increased from 100 kN/m^2 to 120 kN/m^2, then we would have an average upward pressure on the aerofoil of 120 kN/m^2. The actual average lift obtained from an aeroplane wing is from about $\frac{1}{2}$ up to 5 kN/m^2. Take a piece of cardboard of about 100 cm^2, or 1/10th of a square metre, and place a weight of 100 N on it; lift this up and it will give you some idea of the average lift provided by one-tenth of a square metre of aeroplane wing, and the type of load that has to be carried by the skin. You will not want to repeat the experiment with more than 10 000 N on the cardboard!

The reason why the pressure distribution diagram has not been completed round the leading edge is because the changes of pressure are very sudden in this region and cannot conveniently be represented on a diagram. The increased pressure on the underside continues until we reach a point head-on into the wind where the air is brought to rest and the increase of pressure is $\frac{1}{2}\rho V^2$, or q, as recorded on a pitot tube. The point at which this happens is called the **stagnation point,** and its position round the leading edge varies slightly as the angle of attack of the aerofoil is changed but is always just behind the nose on the underside of positive angles of attack. After the stagnation point there is a very sudden drop to zero, followed by an equally sudden change to the decreased pressure of the upper surface, and rather surprisingly on the nose.

CENTRE OF PRESSURE

The second thing that we learn from the pressure distribution diagram – namely, that both decreases and increases of pressure are greatest near the leading edge of the aerofoil – means that if all the distributed forces due to pressure were replaced by a **single resultant force**, this single force would act less than halfway back along the chord. The position on the chord at which this resultant force acts is called the **centre of pressure** (Fig. 3.9). The idea of a centre of pressure is very similar to that of a centre of gravity of a body whose weight is unevenly distributed, and it should therefore present no difficulty to the student who understands ordinary mechanics.

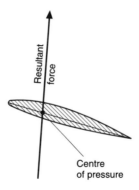

Fig. 3.9   Centre of pressure

To sum up, we may say that we have a **decreased pressure above the aerofoil and an increased pressure below, that the decrease of pressure above is greater than the increase below, and that in both cases the effect is greatest near the leading edge** (Fig. 3.8).

All this is important when we consider the structure of the wing; for instance, we shall realise that the top surface or 'skin' must be held down on to the ribs, while the bottom skin will simply be pressed up against them.

TOTAL RESULTANT FORCE ON AN AEROFOIL

If we add up the distributed forces due to pressure over an aerofoil, and replace it by the total resultant force acting at the centre of pressure, we find that this force is not at right angles to the chord line nor at right angles to the flight direction. Near the tips of swept wings it can sometimes be inclined forward relative to the latter line due to rather complicated three dimensional effects, but over most of the wing, and on average, it must always be inclined backwards, otherwise we would have a forwards component, or negative drag, and hence perpetual motion.

Although the force must on average be inclined backwards relative to the flight direction as in Fig. 3.10 it can often be inclined forwards relative to the chord line normal. Figure 3.10 illustrates the situation. You will see from this figure that there can be a component of the force that is trying to bend the wings forward. This may come as a surprise, because you might have expected that the wings would always be bent rearwards.

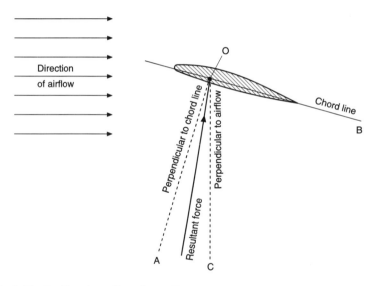

Fig. 3.10   Inclination of resultant force

MOVEMENT OF CENTRE OF PRESSURE

Pressure plotting experiments also show that as the angle of attack is altered the distribution of pressure over the aerofoil changes considerably, and in consequence there will be a **movement of the centre of pressure**. The position of the centre of pressure is usually defined as being a certain proportion of the chord from the leading edge. Figure 3.11 illustrates typical pressure distribution over an aerofoil at

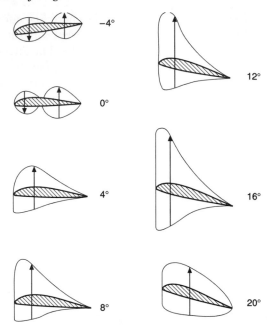

Fig. 3.11   How the lift distribution changes with angle of attack

varying angles of attack. In these diagrams only the lift component of the total pressure has been plotted – the drag component has hardly any effect on the position of the centre of pressure. It will be noticed that at a negative angle, and even at 0°, the pressure on the upper surface near the leading edge is increased above normal, and that on the lower surface is decreased; this causes the loop in the pressure diagram, which means that this portion of the aerofoil is being pushed **downwards**, while the rear portion is being pushed **upwards, so that the whole aerofoil tends to turn over nose first**.

So, **even at the angle of zero lift**, when the upward and downward forces are equal, **there is a nose–down pitching movement on the aerofoil**; as will be seen later this is a matter of considerable significance. Putting it another way, at these negative angles the centre of pressure is a long way back – the only place where we could put one force which would have the same moment or turning effect as the distributed pressure would be a long way behind the trailing edge, in fact at zero lift it could not provide a pitching moment at all unless it were an infinite distance back – **which is absurd**. Perhaps a more sensible way of putting it is to say that there is **a couple** acting on the aerofoil, and **a couple has no resultant** and **has the same moment about any point** (see Chapter 1).

As the angle of attack is increased up to 16°, the centre of pressure gradually moves forward until it is less than one-third of the chord from the leading edge; above this angle it begins to move backwards again.

Now during flight, for reasons which we shall see later, the angle of attack is usually between 2° and 8° and is very rarely below 0° or above 16°. So, **for the ordinary angles of flight, as the angle of attack of the aerofoil is increased, the centre of pressure tends to move forward**.

Lift a poker at its centre of gravity and it will lie horizontal; move the position at which you lift it forwards towards the knob and the rear end of the poker will drop: this is because the centre of lift has moved forwards as compared with the centre of gravity. Therefore if the aerofoil is in balance or 'trimmed' at one angle of attack, so that the resultant force passes through the centre of gravity, then the forward movement of the centre of pressure on the aerofoil as the angle of attack is increased will tend to drop still farther the trailing edge of the aerofoil; in other words, the angle of attack will increase even more, and this will in turn cause the centre of pressure to move farther forward, and so on. This is called **instability**, and it is one of the problems of flight.

If we were to take the wing off a model aeroplane and try to make it glide without any fuselage or tail, we would find that it would either turn over nose first or its nose would go up in the air and it would turn over on to its back. This is because the wing is unstable, and although we might be able to weight it so that it would start on its glide correctly, it would very soon meet some disturbance in the air which would cause it to turn over one way or the other.

Curiously enough, in the case of a **flat plate, an increase of the angle of attack over the same angles causes the centre of pressure to move backwards**; this tends to dip the nose of the plate back again to its original position, and so makes the flat plate **stable**. For this reason it is possible to take a **flat** piece of stiff paper or cardboard, and, after properly weighting it, to make it glide across the room. If it meets a disturbance the centre of pressure moves in such a way as to correct it. Note that the flat piece of paper will only glide if it is weighted so that the centre of gravity is roughly one-third of the chord back from the leading edge. If it is not weighted the centre of pressure will always be in front of the centre of gravity, and this will cause the piece of paper to revolve rapidly.

The unstable movement of the centre of pressure is a disadvantage of the ordinary curved aerofoil, and in a later chapter we shall consider the steps which are taken to counteract it. Attempts have been made to devise aerofoil shapes which have not got this unpleasant characteristic, and it has been found possible to design an aerofoil in which the centre of pressure remains practically stationary over the angles of attack used in ordinary flight. The chief feature in such aerofoils is that the under-surface is convex, and that there is sometimes a reflex curvature towards the trailing edge (see Fig. 3.12); nearly all modern aerofoil sections have in fact got convex camber on the lower surface. Unfortunately, attempts to improve the stability of the aerofoil may often tend to spoil other important characteristics.

Fig. 3.12   Reflex curve near trailing edge

LIFT, DRAG AND PITCHING MOMENT OF AN AEROFOIL

Now the ultimate object of the aerofoil is to obtain the lift necessary to keep the aeroplane in the air; in order to obtain this lift it must be propelled through the air at a definite **velocity** and it must be set at a definite **angle of attack** to the flow of air past it. We have already discovered that we cannot obtain a purely vertical force on the aerofoil; in other words, **we can only obtain lift at the expense of a certain amount of drag**. The latter is a necessary evil, and it must be reduced to the minimum so as to reduce the power required to pull the aerofoil through the air, or alternatively to increase the velocity which we can obtain from a given engine power. Our next task, therefore, is to investigate how much lift and how much drag we shall obtain from different shaped aerofoils at various angles of attack and at various velocities. The task is one of appalling magnitude; there is no limit to the number of aerofoil shapes which we might test, and in spite of thousands of experiments carried out in wind tunnels and by full-scale tests in the air, it is still impossible to say that we have discovered the best-shaped aerofoil for any particular purpose. However, modern theoretical methods make it possible to predict the behaviour of aerofoil sections. Such methods can even be used to design aerofoils to give specified characteristics.

In wind-tunnel work it is the usual practice to measure lift and drag separately, rather than to measure the total resultant force and then split it up into two components. The aerofoil is set at various angles of attack to the airflow, and, the lift, drag and pitching moment are measured on a balance.

The results of the experiments show that within certain limitations the lift, drag and pitching moment of an aerofoil depend on –

(*a*) **The shape of the aerofoil.**
(*b*) **The plan area of the aerofoil.**
(*c*) **The square of the velocity.**
(*d*) **The density of the air.**

Notice the similarity of these conclusions to those obtained when measuring drag, and in both cases there are similar limitations to the conclusions arrived at.

The reader should notice that whereas when measuring drag we considered the **frontal area** of the body concerned, on aerofoils we take the **plan area**. This is more convenient because the main force with which we are concerned, i.e. the lift, is at right angles to the direction of motion and very nearly at right angles to the aerofoils themselves, and therefore this force will depend on the plan area rather than the front elevation. The actual plan area will alter as the angle of attack is changed and therefore it is more convenient to refer results to the **maximum plan area** (the area projected on the plane of the chord), so that the area will remain constant whatever the angle of attack may be. Unfortunately it is customary to use the same symbol S both for the plan area of a wing and the frontal area of any other body.

In so far as the above conclusions are true, we can express them as formulae in the forms –

$$\textbf{Lift} = C_L \cdot \tfrac{1}{2}\rho V^2 \cdot S \text{ or } C_L \cdot q \cdot S$$
$$\textbf{Drag} = C_D \cdot \tfrac{1}{2}\rho V^2 \cdot S \text{ or } C_D \cdot q \cdot S$$
$$\textbf{Pitching Moment} = C_M \cdot \tfrac{1}{2}\rho V^2 \cdot Sc \text{ or } C_M \cdot q \cdot Sc$$

Since the pitching moment is a moment, i.e. **a force × distance**, and since $\tfrac{1}{2}\rho V^2 \cdot S$ represents a force, it is necessary to introduce a length into the equation – this is in the form of the chord, c, measured in metres.

The pitching moment is positive when it tends to push the nose upwards, negative when the nose tends to go downwards – as at zero lift.

The symbols C_L, C_D and C_M are called the **lift coefficient, drag coefficient** and **pitching moment coefficient** of the aerofoil respectively; they depend on the shape of the aerofoil, and they alter with changes in the angle of attack. The air density is represented by ρ in kilograms per cubic metre, S is the plan area of the wing in square metres, V is the air speed, in metres per second, c the chord of the aerofoil in metres; the method of writing the formulae in terms of $\tfrac{1}{2}\rho V^2$, or q, has already been explained in Chapter 2.

AEROFOIL CHARACTERISTICS

The easiest way of setting out the results of experiments on aerofoil sections is to draw curves showing how –

(*a*) **the lift coefficient,**
(*b*) **the drag coefficient,**
(*c*) **the ratio of lift to drag,** and
(*d*) **the position of the centre of pressure,** or **the pitching moment coefficient,**

alter as the **angle of attack** is increased over the ordinary angles of flight.

Typical graphs are shown in Figs 3.13, 3.15, 3.16 and 3.17. These do not refer to any particular aerofoil; they are intended merely to show the type of curves obtained for an ordinary general purpose aerofoil.

In Appendix 1 at the end of the book, tables are given showing the values of C_L, C_D, L/D, position of the centre of pressure, and C_M, for a few well-known aerofoil sections. The reader is advised to plot the graphs for these sections, and to compare them with one another (see examples Nos. 104 to 107 in Appendix 3). In this way he will be enabled to understand much more clearly the arguments followed in the remaining portion of this chapter.

It is much more satisfactory to plot the **coefficients** of lift, drag and pitching moment rather than the total lift, drag and pitching moment, because the coefficients are practically independent of the air density, the scale of the aerofoil and the velocity used in the experiment, whereas the total lift, drag and moment depend on the actual conditions at the time of the experiment. In other words, suppose we take a particular aerofoil section and test it on different scales at different velocities in various wind tunnels throughout the world, and also full-scale in actual flight, we should in each case obtain the same curves showing how the coefficients change with angle of attack.

It must be admitted that, in practice, the curves obtained from these various experiments do not exactly coincide; this is because the theories which have led us to adopt the formula lift $= C_L \cdot \frac{1}{2}\rho V^2 \cdot S$ are not **exactly** true for very much the same reasons as those we mentioned when dealing with drag – for instance, scale effect and the interference of wind-tunnel walls. As a result of the large number of experiments which have been performed, it is possible to make allowances for these errors and so obtain good accuracy whatever the conditions of the experiment.

Now let us look at the curves to see what they mean, for a graph which is properly understood can convey a great deal of information in a compact and practical form.

LIFT CURVE

Let us first see how the **lift coefficient** changes with the angle of attack (Fig. 3.13).

We notice that **when the angle of attack has reached 0° there is already a definite lift coefficient and therefore a definite lift;** this is a property of most cambered aerofoils. A flat plate, or a symmetrical aerofoil, will of course give no lift when there is no angle of attack.

Then **between 0° and about 12° the graph is practically a straight line, meaning that as the angle of attack increases there is a steady increase in the lift;** whereas above 12°, although the lift

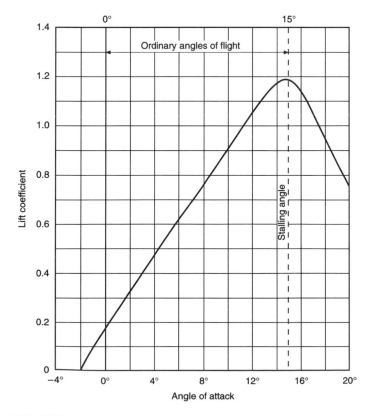

Fig. 3.13 Lift curve

still increases for a few degrees, the increase is now comparatively small and the graph is curving to form a top, or maximum point.

At about 15° the lift coefficient reaches a maximum, and above this angle it begins to decrease, the graph now curving downwards.

STALLING OF AEROFOIL

This last discovery is perhaps the most important factor in the understanding of the why and wherefore of flight. It means that whereas at small angles any increase in the angle at which the aerofoil strikes the air will result in an increase in lift, **when a certain angle is reached any further increase of angle will result in a loss of lift.**

This angle is called the **stalling angle of the aerofoil**, and, rather curiously, perhaps, we find that the **shape** of the aerofoil makes little difference to the **angle** at which this stalling takes place, although it may affect considerably the amount of lift obtained from the aerofoil at that angle.

Now, what is the cause of this comparatively sudden breakdown of lift? The student will be well advised to take the first available opportunity of watching, or trying for himself, some simple experiment to see what happens. Although, naturally, the best demonstration can be given in wind tunnels with proper apparatus for the purpose, perfectly satisfactory experiments can be made by using paper or wooden model aerofoils and inserting them in any fairly steady flow of air or water, or moving them through air or water. The movement of the fluid is emphasised by introducing wool streamers or cigarette smoke in the case of air and coloured streams in the case of water.

Contrary to what might be expected, **the relative speed at which the aerofoil moves through the fluid makes very little difference to the angle at which stalling takes place; in fact, an aerofoil stalls at a certain angle, not at a certain speed.** (It is not correct to talk about the stalling speed of an **aerofoil**, but it will be seen in a later chapter why we talk about the stalling **speed** of an **aeroplane**.) Now what happens? While the angle at which the aerofoil strikes the fluid is comparatively small, the fluid is deflected by the aerofoil, and the flow is of a **streamline** and steady nature (compare Fig. 3.4); but suddenly, when the critical angle of about 15° is reached, there is a complete change in the nature of the flow. The airflow breaks away or separates from the top surface forming vortices similar to those behind a flat plate placed at right angles to the wind; there is therefore very little lift. Some experiments actually show that the fluid which has flowed beneath the under-surface doubles back round the trailing edge and proceeds to flow forward over the upper surface. In short, the streamline flow has broken down and what is called separation or 'stalling' has taken its place, with consequent loss in lift (Fig. 3.14).

Direction of airflow

Fig. 3.14 Stalling of an aerofoil

Anyone who has steered a boat will be familiar with the same kind of phenomenon when the rudder is put too far over, and yachtsmen also experience 'stalling' when their sails are set at too large an angle to the relative wind. There are, in fact, many examples of stalling in addition to that of the aerofoil.

What happens is made even more clear if we look again at the results of pressure plotting (Fig. 3.11). We notice that up to the critical angle considerable suction has been built up over the top surface, especially near the leading edge, whereas when we reach the stalling angle the suction near the leading edge disappears, and this accounts for the loss in lift, because the pressure on other parts of the aerofoil remains much the same as before the critical angle.

Some students are apt to think that all the lift disappears after the critical angle; this is not so, as will easily be seen by reference to either the lift curve or to the pressure plotting diagrams. The aerofoil will, in fact, give some lift up to an angle of attack of 90°. Modern interceptor aircraft are sometimes flown at very high angles of attack during violent manoeuvres, so the upper portion of the graph is nowadays quite important.

The stalling angle, then, is that angle of attack at which the lift coefficient of an aerofoil is a maximum, and beyond which it begins to decrease owing to the airflow becoming separated instead of streamlined.

THE DRAG CURVE

Now for the **drag coefficient curve** (Fig. 3.15). Here we find much what we might expect. The drag is least at about 0°, or even a small negative angle, and increases on both sides of this angle; up to about 6°, however, the increase in drag is not very rapid, then it gradually becomes more and more rapid, especially after the stalling angle when the airflow separates.

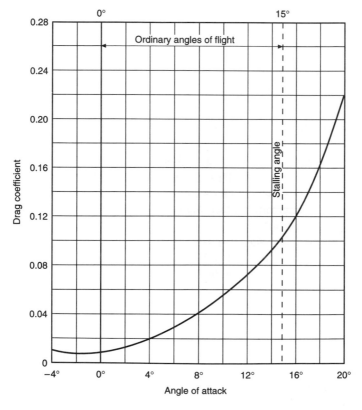

Fig. 3.15   Drag curve

THE LIFT/DRAG RATIO CURVE

Next we come to a very interesting curve (Fig. 3.16), that which shows **the relation between the lift and the drag** at various angles of attack.

In a former paragraph we came to the conclusion that we want as

Fig. 3.16   Lift/drag curve

much lift, but as little drag, as it is possible to obtain from the aerofoil. Now from the lift curve we find that we shall get most lift at about 15°, from the drag curve least drag at about 0°, but both of these are at the extreme range of possible angles, and at neither of them do we really get the best conditions for flight, i.e. **the best lift in comparison to drag, the best lift/drag ratio.**

If the reader has available the lift curve and the drag curve for any aerofoil, he can easily plot the lift/drag curve for himself by reading C_L off the lift curve at each angle and dividing it by the C_D at the same angle. It should be noted that it makes no difference whether we plot L/D or C_L/C_D, as both will give the same numerical value, since $L = C_L \cdot \frac{1}{2}\rho V^2 . S$ and $D = C_D \cdot \frac{1}{2}\rho V^2 . S$.

We find that **the lift/drag ratio increases very rapidly up to about 3° or 4°**, at which angles the lift is nearly 24 times the drag (some aerofoils give an even greater maximum ratio of lift to drag); the ratio then gradually falls off because, although the lift is still increasing, the drag is increasing even more rapidly, until at the stalling angle the lift may be only 10 or 12 times as great as the drag, and after the stalling angle the ratio falls still further until it reaches 0 at 90°.

The chief point of interest about the lift/drag curve is the fact that **this ratio is greatest at an angle of attack of about 3° or 4°**; in other words, it is **at this angle that the aerofoil gives its best all-round results** – i.e. it is most able to do what we chiefly require of it, namely to give as much lift as possible consistent with a small drag.

THE CENTRE OF PRESSURE AND MOMENT COEFFICIENT

Lastly, let us examine the curves (Fig. 3.17) which show **how the centre of pressure moves**, and **what happens to the pitching moment coefficient**, as the angle of attack is increased.

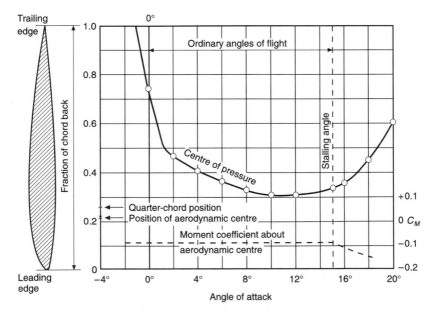

Fig. 3.17   Centre of pressure and moment coefficient curves

The centre of pressure curve merely confirms what we have already learnt about the movement of the centre of pressure on an ordinary aerofoil. After having been a long way back at negative angles, at 0° it is about 0.70 of the chord from the leading edge, at 4° it is 0.40 of the chord back, and at 12° 0.30 of the chord; in other words, the centre of pressure gradually moves forward as the angle is increased over the ordinary angles of flight; and this tends towards instability. After 12° it begins to move back again, but this is not of great importance since these angles are not often used in flight.

It is easy to understand the effect of the movement of the centre of pressure, and for that reason it has perhaps been given more emphasis in this book than it would be in more advanced books on the subject.

It is important to remember that **the pitching moment, and its coefficient**, depend not only on the lift (or more correctly on the resultant force) and on the position of the centre of pressure, but also **on the point about which we are considering the moment – which we shall call the reference point.** There is, of course, **no moment** about the centre of pressure itself – that, after all, is the meaning of centre of pressure – but, as we have seen, **the centre of pressure is not a fixed point.** If we take as our point of reference **some fixed point** on the chord we shall find that the pitching moment – which was already slightly nose-down (i.e. slightly negative) at the angle of zero lift **– increases or decreases as near as matters in proportion to the angle of attack**, i.e. the graph is a straight line, like that of the lift coefficient, over the ordinary angles of flight. About the leading edge, for instance, it becomes more and more nose-down as the angle is increased; but about a point near the trailing edge, although starting at the same slightly nose-down moment at zero lift, it becomes less nose-down, and finally nose-up, with increase of angle (Fig. 3.18).

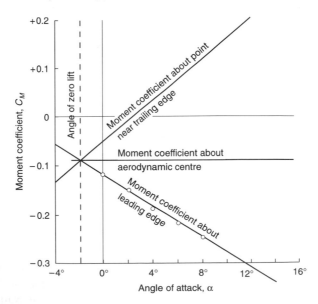

Fig. 3.18 Moment coefficient about different reference points

The reader may be surprised at the increasing nose-down moment about the leading edge, because is not the centre of pressure moving forward? Yes, but the movement is small and the increasing lift has more effect on the pitching moment. The intelligent reader may be even more surprised to hear that an increasing nose-down tendency is a requirement for the pitching stability of the aircraft, for have we not said that the movement of the centre of pressure was an unstable one? Yes, this is a surprising subject, but the answer to the apparent paradox emphasises once again the importance of the point of reference; **in considering the stability of the whole aircraft our point of reference must be the centre of gravity,** and the centre of gravity is always, or nearly always, behind the leading edge of the wing, so the change of pitching moment with angle of attack is more like that about the trailing edge – which is definitely unstable.

AERODYNAMIC CENTRE

But something else of considerable importance arises from the differing effects of different reference points. For if **about the leading edge there is a steady increase,** and **about a point near the trailing edge a steady decrease** in the nose-down pitching moment, there must be **some point** on the chord about which there is **no change in the pitching moment as the angle of attack is increased,** about which the moment remains at the small negative nose-down value that it had at the zero lift angle (Figs 3.17 and 3.18).

This point is called the aerodynamic centre of the wing.

So we have two possible ways of thinking about the effects of increase of angle of attack on the pitching moment of an aerofoil, or later of the whole aeroplane; one is to think of **the lift changing, and its point of application (centre of pressure) changing;** the other is to think of **the point of application (aerodynamic centre) being fixed, and only the lift changing** (Fig. 3.19). Both are sound theoretically; the conception of a moving centre of pressure may sound easier at first, but for the aircraft as a whole it is simpler to consider the lift as always acting at the aerodynamic centre. In both methods we really ought to consider the total force rather than just the lift, but the drag is small in comparison and, for most purposes, it is sufficiently accurate to consider the lift alone.

At subsonic speeds the aerodynamic centre is usually **about one-quarter of the chord from the leading edge,** and theoretical considerations confirm this. In practice, however, it differs slightly according to the aerofoil section, usually being ahead of the quarter-

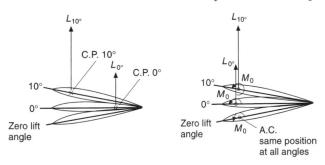

Fig. 3.19   Centre of pressure and aerodynamic centre
Two ways of looking at it.

chord point in older type sections, and slightly aft in more modern low drag types.

The graph in Fig. 3.18 (it can hardly be called a curve) shows how nearly the moment coefficient, about the aerodynamic centre, remains constant on our aerofoil at its small zero-lift negative value of about −0.09. This is further confirmed by the figures of C_M given in Appendix 1 for a variety of aerofoil shapes.

The graphs tell us all we want to know about a particular wing section; they give us the '**characteristics**' of the section, and from them we can work out the effectiveness of a wing on which this section is used.

For example, **to find the lift, drag and pitching moment** per unit span (about the aerodynamic centre) of an aerofoil of this section, of chord 2 metres at 6° angle of attack, and flying at 100 knots at standard sea-level conditions.

From Figs 3.13, 3.15 and 3.17, we find that at 6° –

$$C_L = 0.6$$
$$C_D = 0.028$$
$$C_M = -0.09 \text{ about aerodynamic centre}$$
$$100 \text{ knots} = 51.6 \text{ m/s}$$

Since $\frac{1}{2}\rho V^2$ (or q) is common to the lift, drag and moment formulae, we can first work out its value –

$$q = \tfrac{1}{2}\rho V^2 = \tfrac{1}{2} \times 1.225 \times 51.6 \times 51.6 = 1631 \text{ N/m}^2$$
So **lift** $= C_L \cdot q \cdot S = 0.6 \times 1631 \times 2 = \mathbf{1957\,N}$
 drag $= C_D \cdot q \cdot S = 0.028 \times 1631 \times 2 = \mathbf{91.3\,N}$

pitching moment $= C_M \cdot q \cdot Sc = -0.09 \times 1631 \times 20 \times 2$
$$= \mathbf{-5872\,N\text{-}m}$$

But where is the aerodynamic centre on this aerofoil?

At zero lift there is only a pure moment, or couple, acting on the aerofoil, and since the moment of a couple is the same about any point, this moment, and its coefficient, must be equal to that about the aerodynamic centre, which we shall call $C_{M \cdot AC}$ (sometimes written as C_{MO}), and this by definition will remain the same whatever the angle of attack.

For all practical purposes we can assume that the aerodynamic centre is on the chord line, though it may be very slightly above or below. So let us suppose that it is on the chord line, and at distance x from the leading edge, and that the angle of attack is $\alpha°$ (Fig. 3.20).

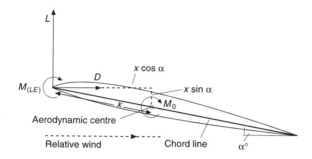

Fig. 3.20   To find aerodynamic centre

The moment about the aerodynamic centre, i.e. $C_{M \cdot AC} \cdot q \cdot Sc$, will be equal to the moment about the leading edge (which we will call $C_{M \cdot LE} \cdot q \cdot Sc$) plus the moments of L and D about the aerodynamic centre; the leverage being $x \cos \alpha$ and $x \sin \alpha$ respectively.

So

$$C_{M \cdot AC} \cdot q \cdot Sc = C_{M \cdot LE} \cdot q \cdot Sc + C_L \cdot q \cdot S \cdot x \cdot \cos \alpha + C_D \cdot q \cdot S \cdot x \cdot \sin \alpha$$

and, dividing all through by $q \cdot S$,

$$C_{M \cdot AC} \cdot c = C_{M \cdot LE} \cdot c + C_L \cdot x \cdot \cos \alpha + C_D \cdot x \cdot \sin \alpha$$
$$\therefore x = c \cdot (C_{M \cdot AC} - C_{M \cdot LE})/(C_L \cos \alpha + C_D \sin \alpha)$$

or, expressed as a fraction of the chord,

$$x/c = (C_{M \cdot AC} - C_{M \cdot LE})/(C_L \cos \alpha + C_D \sin \alpha)$$

But the moment coefficient about the leading edge for this aerofoil at 6° is −0.22 (see Fig. 3.18), and $C_{M \cdot AC}$ is −0.09 (Fig. 3.17),

$$C_L = 0.6, \cos 6° = 0.994, C_D = 0.028, \sin 6° = 0.10$$

So

$$x/c = (-0.09 + 0.22)/(0.6 \times 0.994 + 0.028 \times 1.10)$$
$$= 0.13/(0.60 + 0.003)$$
$$= \mathbf{0.216}$$

which means that **the aerodynamic centre is 0.216 of the chord, or 0.432 metres, behind the leading edge**, and so in this instance is forward of the quarter-chord (0.25) point.

Notice that at small angles, such as 6°, cos α is approx 1, sin α is nearly 0, so we can approximate by forgetting about the drag and saying that $x/c = (C_{M.AC} - C_{M.LE})/C_L$ approx.

About the centre of pressure there is no moment, so

$$(\text{Distance of C.P. from L.E.})/c = -C_{M.LE}/(C_L \cos \alpha + C_D \sin \alpha)$$
$$= -C_{M.LE}/C_L \text{ approx}$$
$$= +0.22/0.60 = \mathbf{0.37}$$

thus confirming the position of the C.P. as shown in Fig. 3.17.

All this has been explained rather fully at this stage; its real significance in regard to the stability of the aircraft will be revealed later.

THE IDEAL AEROFOIL

But what characteristics do we want in the ideal aerofoil section? We cannot answer that question fully until a later stage, but briefly we may say that we need the following –

1. A High Maximum Lift Coefficient. In other words, the top part of the lift curve should be as high as possible. In our imaginary aerofoil it is only about 1.18, but we would like a maximum of 1.6 or even more. Why? Because we shall find that the higher the maximum C_L that we can obtain, the lower will be the landing speed of the aeroplane, and nothing will contribute more towards the safety of an aircraft than that it shall land at a low speed.

2. A Good Lift/Drag Ratio. If we look again at Fig. 3.16, we can see that at a particular angle of attack, the lift/drag ratio of the aerofoil has a maximum value. This ratio does not occur at the angle of attack for minimum drag (Fig. 3.15) or at that for maximum lift coefficient (Fig. 3.13), but somewhere in between. Why is this ratio important? Because to get the smallest possible resistance to motion for a given weight we must operate at this angle of attack, and the higher the maximum lift/drag ratio, the smaller the air resistance that will be experienced.

The real importance of both high lift/drag ratio and high $C_L^{3/2}/C_D$ discussed below will become clearer when we talk about aircraft performance (Chapter 7). Let us just note here that both are important from the point of view of aerofoil design.

3. **A High Maximum Value of** $C_L^{3/2}/C_D$. The power required to propel an aeroplane is proportional to drag $\times$ velocity, i.e. to DV. For an aeroplane of given weight, the lift for level flight must be constant (being equal to the weight). If L is constant, D must vary inversely as L/D (or C_L/C_D). From the formula $L = C_L \cdot \frac{1}{2}\rho V^2 S$ it can be seen that if L, ρ and S are constant (a reasonable assumption), then V is inversely proportional to $\sqrt{C_L}$ (or $C_L^{1/2}$). Thus power required is proportional to DV, which is inversely proportional to $(C_L/C_D) \times C_L^{1/2}$, i.e. to $C_L^{3/2}/C_D$. In other words, the greater the value of $C_L^{3/2}/C_D$, the less the power required, and this is especially important from the point of view of climbing and staying in the air as long as possible on a given quantity of fuel and as we have seen, getting the best economy from a piston-engined aircraft. If the reader likes to work out the value of this fraction for different aerofoils at different angles, and then compares the best value of each aerofoil, he will be able to decide the best aerofoil from this point of view.

4. **A Low Minimum Drag Coefficient.** If high top speed rather than economical cruise is important for an aircraft, then we will need low drag at small lift coefficient, and hence small angles of attack. The drag coefficient at these small angles of attack will be related to the **minimum drag coefficient** (Fig. 3.15).

5. **A Small and Stable Movement of Centre of Pressure.** The centre of pressure of our aerofoil moves between 0.75 and 0.30 of the chord during ordinary flight; we would like to restrict this movement because if we can rely upon the greatest pressures on the wing remaining in one fixed position we can reduce the weight of the structure required to carry these pressures. We would also like the movement to be in the stable rather than in the unstable direction.

Looking at this another way: as we have explained, the moment coefficient at zero lift is slightly negative on most aerofoils, and about the leading edge becomes more nose-down as the angle of attack is increased, and this tends towards stability. Yes, but **our real reference point should be about the centre of gravity** and, as we have also explained, this is usually not only behind the leading edge but also behind the aerodynamic centre, and may even be behind the trailing edge. So, in fact, this is not what we want for stability about the centre of gravity. On the contrary, we would prefer the exact opposite, i.e. a slight positive (nose-up) moment coefficient at zero lift, and this decreasing to negative as the angle of attack is increased. Most aerofoil sections do not give this; but later we shall find that there are means of achieving it.

6. Sufficient Depth to enable Good Spars to be Used. Here we are up against an altogether different problem. Inside the wing must run the spars, or other internal members, which provide the strength of the structure. Now the greater the depth of a spar, the less will be its weight for a given strength. We must therefore try to find aerofoils which are deep and which at the same time have good characteristics from the flight point of view.

COMPROMISES

So much for the **ideal aerofoil**. Unfortunately, as with most ideals, we find that no practical aerofoil will meet all the requirements. In fact, attempts to improve an aerofoil from one point of view usually make it worse from other points of view, until we are forced either to go all out for one characteristic, such as maximum speed, or to take a happy mean of all the good qualities – in other words, to make a **compromise**, and all compromises are bad! It is perhaps well that we have introduced the word 'compromise' at this stage, because **the more one understands about aeroplanes the more one realises that an aeroplane is from beginning to end a compromise.** We want an aeroplane which will do this, we want an aeroplane which will do that, we cannot get an aeroplane which will do both this and that, therefore we make an aeroplane which will half do this and half do that – a 'half and half affair', not a 'regular right down aeroplane'. **And of all the compromises which go to make up that final great compromise, the finished aeroplane, the shape of the aerofoil is the first, and perhaps the greatest, compromise.**

CAMBER

How can we alter the shape of the aerofoil section in an attempt to obtain better results?

The main changes that we can make are in **the curvature, or camber, of the centre line**, i.e. the line equidistant from the upper and lower surfaces, and in **the position of the maximum camber along the chord**.

In symmetrical sections, some of which have been very successful, there is of course no camber of the centre line; other sections have centre line cambers of up to 4 per cent or more of the chord.

Generally speaking we get good all-round characteristics and a smooth stall when the maximum camber is situated about 40 per cent of the chord back. Aerofoils with the maximum camber well forward, say at 15 per cent to 20 per cent of the chord, may have low drag but are apt to have poor stall characteristics – a rather sudden breakaway of the airflow.

The other main features that can be varied are **the maximum thickness, the variation of thickness along the chord** and **the position of maximum thickness** – not necessarily the same as that of maximum camber.

There is considerable variation of maximum thickness (Fig. 3.21)

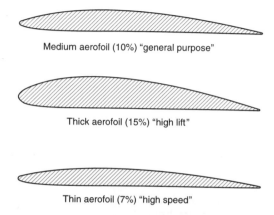

Medium aerofoil (10%) "general purpose"

Thick aerofoil (15%) "high lift"

Thin aerofoil (7%) "high speed"

Fig. 3.21   Thickness of aerofoil sections

even in commonly used aerofoils, from very thin sections with about 6 per cent of the chord to thick sections of 18 per cent or more. Reasonably thick sections are best at low speed, and for pure weight carrying, thin ones for high speed. Remember that it is **the thickness compared with the chord that matters,** thus the Concorde with its large chord of nearly 30 metres achieves a remarkable thickness/chord ratio of 3 per cent.

The greater the camber of the centre line the more convex will be the upper surface, while the lower surface may be only slightly convex, flat or even slightly concave (though this is rare in modern types). Sometimes there is a reflex curve of the centre line towards the trailing edge (Fig. 3.12); this tends to reduce the movement of the centre of pressure and makes for stability.

LAMINAR FLOW AEROFOILS

The attainment of really high speeds, speeds approaching and exceeding that at which sound travels in air, has caused new

problems in the design and in the flying of aeroplanes. **Not the least of these problems is the shape of the aerofoil section.**

Speed is a comparative quantity and the term 'high speed' is often used rather vaguely; in fact, the problem changes considerably at the various stages of high speed. In general, we may say that we have so far been considering aerofoil sections that are suitable for speeds up to 400 or 500 km/h (say 220 to 270 knots) – and we must remember that although these speeds have now been far exceeded they can hardly be considered as dawdling. Furthermore all aeroplanes, however fast they may fly, must pass through this important region. At the other end of the scale are speeds near and above the so-called 'sound barrier', shall we say from 800 km/h (430 knots) up to – well, what you will! The problems of such speeds will be dealt with in later chapters. Notice that there is a gap, from about 500 to 800 km/h (say 270 to 430 knots), and this gap has certain problems of its own; among other things, **it is in this region that the so-called laminar flow aerofoil sections have proved of most value.**

The significance of the boundary layer was explained in Chapter 2. Research on the subject led to the introduction of the **laminar flow or low drag aerofoil,** so designed as to maintain laminar flow over as much of the surface as possible. By painting the wings with special chemicals the effect of turbulent flow in the boundary layer can be detected and so the transition point, where the flow changes from laminar to turbulent, can actually be found both on models and in full-scale flight. Experiments on these lines have led to the conclusion that **the transition point occurs where the airflow over the surface begins to slow down, in other words at or slightly behind the point of maximum suction.** So long as the velocity of airflow over the surface is increasing the flow in the boundary layer remains laminar, so it is necessary to maintain the increase over as much of the surface as possible. The aerofoil that was evolved as a result of these researches (Fig. 3.22) **is thin, the leading edge is more pointed than in the older conventional shape, the section is nearly symmetrical** and, most important of all, **the point of maximum camber (of the centre line) is much farther back than usual,** sometimes as much as 50 per cent of the chord back.

The pressure distribution over these aerofoils is more even, and the airflow is speeded up very gradually from the leading edge to the point of maximum camber.

Fig. 3.22   Laminar flow aerofoil section

There are, of course, snags – and quite a lot of them. It is one thing to design an aerofoil section that has the desirable characteristics at a small angle of attack, but what happens when the angle of attack is increased? As one would expect, the transition point moves rapidly forward! It has been found possible, however, to design some sections in which the low drag is maintained over a reasonable range of angles. Other difficulties are that the behaviour of these aerofoils near the stall is inferior to the conventional aerofoil and the value of $C_L$max is low, so stalling speeds are high. Also, the thin wing is contrary to one of the characteristics we sought in the ideal aerofoil.

But by far the most serious problem has been that wings of this shape are very sensitive to slight changes of contour such as are within the tolerances usually allowed in manufacture. The slightest waviness of the surface, or even dust, or flies, or raindrops that may alight on the surface, especially near the leading edge and, worst of all, the formation of ice – any one of these may be sufficient to cause the transition point to move right up to the position where the irregularity first occurs, thus causing all the boundary layer to become turbulent and the drag due to skin friction to be even greater than on the conventional aerofoil. This is a very serious matter, and led to the tightening up of manufacturing and maintenance tolerances.

Another and more drastic method of controlling the boundary layer is to provide a source of suction near the trailing edge, with the object of **'sucking the boundary layer away'**.

This has the advantage that a much thicker wing section can be used (Fig. 3.23). The practical difficulty is in the power and weight involved

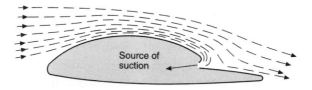

Fig. 3.23   Control of boundary layer by suction

in providing a suitable source of suction. An alternative, and one easier to provide, is a discharge of air backwards from a similar position, thus **'blowing away the boundary layer'** (Fig. 3.24). On aircraft with jet engines one way of doing this is to divert a portion of the jet stream for this purpose.

Fig. 3.24   **Control of boundary layer by pressure**

DESIGN AND NOMENCLATURE OF AEROFOIL SECTIONS

In the early days, in fact until the late 1930s, very few aerofoil shapes were suggested by theory; the usual method was to sketch out a shape by eye, give it a thorough test and then try to improve on it by slight modifications. As a result of this method we had a mass of experimental data obtained under varying conditions in the various wind tunnels of the world. The results were interpreted in different ways, and several systems of units and symbols were used, so that it was difficult for the student or aeronautical engineer to make use of the data available.

It is true that this hit-and-miss method of aerofoil design produced a few excellent sections but it was gradually replaced by more systematic methods. The first step in this direction was to design and test a 'family' of aerofoils by taking a standard symmetrical section and altering the curvature, or camber, of its centre line. An early example of this system was the series beginning with RAF 30, a symmetrical section from which RAF 31, 32, 33, etc., were evolved by curving the centre line in various ways and according to a definite plan. Then RAF 40 was used as the basic section of a new family, and so on. RAF referred to the Royal Aircraft Factory at Farnborough, now the Defence Research Agency (DRA). In Germany similar investigations were made with series named after the Gottingen Laboratory, and in America with the Clark Y series.

Later sections have been based on theoretical calculations but, whatever the basis of the original design, **we still rely on wind tunnel tests to decide the qualities of the aerofoil.**

The naming and numbering of sections has also been rather haphazard. At first the actual number, such as RAF 15, meant nothing except perhaps that it was the 15th section to be tried. But the National Advisory Committee for Aeronautics in America soon attempted to devise **a system whereby the letters and numbers denoting the aerofoil section served as a guide to its main features**; this meant that we could get a good idea of what the section was like simply from its number. Unfortunately the system has been changed from time to time, and this has caused confusion; while the modern tendency to have more figures and letters in a number has resulted in such complication that the student finds it more difficult to get information about the section from the number than he did with some of the earlier ones. However since NACA sections, or slight modifications of them, are now used by nearly every country in the world, the reader may be interested in getting some idea of the systems.

The geometric features that have most effect on the qualities of an aerofoil section are –

(*a*) **the camber of the centre line;**
(*b*) **the position of maximum camber;**
(*c*) **the maximum thickness, and variation of thickness** along the chord;

and, perhaps rather surprisingly –

(*d*) **the radius of curvature of the leading edge;**
(*e*) **whether the centre line is straight, or reflexed** near the trailing edge; and **the angle between the upper and lower surfaces at the trailing edge.**

The NACA sections designed for comparatively low speed aircraft are based on either the four- or five-digit system; laminar flow sections for high subsonic speeds on the 6, 7 or 8 systems (the 6, 7 or 8 being the first figure, not the number of digits).

In each system there are complicated formulae for the thickness distribution, the radius of the leading edge and the shape of the centre line, but we need not worry about these; what is easier to understand is the meaning of the digits or integers, for instance, in the **four-digit** system –

(*a*) the **first digit** gives the **maximum camber** as a percentage of the chord;
(*b*) the **second digit** gives the **position of the maximum camber,** i.e. distance from the leading edge, in **tenths** of the chord;
(*c*) the **third** and **fourth digits** indicate the **maximum thickness** as a percentage of the chord.

Thus NACA 4412 has a maximum camber of 4 per cent of the chord, the position of this maximum camber is 40 per cent of the chord back, and the maximum thickness is 12 per cent of the chord. In a symmetrical section there is of course no camber so the first two digits will be zero; thus NACA 0009 is a symmetrical section of 9 per cent thickness.

Notice that these are all geometric features of the section, but in later systems attempts are made to indicate also some of the aerodynamic characteristics, for instance, in the **five-digit** system –

(*a*) the **'design lift coefficient'** (in tenths) is three-halves of the **first digit**;

(*b*) the **second** and **third digits** together indicate **twice the distance back of the maximum camber,** as a percentage of the chord;

(*c*) and the **last two** once again the **maximum thickness**.

The 'design lift coefficient' is the lift coefficient at the angle of attack for normal level flight, usually at about 2° or 3°.

Most of these sections have a 2 per cent camber, and in fact there is some relationship between the design lift coefficient and the maximum camber which has sometimes led to confusion about the meaning of the first digit; also the point of maximum camber is well forward at 15 per cent, 20 per cent or 25 per cent of the chord (which accounts for the doubling of the second and third digits to 30, 40 or 50). In fact the most successful, and so the most common of these sections, begins with the digits 230, followed by the last two indicating the thickness. Thus NACA 23012, as used on the Britten-Norman Islander (Fig. 4D), has a design lift coefficient of 0.3 (it also has 2 per cent camber), the maximum camber is at 15 per cent of the chord, while the maximum thickness is 12 per cent.

The forward position of the maximum camber in the five-digit sections results in **low drag, but poor stalling characteristics,** which explains why, when these sections are used near the root of a wing, they are often changed to a four-digit one (which gives a smooth stall) near the tip.

It should be noted that **the position of maximum thickness** (not indicated in either of these systems) **is not necessarily the same as that of maximum camber,** and in one British system eight digits were used so that this too could be indicated; two pairs of digits gave the thickness and its position, two other pairs the maximum camber and its position. Figure 3.25 illustrates 1240/0658 based on this system. For a

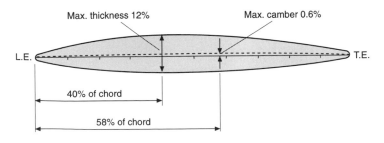

Fig. 3.25   Aerofoil section 1240/0658

symmetrical section the last four figures are omitted since they would all have been zero.

The reader may like to sketch for himself such sections as NACA 4412 and 23012, but he will have to judge the position of maximum thickness by eye.

In the NACA 6, 7 and 8 series, as in nearly all the NACA series, the last two digits again indicate the percentage thickness, but the other figures, letters, suffixes, dashes and brackets become so complicated that it is necessary to refer to tables. Most of these sections are particularly good for high subsonic speeds.

ASPECT RATIO

We have so far only considered aerofoils from the point of view of their cross-section, and we must now consider the effect of the **plan form**. Suppose we have a rectangular wing of 12 m² plan area; it could be of 6 m span and 2 m chord, or 8 m span and 1.5 m chord, or even 16 m span and 0.75 m chord. In each case the cross-sectional shape may be the same although, of course, to a different scale, depending on the chord. Now according to the conclusions at which we have already arrived, the lift and drag are both proportional to the area of the wing, and therefore since all of these wings have the same area they should all have the same lift and drag. Experiments, however, show that this is not exactly true and indicate **a definite, though small, advantage to the wings with larger spans**, both from the point of view of lift and lift/drag ratio.

The ratio span/chord is called **aspect ratio** (Fig. 3.26), and the aspect ratios of those wings which we have mentioned are therefore 3, 5.33 and 21.33 respectively, and the last one, with its 'high aspect ratio', gives the best results (at any rate at subsonic speeds which is what we

Fig. 3.26   Aspect ratio
The area of each wing is 12 m².

are concerned with in this chapter). Why? It is a long story, and some of it is beyond the scope of this book; but the reader has the right to ask for some sort of explanation of one of the most interesting and, in some ways, one of the most important, problems of flight. So here goes!

INDUCED DRAG

Experiments with smoke or streamers show quite clearly that the **air flowing over the top surface of a wing tends to flow inwards** (Fig. 3.27). This is because the decreased pressure over the top surface is less

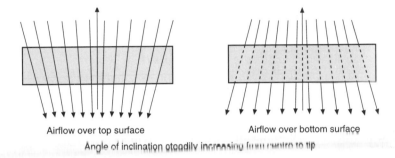

Airflow over top surface   Airflow over bottom surface

Angle of inclination steadily increasing from centre to tip

Fig. 3.27   The cause of trailing vortices

than the pressure outside the wing tip. Below the under-surface, on the other hand, the **air flows outwards,** because the pressure below the wing is greater than that outside the wing tip. Thus there is a continual spilling of the air round the wing tip, from the bottom surface to the top. Perhaps the simplest way of explaining why a high aspect ratio is better than a low one is to say that the higher the aspect ratio the less is the **proportion** of air which is thus spilt and so is ineffective in providing lift – the less there is of what is sometimes called 'tip effect' or 'end effect'.

When the two airflows, from the top and bottom surfaces, meet at the trailing edge they are flowing at an angle to each other and cause **vortices rotating clockwise (viewed from the rear) from the left wing,** and **anti-clockwise from the right wing.** All the vortices on one side tend to join up and form one large vortex which is shed from each wing tip (Fig. 3.28). These are called **wing-tip vortices.**

All this is happening every time and all the time an aeroplane is flying, yet some pilots do not even know the existence of such vortices. Perhaps it is just as well, perhaps it is a case of ignorance being bliss. In earlier editions of this book it was suggested that if only pilots could **see**

Fig. 3.28   Trailing vortices which become wing-tip vortices

the vortices, how they would talk about them! Well, by now most pilots **have seen** the vortices or, to be more correct, the central core of the vortex, which is made visible by the condensation of moisture caused by the decrease of pressure in the vortex (Figs 3A and 3B). These visible (and sometimes audible!) trails from the wing tips should not be confused with the vapour trails caused by condensation taking place in the exhaust gases of engines at high altitudes (Fig. 3C).

Now if you consider which way these vortices are rotating you will realise that there is an upward flow of air outside the span of the wing and **a downward flow of air behind the trailing edge of the wing itself**. This means that the net direction of flow past a wing is pulled downwards. Therefore the lift – which is at right angles to the airflow –

Fig. 3A   Wing-tip vortices
(By courtesy of the former British Aircraft Corporation, Preston)
Vortices on the TSR2; the low pressure at the core of the vortex causes a local condensation fog on a damp day.

Fig. 3B   Rolling up of vortices on TSR2
(By courtesy of the former British Aircraft Corporation, Preston)
A unique demonstration of the rolling up process; the wing tip and flap tip are each shedding vortices that are strong enough to cause condensation, and the pair roll around one another.

is slightly **backwards,** and thus contributes to the drag (Fig. 3.29). **This part of the drag is called induced drag.**

In a sense, **induced drag is part of the lift**; so long as we have lift we must have induced drag, and we can never eliminate it altogether however cleverly the wings are designed. **But the greater the aspect ratio,** the less violent are the wing-tip vortices, and **the less the induced drag.** If we could imagine a wing of infinite aspect ratio, the air would flow over it without any inward or outward deflection, there would be no wing-tip vortices, no induced drag. Clearly such a thing is impossible in practical flight, but it is interesting to note that an aerofoil in a wind tunnel may approximate to this state of affairs if it extends to the wind-tunnel walls at each side, or outside the jet stream in an open jet type of tunnel. The best we can do in practical design is to make the aspect ratio as large as is practicable. Unfortunately a limit is soon reached – **from the structural point of view.** The greater the span,

Fig. 3C   Exhaust trails
(By courtesy of General Dynamics Corporation, USA)
A modified B36 with 6 piston engines and 4 jet engines, and carrying an atomic
reactor as an experiment on the shielding of crews and aircraft components.

the greater must be the wing strength, the heavier must be the structure,
and so eventually the greater weight of structure more than
counterbalances the advantages gained. Again it is a question of
compromise. In practice, aspect ratios for flight at subsonic speeds
vary from 6 to 1 up to about 10 to 1 for ordinary aeroplanes, but
considerably higher values may be found on sailplanes, and even more
in man-powered aircraft, where aerodynamic efficiency must take
precedence over all other considerations (see Fig. 3D) and very low
values for flight at transonic and supersonic speeds (see Fig. 3E).

Fig. 3.29   Induced drag

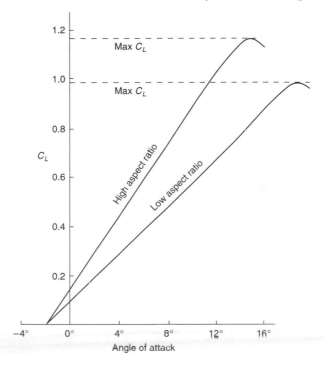

Fig. 3.30   How aspect ratio affects the lift curve

Fig. 3.30 shows how aspect ratio affects the lift curve, not only in the maximum value of C_L but in the slope of the curve, the stalling angle actually being higher with low values of aspect ratio. Notice that the angle of no lift is unaffected by aspect ratio.

The theory of induced drag can be worked out mathematically (in fact Dr. Lanchester worked it out before the Wright Brothers flew), and experiment confirms the theoretical results. The full calculation involved would be out of place in a book of this kind, but the answer is quite simple and the reader may like to know it, especially since it helps to give a clearer impression of the significance of this part of the drag.

The coefficient of induced drag is found to be $C_L{}^2/\pi A$ for elliptical wings, where A is the aspect ratio and C_L the lift coefficient (the value for tapered wings may be 10 per cent to 20 per cent higher, depending on the degree of taper). This means that the actual drag caused by the vortices is $(C_L{}^2/\pi A) \cdot \frac{1}{2}\rho V^2 \cdot S$, but since the $\frac{1}{2}\rho V^2 \cdot S$ applies to all aerodynamic forces, it is sufficient to consider the significance of the coefficient, $C_L{}^2/\pi A$. In the first place, the fact that A is underneath in the fraction confirms our previous statement that the greater the aspect

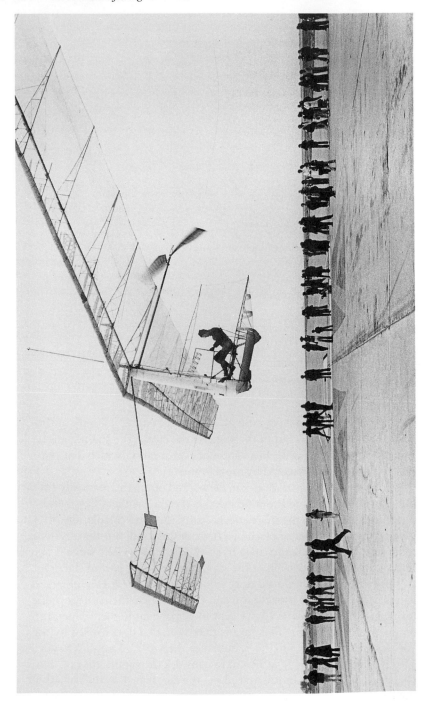

ratio, the less the induced drag; but it tells us even more than this, for it shows that it is a matter of simple proportion: if the aspect ratio is doubled, the induced drag is halved. The significance of the C_L^2 is perhaps not quite so easy to understand. C_L is large when the angle of attack is large, that is to say when the speed of the aircraft is low; so induced drag is relatively unimportant at high speed (probably less than 10 per cent of the total drag), more important when climbing (when it becomes 20 per cent or more of the total) and of great importance for taking off (when it may be as high as 70 per cent of the total). In fact, the **induced drag is inversely proportional to the square of the speed**, whereas all the remainder of the drag is **directly proportional** to the square of the speed.

It is easy to work out simple examples on induced drag, e.g. –

A monoplane wing of area 36 m² has a span of 15 m and chord of 2.4 m. What is the induced drag coefficient when the lift coefficient is 1.2?

$$\text{Aspect ratio} = A = 15/2.4 = 6.25$$

$$\text{Induced drag coefficient} = C_L^2/\pi A = 1.2^2/6.25\pi = 0.073$$

Perhaps this does not convey much to us, so let us work out the actual drag involved, assuming that the speed corresponding to a C_L of 1.2 is 52 knots, i.e. 96 km/h (26.5 m/s), and that the air density is 1.225 kg/m³.

$$
\begin{aligned}
\text{Induced drag} &= (C_L^2/\pi A) \cdot \tfrac{1}{2}\rho V^2 \cdot S \\
&= 0.073 \times \tfrac{1}{2} \times 1.225 \times 26.5^2 \times 36 \\
&= 1130\,\text{N}
\end{aligned}
$$

Let us take it even one step further and find the power required to overcome this induced drag –

$$\text{Power} = DV = 1130 \times 26.5 = 30\,\text{kW (about 40 horse-power)}.$$

This example will help the reader to realise that induced drag is something to be reckoned with; he is advised to work out for himself similar examples, which will be found at the end of the book.

CIRCULATION

An interesting way of thinking about the airflow over wings and wing-tip vortices is the theory of **circulation** as propounded by Lanchester.

Fig. 3D   High aspect ratio (opposite)
(By courtesy of Paul MacCready)
The Gossamer Condor. Flight on one man-power requires a very high value of lift/drag.

The fact that the air is speeded up over the upper surface, and slowed down on the under surface of a wing, can be considered **as a circulation round the wing superimposed upon the general speed of the airflow** (this does **not** mean that particles of air actually travel round the wing). **This circulation is, in effect, the cause of lift.** If we now consider this circulation as slipping off each wing tip, and continuing downstream, we have the wing-tip vortices; and they rotate, as already established, downwards behind the wing and upwards outside the wing tips.

But this is not all. When the wing starts to move, or when the lift is increased, **the wing sheds and leaves behind a vortex rotating in the opposite direction to the circulation round the wing** – sometimes called the **starting vortex** – so there is a complete system of vortices, round the wing, then the wing-tip vortices, and finally the starting vortex. The wing-tip vortices and the starting vortex are gradually damped out with time – owing to viscosity – but the exertion of engine power (which ultimately is what creates the vortices, and so the lift and induced drag) keeps renewing the circulation round the wing, and the wing-tip vortices which result from it.

This is not just a theory; the flow over the wing can be clearly seen in experiments, as can the wing-tip vortices, while the starting vortex is easily demonstrated by starting to move a model wing, or even one's hand, through water. But perhaps the most extraordinary example of the reality of the effect of aspect ratio on circulation and wing-tip vortices is that by clever formation flying of say three or five aircraft, with the centre one leading, and the outer ones with their wing tips just behind the opposite wing tips of the leading aircraft, it is possible to achieve something of the same result (which is illustrated in flying for maximum range) as with an aircraft of three or five times the span! This is hardly a practical proposition for flying across the Atlantic, but it has been illustrated by careful experiment, and geese and other birds used the technique long before we discovered it!

TAPER AND SHAPE OF WING TIPS

In addition to changes of aspect ratio, the plan form of the wing may be tapered from centre to wing tip; this is often accompanied by a taper in

Fig. 3E Low aspect ratio (opposite)
(By courtesy of the Grumman Corporation, USA)
For high-speed flight, the wings of the F14 are swung back producing an aspect ratio of less than 1. For low-speed flight they can be swung forward giving a higher aspect ratio.

Wing tapered in elevation

Wing tapered in plan

Fig. 3.31   Tapered wing

the depth of the aerofoil section (Fig. 3.31) and also by a 'wash-out', or decrease of angle of incidence, towards the wing tip – sometimes too a different aerofoil section is used near the tips. The tapered wing has advantages both from the structural and aerodynamic points of view. This is a feature in which we were slow to accept the teachings of nature, for the wings of most birds have a decided taper. Where the chord is not constant along the span, the numerical value of the aspect ratio is usually taken as the fraction (span/mean chord), or $span^2$/area.

Taper in plan form means a sweepback of the leading edge, or a sweepforward of the trailing edge, or both. Considerable sweepback of the whole wing is sometimes used, but this is usually more for consideration of stability or for very high-speed flight, and discussion of the problem from these points of view is deferred to later chapters.

VARIABLE CAMBER

Many attempts have been made to provide aerofoils with some kind of variable camber so that the pilot might be able to alter his aerofoil from a high-lift type to a high-speed type at will. Owing to the tremendous advantage to be gained by such a device, it is not surprising to find that much ingenuity has been expended, many patents have been taken out, and it is not easy to compare the rival merits of the various slots, flaps, slotted flaps, and so on. Figure 3.32 shows some of the devices with the increase in maximum lift claimed for each, but we must not take these figures as the only guide to the usefulness or otherwise of each device, because there are other points to be considered besides maximum lift. For instance, we may want a good **ratio of maximum** C_L **to minimum** C_D (which indicates **a good speed range**), or an increase in drag as well as lift, the flaps acting as an **air brake**, which may be useful in increasing the gliding angle (explained later). Another important consideration is the simplicity of the device; anything which needs complicated operating mechanism will probably mean more weight, more controls for the pilot to work, something more to go wrong.

FLAPS AND SLOTS

Although there is a large variety of high-lift devices nearly all of them can be classed as either **slots** or **flaps** – or a combination of the two (Fig. 3.32).

Slots may be subdivided into –

(*a*) **Fixed slots.**
(*b*) **Controlled slots.**
(*c*) **Automatic slots.**
(*d*) **Blown slots.**

Flaps may be subdivided into –

(*a*) **Camber flaps.**
(*b*) **Split flaps.**
(*c*) **Slotted flaps.**
(*d*) **Lift flaps.**
(*e*) **Blown flaps.**
(*f*) **Jet flaps.**
(*g*) **Nose flaps.**
(*h*) **Spoilers.**
(*i*) **Lift dumpers.**
(*j*) **Air brakes.**

We can also classify the effects of both slots and flaps on the characteristics of an aerofoil by saying that their use may cause one or more of the following –

High-lift devices	Increase of maximum lift	Angle of basic aerofoil at max. lift	Remarks
Basic aerofoil	–	15°	Effects of all high-lift devices depend on shape of basic aerofoil.
Plain or camber flap	50%	12°	Increase camber. Much drag when fully lowered. Nose-down pitching moment.
Split flap	60%	14°	Increase camber. Even more drag than plain flap. Nose-down pitching moment.
Zap flap	90%	13°	Increase camber and wing area. Much drag. Nose-down pitching moment.
Slotted flap	65%	16°	Control of boundary layer. Increase camber. Stalling delayed. Not so much drag.
Double-slotted flap	70%	18°	Same as single-slotted flap only more so. Treble slots sometimes used.
Fowler flap	90%	15°	Increase camber and wing area. Best flaps for lift. Complicated mechanism. Nose-down pitching moment.

Fig. 3.32   High lift devices
Note. Since the effects of these devices depend upon the shape of the basic aerofoil, and the exact design of the devices themselves, the values given can only be considered as approximations. To simplify the diagram the aerofoils and the flaps have been set at small angles, and not at the angles giving maximum lift.

High-lift devices	Increase of maximum lift	Angle of basic aerofoil at max. lift	Remarks
Double-slotted Fowler flap	100%	20°	Same as Fowler flap only more so. Treble slots sometimes used.
Krueger flap	50%	25°	Nose-flap hinging about leading edge. Reduces lift at small deflections. Nose-up pitching moment.
Slotted wing	40%	20°	Controls boundary layer. Slight extra drag at high speeds.
Fixed slot	50%	20°	Controls boundary layer. Extra drag at high speeds. Nose-up pitching moment.
Movable slat	60%	22°	Controls boundary layer. Increases camber and area. Greater angles of attack. Nose-up pitching moment.
Slat and slotted flap	75%	25°	More control of boundary layer. Increased camber and area. Pitching moment can be neutralized.
Slat and double-slotted Fowler flap	120%	28°	Complicated mechanisms. The best combination for lift; treble slots may be used. Pitching moment can be neutralized.
Blown flap	80%	16°	Effect depends very much on details of arrangement.
Jet flap	60%	?	Depends even more on angle and velocity of jet.

(*a*) **Increase of Lift.**
(*b*) **Increase of Drag.**
(*c*) **Change of Stalling Angle.**
(*d*) **Decrease of Lift.**
(*e*) **Change of Trim.**

SLOTS

If a small auxiliary aerofoil, called a slat, is placed in front of
the main aerofoil, with a suitable gap or slot in between the two
(Fig. 3F), the maximum lift coefficient of the aerofoil may be
increased by as much as 60 per cent (Fig. 3.33). Moreover the
stalling angle may be increased from 15° to 22° or more, not always
an advantage as we shall discover when we consider the problems of
landing. An alternative to the separate slat, simpler but not so effective,
is to cut one or more slots in the basic aerofoil itself, forming as
it were a slotted wing.

The reason behind these results is clearly shown in Fig. 3.34. Stalling
is caused by the breakdown of the steady streamline airflow. On a slotted

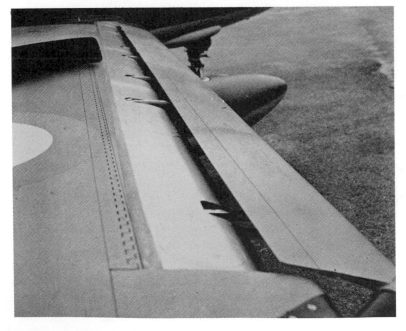

Fig. 3F Leading edge slat and slot
(By courtesy of Fiat Aviazione, Torino, Italy)

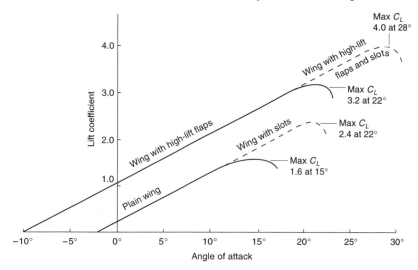

Fig. 3.33   **Effect of flaps and slots on maximum lift coefficient and stalling angle**

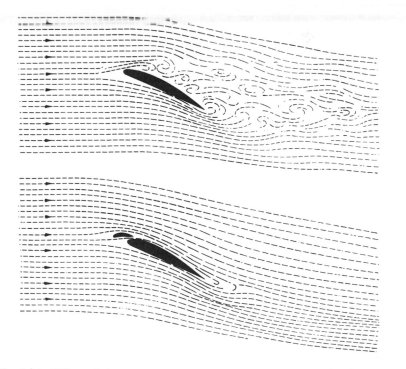

Fig. 3.34   **Effect of slot on airflow over an aerofoil at large angle of attack**

wing the air flows through the gap in such a way as to keep the airflow smooth, following the contour of the surface of the aerofoil, and continuing to provide lift until a much greater angle is reached. Numerous experiments confirm this conclusion. It is, in effect, **a form of boundary layer control** as described earlier.

The extra lift enables us to obtain a lower landing or stalling speed, and this was the original idea. If the slots are permanently open, i.e. **fixed slots**, the extra drag at high speed is a disadvantage, so most slots in commercial use are **controlled slots**, that is to say, the slat is moved backwards and forwards by a control mechanism; and so can be closed for high-speed flight and opened for low speeds. In the early days experiments were made which revealed that, if left to itself, the slat would move forward of its own accord. So **automatic slots** came into their own; in these the slat is moved by the action of air pressure, i.e. by making use of that forward and upward suction near the leading edge. Figure 3.35 shows how the force on the slat inclines forward as the

| 4° | 12° | 16° |

Fig. 3.35   Direction of force on slat at varying angles of attack

stalling angle is reached. The opening of the slot may be delayed or hastened by 'vents' at the trailing or leading edge of the slat respectively (Fig. 3.36), and there may be some kind of spring or tensioning device to prevent juddering, which may be otherwise likely to occur.

Fig. 3.36   Effect of vents on opening of automatic slots

Before leaving the subject of slots – for the time being, at any rate – there are a couple of interesting points which may be worth mentioning. Firstly, the value of the slot in maintaining a smooth airflow over the top surface of the wing can be materially enhanced by blowing air through the gap between slat and wing; this may be called a **blown slot**. Secondly, what might be called **the 'slot idea' may be extended to other parts of the aircraft**. Specially shaped cowlings can be used to smooth the airflow over an engine, and fillets may be used at exposed joints, and other awkward places, to prevent the airflow from becoming turbulent.

FLAPS

The history of flaps is longer, and just as varied, as that of slots. The **plain** or **camber flap** works on the same principle as an aileron or other control surface; it is truly a 'variable camber'. Such flaps were used as early as the 1914-1918 war, and the original idea was the same as with slots, to decrease landing speed with flaps down, and retain maximum speed with flaps up. Their early use was almost exclusively for deck-landing purposes. It seemed at first as though the invention of slots, which followed a few years after that war, might sound the death-knell of flaps. Far from it – if anything it has been the other way round, for flaps have become a necessity on modern aircraft. **Flaps, like slots, can increase lift** – honours are about even in this respect so far as the plain (or camber) flap, or split flap is concerned. But these flaps **can also increase drag** – not, like slots, at high speed when it is not wanted, but at low speed when it is wanted. But the main difference between the effects of flaps and slots is shown in Fig. 3.33; from this it will be seen that whereas **slots merely prolong the lift curve** to higher values of the maximum lift coefficient, when the angle of attack of the main portion of the aerofoil is beyond the normal stalling angle, the high-lift type of flap **increases the lift coefficient available throughout the whole range of angles of attack**.

However it is no longer appropriate to compare the relative merits of slots and flaps because in modern aircraft it is usual to **combine the two** in some form or other; and in this way to get the best of both devices (Fig. 3G). There are a large number of possible combinations, but Fig. 3.32 is an attempt to sum up the main varieties, and to describe the effect they have on the maximum lift coefficient, on the angle of the main aerofoil when maximum lift is obtained, why they improve the lift, what effects they have on the drag, how they affect the pitching moment, and so on.

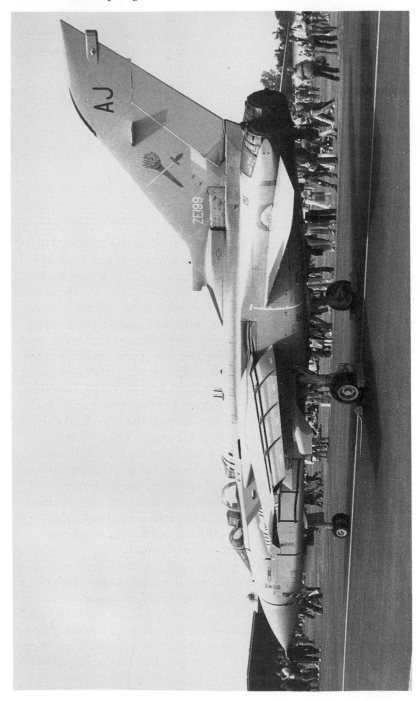

From this figure it will be seen that the simpler flaps such as the **camber flap, split flap** and **single slotted flap** give a **good increase in maximum lift coefficient at a reasonable angle of attack** of the main aerofoil, and therefore a reasonable attitude of the aeroplane for landing; they **also increase drag** which is an advantage in the approach and landing.

The more complicated types such as the **Zap** and **Fowler flap**, and the **double- or treble-slotted flap**, give an **even greater** increase in maximum lift coefficient, but still at a reasonable angle of attack; while the even more complicated combinations of slots and flaps give **yet greater maximum lift coefficients**, but usually at larger angles of attack, and of course at the expense of considerable complication (Fig. 3H).

Blown and **jet flaps** are in a class of their own since they depend on power to produce the blowing, and this may be a serious disadvantage in the event of power failure. **The true jet flap isn't a flap at all**, but simply an efflux of air, or a jet stream in the form of a sheet of air ejected under pressure at or near the trailing edge of the aerofoil. This helps to control the boundary layer, and if the sheet of air can be deflected the reaction of the jet will also contribute directly to the lift.

The **Krueger** and other types of **nose flap** are used mainly for increasing lift for landing and take-off on otherwise high-speed aerofoils.

Spoilers, air brakes, dive brakes, lift dumpers and suchlike are a special category in that their main purpose is to increase drag, or to destroy lift, or both; moreover, they need not necessarily be associated with the aerofoils (Fig. 3K). They are used for various purposes on different types of aircraft; to spoil the L/D ratio and so steepen the gliding angle on high-performance sailplanes and other 'clean' aircraft; to check the speed before turning or manoeuvring; to assist both lateral and longitudinal control; to 'kill' the lift and provide a quick pull-up after landing; and on really high-speed aircraft to prevent the speed from reaching some critical value as in a dive. They will be considered later as appropriate to their various functions.

Fig. 3G Flaps and slats (opposite)
Double-slotted flaps and leading edge slats are used on the Tornado. Because the flaps extend across the entire span, there is no room for ailerons, instead, the slab tailplane surfaces can move differentially as well as collectively, and this 'taileron' serves both for roll and pitch control.

Fig. 3H   Multi-element slotted flaps
Three-element slotted Fowler-type flaps extend rearwards and down as this Boeing
737 prepares to land.

ICING

All these mechanical devices are designed to vary the characteristics of
an aerofoil according to our needs, but there is one important form of
variable camber which is the work of nature and over which we have
little control, namely **the formation of ice**. Brief mention has already
been made of this problem in connection with laminar flow aerofoils,
but the effects of icing may be far wider than this, affecting as they do,
not only the wings, but many parts of the aircraft, the engine intakes and
even the propeller. Icing conditions can arise in various conditions of
atmospheric humidity and temperature, but they become worse **at
regions of low pressure** such as on the upper surface of wings near the
leading edge, and at engine intakes – **just the places where any
alteration of contour can be most serious**. Apart from the effect on
shape, **the actual weight** of accumulation of ice can be considerable
and this alone has been the cause of accidents, as has also the breaking
off of lumps of ice which may enter the engine or strike other parts of
the aircraft.

Many methods both of prevention and cure have been used to combat
the ice problem, and they may be divided into three main categories –
mechanical methods (such as rubber overshoes alternately inflated
and deflated) designed to break up the ice; **heating methods** (using the

Fig. 3K   Speed brakes
Speed brakes on the wings of the last Vulcan bomber (now sadly retired). The cables
of a braking parachute can also just be seen trailing from the rear.

heat of the engines or separate heaters) designed to melt the ice on the
leading edges of the wings, fins, engine intakes, etc.; and the use of
special anti-icing fluid (about the only method suitable for propellers
where it is flung out from the hub). All these, necessary though they
may be, mean extra weight and complication, and some of them absorb
part of the engine power.

CAN YOU ANSWER THESE?

If you understand aerofoils you have broken the back of the problems of
flight – so test yourself with the following questions.

1. How does the pressure distribution over an aerofoil change as we
   increase the angle of attack from negative values to beyond the
   stalling angle?
2. What is meant by the centre of pressure of an aerofoil?
3. Why is it more convenient to speak of the lift coefficient and drag
   coefficient rather than the lift and drag of an aerofoil?
4. What is meant by the aerodynamic centre of an aerofoil section?
5. What do you understand by the stalling angle of an aerofoil? Why
   should one not talk about the stalling speed of an aerofoil?
6. What is aspect ratio and what is its significance?

CHAPTER 4

THRUST

INTRODUCTION

In Chapter 2 we made a study of drag – the force that tries to hold the aeroplane back. In this chapter we shall deal with **thrust** – the force that opposes drag and keeps the aeroplane going forward. **In steady level flight the thrust must be equal to the drag, in order to accelerate the aeroplane it must be greater than the drag, and in climbing it must also be greater than the drag because it will have to support some proportion of the weight.** The actual conditions of balance of the forces will be dealt with in the next chapter; it is sufficient at this stage to realise that we must provide the aeroplane with considerable thrust, and that the performance that we can achieve from the aeroplane will be largely dependent upon the amount of thrust that we can provide.

Once the aeroplane is clear of the ground, the only reasonable way of obtaining thrust is to push air or something else, backwards and to rely on the reaction to push the aeroplane forwards. This is, in fact, what is done, and to save complication the same system is usually used while still on the ground although it must be admitted that it is not always a very efficient system for this purpose. The precise physical process by which this reaction is produced and transmitted to the aircraft depends on the type of propulsion system used.

The **thrust-provider**, of whatever kind it may be, must be supplied with **energy**. This will usually be in the form of **a fuel, which is fed into some kind of 'engine' where, in burning, its chemical energy is changed into thermal energy, which in turn is converted into the mechanical work done in propelling the aeroplane against the drag.** Methods of providing thrust differ only in the way in which these various conversions are effected, and in the efficiency of the conversion, that is to say in the proportion of useful work got out to the energy supplied.

THE ENGINE

This is a book on *Mechanics of Flight*, and it would be out of place to go into any details of thermodynamics or aero-engines. When considering propulsion, however, it is difficult to know where to draw the line, and it is becoming even more difficult as speeds increase and new methods of providing thrust are evolved. The propeller coupled to the reciprocating or piston engine has had a long innings – and it is not out yet. In this system there is a clear-cut boundary line between engine and air-frame in the form of the propeller shaft. The propeller is a problem in mechanics of flight, an aerodynamic problem; the engine is a thermodynamic problem. But the **ramjet**, is almost entirely an aerodynamic problem; in a sense no engine is involved. The various jet and rocket systems come between the two extremes.

THE RAMJET

Let us begin at the end, as it were, by considering for a moment the **ramjet** – originally called an **aero-thermodynamic-duct** (a long name that was quite out of keeping with the extreme simplicity of the device). The ramjet is simply a duct, or tube, of special shape, which faces the airflow caused by the motion of the aircraft through the air (Fig. 4.1). It relies on the forward speed, or ram effect, to collect and compress the air which then flows over some source of heat – the Germans originally tried a coal-burning brazier! – from which it gains energy and so flows out of the duct at a higher speed than that at which it entered. There are no moving parts, neither reciprocating nor rotary, no need for lubrication of any kind, nothing in fact except a tube, a

Fig. 4.1 Principle of the ramjet

source of heat and a fuel. Once the thing is going there is sufficient heat generated to keep the fuel burning without even an ignition device or source of heat.

So simple. What's the snag?

Well, of course, there is one, and it is rather obvious. **The ramjet will only work when it is going,** and moreover it will only work really well when it is going rather fast, say 1600 km/h, 1000 knots, or so! **It gives no thrust at all at no forward speed,** so we cannot start. The only hope for the ramjet, therefore, is to use an auxiliary means of propulsion, possibly rockets, for starting and for low speeds, and to rely on the ramjet when high speed has been obtained – this means, of course, a sacrifice of its greatest virtue, simplicity.

Ramjets are not just an impossible ideal (see Fig. 4A). **They do work, they have been used in practice, and we shall certainly hear more of them.** The real point of introducing the ramjet at this stage, is to show the tendency, which is towards what might be called an aerodynamic engine, or an engine that is part of the aircraft.

Fig. 4A   Ramjet propulsion
The BAC Bloodhound missile used two ramjets: the large tubular shapes with silvery ends which can be seen above and below the body. The smaller tubes are solid-fuel rockets which were used to boost the missile to high speed; they were jettisoned once the ramjets had taken over.

JET PROPULSION

It is but a step backwards from the ramjet to what is commonly called **jet propulsion**. The ramjet, of course, relies on jet propulsion, nothing more nor less, and perhaps in its purest form; but so does a rocket and so does a propeller; they are all jet propulsion, they all provide thrust by reaction to throwing a jet backwards. It has become conventional, however, to think of jet propulsion in terms of gas-turbine-cum-jet, or what is now usually called a **turbojet** (or **turbofan** or **by-pass** engine when the fan is large and acts as a ducted propeller). Figure 4.2 illustrates the pure turbojet system by means of a diagrammatic sketch. **The air is collected, partly by ram effect as in the ramjet, compressed by a compressor, fired and burned as in all types of heat engine, it gains energy and momentum and flows out faster than it came in, as in the ramjet.** The jet velocity is of the order of 600 m/s or about 2250 km/h. On its way out it loses some of its energy and momentum in driving the turbine, which in turn drives the compressor which, etc., etc. This engine is simply a ramjet provided with a means of producing thrust – by the turbine and compressor – when there is little or no forward speed. The rotary parts have, of course, introduced a complication, though nothing to compare with a large reciprocating engine, and the whole engine is but a fraction of the weight of a reciprocating engine giving the same power.

While this system is a little more like an engine than the ramjet, and while many problems of thermodynamics are involved, the jet engine is essentially a part of the aircraft. The flow through and over it, the effects of the flow on the blades of the turbine and compressor, the design of the nozzle through which the gases are ejected – all these are aerodynamic problems involving a knowledge of high-speed flight, Mach Numbers, etc. (see later chapters).

The simple type of jet engine shown in Fig. 4.2 has now been largely

Fig. 4.2   Principle of the turbojet

Fig. 4B   Turbojet propulsion
The hot end of a MiG-29. Two Tumansky R-33D low-bypass turbojet engines, each producing 81.4 kN thrust (with reheat). Note the complex variable area outlet nozzles.

replaced on subsonic aircraft by the more efficient high by-pass and fan jet engines described later in this chapter.

ROCKET PROPULSION

It is not easy to know where the rocket should come in this story. It differs from all the other forms of propulsion in that **it does not rely on the air either to provide the oxygen for the combustion of the fuel, or to provide the mass which is thrown out backwards to produce the thrust.** For these reasons it is fundamentally uneconomical and, except for flights of very short duration, its most promising uses are as an auxiliary to other means of propulsion, as a means of testing research models at high speed, for the launching of missiles and space-craft, and for the strange kind of flight of Fig. 4C. It has, of course, the great advantage that **it needs no air in order to function, and so it can propel aircraft or missiles at great altitudes where the air is very thin and drag is very low, or even outside the atmosphere altogether** where there is no air and no drag – and no lift! It is the obvious, and in fact the only reasonable, means of propulsion for launching satellites, and for starting and stopping and steering space-ships.

Fig. 4C   Rocket propulsion
(By courtesy of Bell Aerospace Division of Textron Inc, USA)
Flying the two-man Pogo from the rear platform is the rocket belt operator; on the
front deck is the rocket technician.

A machine gun, firing bullets backwards, is a form of rocket and an
excellent illustration of the principle; but the usual method is to use the
chemical reactions of two fuels, which may be solid, liquid or gaseous, to
create the heat, and to give energy to the gases formed by the
combustion. The gases are then ejected at high speed through a duct
specially shaped rather like a venturi tube (Fig. 4.3).
In principle it is very simple and the engine itself is light in weight,

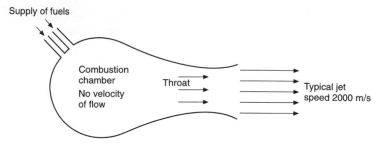

Supply of fuels

Combustion
chamber
No velocity
of flow

Throat

Typical jet
speed 2000 m/s

Fig. 4.3   **Principle of rocket propulsion**

but it requires large quantities of fuel even for short durations and very high temperatures are involved in the combustion chambers. The external shape of the rocket type of engine, and the internal flow, especially through the exit duct, are again aerodynamic problems, while the fuel combustion problem is largely chemical.

Experimental rocket motors have been tested which use the heat generated by a nuclear reactor. However, the radiation problems have so far prevented the use or further development of such motors. Other forms of rocket have been tested in which ionised particles or plasma is accelerated to high speeds by electrostatic or electromagnetic forces.

ENGINE AND PROPELLER PROPULSION

Taking the final step backwards we come to the old and well-tried system of a propeller driven by an internal combustion engine (Figs 4.4 and 4D). Here there is the clear dividing line between the propeller and the engine. We shall consider the propeller in more detail later in this chapter. There are, of course, some problems of airflow even in a reciprocating engine, and we may often use **the ram effect of a forward-facing intake** as an aid to raising the pressure of the incoming air, just as we may use the **backward exhaust as a partial form of jet propulsion**. In the cooling system we may even emulate the ramjet by **collecting the air in ducts, using the otherwise wasted heat of the engine to give it energy, and ejecting it through a venturi tube** – another little bit of jet propulsion. Or, of course, the engine that drives the propeller may itself be a gas turbine, and in this case we can allot almost at will the proportion of the power that we take from the propeller and from the jet respectively; and thus we have, in what is called the **turboprop system** (Fig. 4E), all the advantages – and all the disadvantages – of propellers and jets.

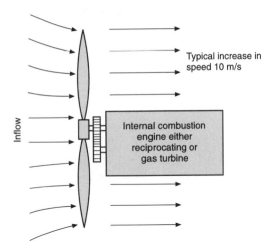

Fig. 4.4   Principle of propeller propulsion

Fig. 4D   Piston engine and propeller
The piston engine and propeller is still the most common arrangement for light general aviation aircraft.

THRUST AND MOMENTUM

All these systems have the common feature that they provide thrust as a result of giving momentum to the air, or other gases. In accordance with

Fig. 4E   Turboprop propulsion
The Lockheed C-130 Hercules with four turboprops. Note the huge wide-chord 'paddle-blade' propellers.

the principles of mechanics the **amount of thrust provided will be equal to the rate at which momentum is given to the air.**

In symbols, **if m kilograms is the mass of air affected per second**, and if it is given an extra velocity of v **metres per second** by the propulsion device, then the **momentum given to the air per second is mv**, so

$$T = mv$$

Now clearly the same thrust could be provided by a large m and a small v, or by a small m and a large v; in other words, by giving a large mass of air a small extra velocity or small mass of air a large extra velocity. Which will work best in practice?

Let us now work a step further in symbols and figures. To make life easy, we will choose to consider the case of a stationary aircraft, perhaps just about to start its take-off run. Now the rate at which kinetic energy is given to m kg of air per second to produce a slipstream or jet speed of v m/s is $\frac{1}{2}mv^2$ watts. So while 1 kg/s given 10m/s has the same rate of momentum change and therefore produces the same thrust as 10 kg/s given 1 m/s, the rate of change of energy of the former is

$$\tfrac{1}{2} \times 1 \times 10^2 = 50 \text{ watts}$$

and of the latter

$\frac{1}{2} \times 10 \times 1^2 = 5$ watts

It is clear therefore that the latter will require less work and that there will be less waste of energy; **in other words, it will be more efficient than the former as a means of producing thrust**.

From this point of view the propeller comes first because it throws back a large mass of air at comparatively low velocity, the jet engine comes next, and the rocket a bad third in that it throws back a very small mass at a very high velocity.

You may wonder therefore why propellers ever started to go out of fashion. The problem is that it is difficult to make them work well at high speed. Since the propeller has a rotational as well as a forward speed, it follows that the blade tips will start to move through the air faster than the speed of sound long before the rest of the aircraft. The occurrence of supersonic flow at the blade tips causes all sorts of problems, and although great advances in propeller design have been made, jet propulsion provides the only practical alternative for high-speed flight.

HIGH BY-PASS AND TURBOFAN ENGINES

In order to make a jet engine more efficient, we need to arrange it so that a larger mass of air is somehow given a smaller increase in speed. The method used is to increase the size of the compressor fan and to allow a proportion of the air to pass round the outside of the engine. The momentum given to this 'by-pass' air contributes to the thrust. There are also a number of secondary advantages, the most significant being a reduction in noise. Another function of the by-pass air is to help cool the engine and to make use of some of the otherwise wasted heat to increase the thrust.

By increasing the amount of by-pass air, the so-called fan jet (see Fig. 4.5) was evolved. The fan is not really part of the gas turbine compressor, and may sometimes be mounted at the rear of the engine.

Attempts to increase the efficiency still further lead to even larger fans until they become ducted propellers, or eventually unducted advanced turboprops, so that after many stages of development we will have come full circle back to the propeller! Lower by-pass engines will still however be required for very high-speed flight.

After so much talk of efficiency it is as well to remember that **efficiency is not everything!** We sometimes want value at all costs rather than value for money. The thrust given by a jet engine is almost independent of speed, while the thrust of a propeller, especially if it is of

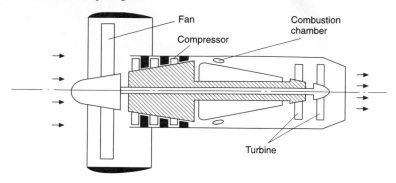

Fig. 4.5   A turbofan engine

fixed pitch, falls off badly both above and below a certain speed. It is
thrust that enables us to fly and gives us performance, and sometimes
we may be more than willing to pay the price provided we get the
thrust.

This seems an appropriate point at which to mention yet another
difference between jet propulsion and propeller propulsion, one that is
related to the fact just mentioned that the thrust of a jet is almost
independent of speed; **so the power developed by a jet engine,** i.e.
thrust × speed, **varies with the speed** and there is no satisfactory way
of measuring it, either on the ground or in flight; when the aircraft is
stationary on the ground, and the engine is running, there is no forward
velocity – **so the power is nil, but the thrust may be considerable,
and can be measured.** That is why the performance capability of a jet
engine – or of a rocket – is given **in terms of thrust and not of
power.** But when an engine drives a propeller, and this applies whether
the engine is of the turbine or piston type, the thrust, as we have said, is
variable, **but the power produced at the propeller shaft may be
considerable** even when the aircraft is stationary, and what is more **it
can be measured** – the propeller acts as a brake on the engine, and the
power is sometimes measured by other kinds of brake, and is sometimes
called **brake power** – so these engines are compared **according to the
power they produce,** and not by the thrust which would be
meaningless.

THE PROPELLER OR AIRSCREW

Of the various systems of propulsion, the propeller has been most used
in the past, and for many types of aircraft it is likely to be a long time in
dying. More and more gas turbines, rather than reciprocating engines,

are being used for driving propellers but that does not in any way affect the aerodynamic problems involved. It is right, therefore, that we should give brief consideration to those problems. Some of them also are common to those of the helicopter, some too to the blades of compressors and fans and turbines, and these are further reasons why we should consider them.

The object of the propeller is to convert the **torque**, or turning effect, given by the power of the engine, into a straightforward pull, or push, called **thrust**.

If an airscrew is in front of the engine it will cause tension in the shaft and so will **pull** the aeroplane – such an airscrew is called a **tractor**. If, on the other hand, it is behind the engine, it will **push** the aeroplane forward, and it is called a **pusher** (Figs 4F and 4G). In Fig. 4H there is an unusual combination of both pusher and tractor propellers.

HOW IT WORKS

Each part of a propeller blade has a cross-section similar to that of an aerofoil; in fact, in some cases exactly the same shape of section has been used for both purposes. The thrust of the propeller is obtained because the chord at each part of the blade is inclined at a small angle (similar to the angle of attack of an aerofoil) to its direction of motion. Since, however, the propeller is both rotating and going forward, the direction of the airflow against the blade will be at some such angle as is shown in Figs 4.6 and 4.7. This will result in lift and drag on the blade section, just as it does on an aerofoil. Actually in a propeller we are not so much concerned with the forces perpendicular and parallel to the airflow, i.e. lift and drag, as the force acting along the axis of the aeroplane (**the thrust force**) and at right angles to the rotation (**the torque force**). So the total force on the blade must be resolved into thrust and torque forces, as in Fig. 4.7. The difference between these and lift and drag is clearly seen by comparing Figs 4.6 and 4.7.

The total torque force on the propeller blades will cause a turning moment or torque which opposes the engine torque, and also **tends to rotate the complete aeroplane in the opposite direction to that in which the propeller is revolving**. When the propeller is revolving at a steady number of revolutions per minute, then the **propeller torque and the engine torque will be exactly equal and opposite**.

HELIX ANGLE AND BLADE ANGLE

Why is the theory of the propeller more involved than that of the aerofoil? Chiefly because the local direction of motion of the blade is

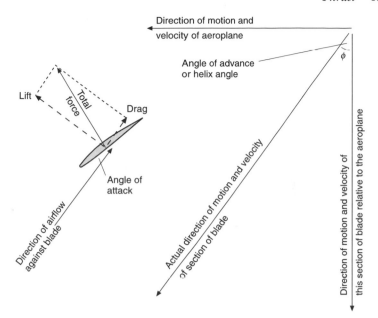

Fig. 4.6   Motion of a section of propeller blade
Showing resolution of total force into lift and drag

along a helix rather than a straight line, and, what is more, every section
of the propeller blade travels on a different helix (Fig. 4.8). The angle
(ϕ) between the resultant direction of the airflow and the plane of
rotation (Fig. 4.6) is called the **angle of advance** or **helix angle**, and it
is a different angle at each section of the blade. The sections near the tip
move on a helix of much greater diameter, and they also move at a much
greater velocity than those near the boss.

Since all the sections must be set at a small **extra** angle to give the
angle of attack, and since for maximum efficiency this extra angle should
be approximately the same at all parts of the blade, it is clear that the
blade angle, or **pitch angle**, must vary like the helix angle from boss
to tip. Figure 4.9 shows a typical variation of blade angle.

The **blade angle** is best defined as **the angle which the chord of
the propeller section at any particular place makes with the
horizontal plane when the propeller is laid flat on its boss on this
horizontal plane, its axis being vertical** (Fig. 4.10). The figure

Fig. 4F   Pusher – old type (opposite)
The Bristol Boxkite. One blade of the small pusher propeller can be seen protruding
behind and below the minimal wooden box fuselage with its barrel-shaped fuel tank.

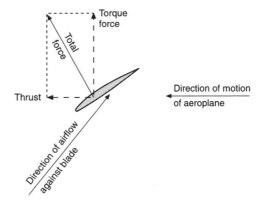

Fig. 4.7 Motion of a section of a propeller blade
Showing resolution of total force into thrust and torque

Fig. 4.8 Helical paths travelled by various sections of propeller blade

shows how the blade angle is made up of the helix angle plus the angle of attack.

ADVANCE PER REVOLUTION

In a propeller, the blade angle at each section is greater than the helix angle and, what is more important, the distance moved forward in one revolution (called the **advance per revolution**) **is not by any means a fixed quantity**, as it depends entirely on the forward speed of the aeroplane.

Fig. 4G Pusher – new type (opposite)
(By courtesy of the Beech Aircraft Corporation, USA)
The Beech Starship. Twin pusher turboprop with many other advanced features including tail-first 'canard' configuration.

Fig. 4H   Pusher and tractor
(By courtesy of Cessna Aircraft Company, USA)
Unique 4/6 seater with two tandem horizontally opposed air-cooled engines, each
driving a 2-blade feathering constant-speed propeller.

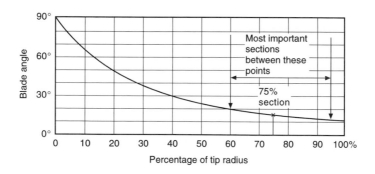

Fig. 4.9   Variation of blade angle

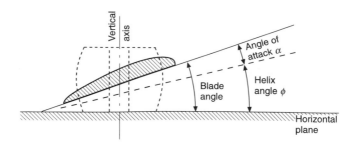

Fig. 4.10   Blade angle

For instance, if an aeroplane is flying at 100 m/s and the propeller is making 1200 rpm – i.e. 20 revs per second – then the advance per revolution will be 100/20 = 5 metres. But the same aeroplane may fly at 80 m/s, with the same revolutions of the propeller, and the advance per revolution will be only 4 metres; while when the engine is run up on the ground and there is no forward motion, the advance per revolution will obviously be 0.

Considering first a fixed-pitch propeller, if the angle of a blade section at a radius of r metres is $\theta°$, and if this particular blade section were to move parallel to its chord – i.e. so that its angle of attack was $0°$ – while at the same time it made one complete revolution, then the distance travelled forward, p metres, would be a definite quantity and would correspond to the pitch of an ordinary screw, the relation $p = 2\pi r \tan \theta$ being true. This is best seen graphically by setting off the blade angle θ from the distance $2\pi r$ drawn horizontally, p being the vertical height. If the same operation is carried out at different distances from the axis of the propeller (Fig. 4.11), it will be found that the value of p is practically the same for all sections of the blade since as the radius r increases there is a corresponding decrease in the blade angle θ, and $2\pi r \tan \theta$ remains constant.

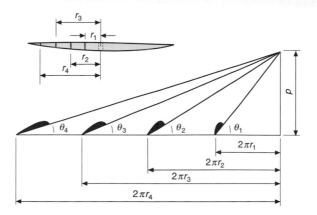

Fig. 4.11   Geometric pitch

PITCH OF A PROPELLER

This quantity, p, is called the **geometric pitch**, since it depends only on the geometric dimensions and not on the performance of the propeller. The value of the geometric pitch of a fixed-pitch propeller may vary from about 1 metre for a slow type of aeroplane to the 5 or 6

Fig. 4I   Fixed pitch propeller
(By courtesy of what was the Fairey Aviation Co Ltd)
Two-blader with very large pitch angle, as used in the Schneider Trophy contest,
1931.

metres that was used on Schneider Trophy and other racing aircraft
(Fig. 4I).

The designer of a propeller may find it convenient to consider the
pitch from a different viewpoint. When the advance per revolution
reaches a certain value, the thrust becomes zero, the reason being that
the angle of attack of each part of the blade has become so small that the
aerofoil section of the blade provides no thrust. (Notice how this
corresponds to the small negative angle at which an aerofoil ceases to
give lift.) The **experimental mean pitch** is defined as **the distance
the propeller will move forward in one revolution when it is
giving no thrust.**

EFFICIENCY

Now the **efficiency** of a propeller, like the efficiency of anything else, is
the ratio of the useful work given out by the propeller to the work put
into it by the engine. Mechanical work done is measured by the force
multiplied by the distance moved, and so when either the force or the
distance is zero, the useful work done is zero, and the efficiency nil.
Thus when the propeller moves forward in each revolution a distance
equal to the experimental pitch, the fact that there is no thrust means
that there is no efficiency. Also, when there is no forward speed, there is
no distance moved, no work done and therefore no efficiency. **Between
these two extremes are the normal conditions of flight.**

It might be thought that the object of the propeller is to give the
maximum thrust (T) with the minimum torque (Q), i.e. to give **the
maximum T/Q ratio.** However, Figs 4.6 and 4.7 show that in order to
get a high value of T/Q, two things are required – **a high value of L/D**
and **a small helix angle.** The high value of L/D is fairly easy, and is

an old problem; what is needed is a good aerofoil section, set at the correct small angle, and this means twisting the blade as already explained. But the provision of a small helix angle is a very different matter – this implies a low forward speed with a relatively high rotational speed. This is turn means that the propeller would be doing a great deal of work against the resistance without getting anywhere, so the efficiency would be very low.

Under conditions of maximum efficiency the advance per revolution is usually considerably less than the experimental pitch. The experimental pitch is sometimes called the **ideal pitch**, while the advance per revolution is the actual **practical pitch**. The difference between the two is called the **slip**, and is usually expressed as a percentage.

The calculation of propeller efficiency is quite straightforward. For example, if the total drag of an aeroplane at 65 m/s is 4.22 kN and the power developed by the engine when the aeroplane is flying at this speed is 336 kW, then –

Work given to propeller per second = 336 000 joules
Work done by propeller per second = 4220 × 65
= 274 300 joules
So efficiency of the propeller =
Work got out/Work put in × 100 per cent
= (274 300/336 000) × 100 per cent
= **81.6 per cent**

This represents the approximate value of the efficiency obtainable from a good propeller, although in some instances it may rise as high as 85 or even 90 per cent. The best efficiency is obtained when the slip is of the order of 30 per cent.

For those who prefer to examine this question in terms of mathematical symbols the efficiency of a propeller can be deduced as follows –

Let v = forward velocity in m/s
T = thrust of propeller in newtons
n = revolutions per second of engine
Q = torque exerted by engine in N-m

Work done by thrust T at v metres per second = Tv joules per second, or watts

Work given by engine = $2\pi Q$ joules per rev
= $2\pi nQ$ joules per second, or watts

Efficiency of propeller = $(Tv/2\pi nQ) \times 100$ **per cent**

TIP SPEED

The power developed by a piston engine depends upon the pressures attained during combustion in the cylinders and on the revolutions per minute. The greatest power in most engines is developed at a fairly high number of revolutions per minute; and if the propeller rotates at the same speed as the engine crankshaft, the **tip speed** of the propeller blades **is liable to approach or exceed the speed of sound** (about 340 m/s in air at ground level, and less at higher altitudes). This causes compressibility effects (see Chapter 11), which, in turn, mean an increase in torque and decrease in thrust; in other words, **a loss of efficiency**. It is clearly of little purpose to design an engine to give high power, if at such power the propeller is to become less efficient and so transfer a lower proportion of the engine power to the aircraft. In the early stages of compressibility some improvement can be effected by changing the blade section near the tip to a thin laminar-flow type and by washing out the blade angle slightly; if this is done the loss is not serious so long as the actual speed of the tip does not exceed the speed of sound. As a further help a reduction gear is often introduced between the engine crankshaft and the propeller; the reduction is not usually very large, perhaps 0.7 or 0.8 to 1, but is just sufficient to reduce the tip speed to a reasonable margin below the speed of sound.

The tip speed, of course, depends not only on the revolutions per minute, but also on the forward speed of the aeroplane and the diameter of the propeller. The high forward speed of modern aeroplanes is such that it is becoming very difficult to keep the tip speed down below the speed of sound, and it would seem that at forward speeds of 350 knots or more some loss in efficiency must be accepted. At 430 knots the loss in efficiency is serious and has spread to a larger proportion of the propeller blades so that it affects not only the tips but what should be the most efficient sections. At this stage there is nothing for it but for the propeller to retire gracefully and hand over supremacy to jet propulsion.

A further objection to high tip speed is that the noise caused by the propeller (incidentally a large proportion of the total noise) is much intensified, especially in the plane in which the propeller is rotating. This can be annoying both outside and inside the aircraft, and in severe cases, structural damage can result.

VARIABLE PITCH

For low-speed aeroplanes the thrust of a fixed-pitch propeller is usually found to be greatest when there is no forward speed, i.e. when the aeroplane is stationary on the ground. The thrust developed under these conditions is called the **static thrust**, and its large value is very useful since it serves to give the aeroplane a good acceleration when starting from rest and thus reduces the run required for taking off. But in high-speed aircraft a fixed-pitch propeller designed for maximum speed would have such a large pitch, and, therefore, such steep pitch angles, that some portions of the blades would strike the air at angles of as much as 70° or more when there is no forward speed, the efficiency and static thrust would be very poor, and great difficulty would be experienced in taking off. The only remedy is variable pitch.

This requirement led to the development of the so-called **constant-speed propeller** (Fig. 4J) in which the pitch is automatically adjusted so that the propeller revolves at a given rate decided by the pilot, and remains at that rate irrespective of throttle opening or manoeuvres of the aeroplane. Thus engine and propeller can work at high efficiency irrespective of conditions, such as take-off, climb, maximum speed, altitude, and so on.

Fig. 4J   Constant-speed propeller
A classic constant-speed propeller design. The hydraulically-actuated speed control unit is housed in the small domed unit on the front of the hub.

It has already been stated that there are problems common to propeller and helicopter blades, and one of these is that of variable pitch, which in helicopters is a virtual necessity as a means of control – more interesting, perhaps, and certainly more recent, is the **application of**

variable pitch to the blades in jet engines, mainly as a means of reducing noise, one of the bugbears of jet engines and indeed of modern flying in general.

An extension of the idea of variable pitch leads to a propeller with the pitch variable not only over the range of blade angles that will be required for normal conditions of flight but **beyond these angles in both directions**.

If the blade can be turned beyond the normal fully-coarse position until the chord lies along the direction of flight, thus offering the minimum resistance, the propeller is said to be **feathered**. This condition is very useful on a multi-engined aircraft for reducing the drag of the propeller on an engine that is out of action. It has another advantage too in that it is a convenient method of stopping the propeller and so preventing it from 'windmilling'; this reduces the risk of further damage to an engine that is already damaged.

The turning of the blade beyond the fully-fine position makes the propeller into an effective **air brake**; it has exactly the opposite effect to feathering by causing the maximum drag, which occurs when the blade angle is approximately $2°$ or $3°$.

If the blade angle is still further reduced, i.e. to negative angles, then instead of allowing the blades to windmill, we can run the engine and produce negative thrust or drag. This produces an **excellent brake for use in slowing up the aeroplane after landing** since it gives a high negative thrust at low forward speeds.

NUMBER AND SHAPE OF BLADES

The propeller must be able to absorb the power given to it by the engine; that is to say, it must have a resisting torque to balance the engine torque, otherwise it will race, and both propeller and engine will become inefficient.

The climbing conditions are particularly difficult to satisfy since high power is being used at low forward speeds; and if we do satisfy these conditions – by any of the methods suggested below – it will be difficult to get efficiency in high-speed flight. Thus, the propeller becomes a **compromise** like so many things in an aeroplane.

The ability of the propeller to absorb power may be increased by –

1. Increasing the blade angle and thus the angle of attack of the blades.
2. Increasing the length of the blades, and thus the diameter of the propeller.
3. Increasing the revolutions per minute of the propeller.

4. Increasing the camber of the aerofoil section of which the blade is made.
5. Increasing the chord (or width) of the blades.
6. Increasing the number of blades.

With so many possibilities one might think that this was an easy problem to solve, but in reality it is one that has caused considerable difficulty. First, the blade angle should be such that the angle of attack is that giving maximum efficiency; there is, therefore, little point in trying to absorb more power if, in so doing, we lose efficiency. The second possibility is to increase the diameter, in other words, to increase the blade aspect ratio, but quite apart from the bogey of tip speed, the aeroplane designer will probably not allow him to spread himself! The third, would mean high tip speed and consequent loss of efficiency. The fourth, as with aerofoils, would simply mean a less efficient section; it would seem, too, that we must face even thinner aerofoil sections to avoid loss of efficiency at high speed. So we are left with the last two, and fortunately they provide some hope. Either will result in an increase in what is called the **solidity** of the propeller. This really means the ratio between that part of the propeller disc which, when viewed from the front, is solid and the part which is just air. **The greater the solidity, the greater the power that can be absorbed.**

Of the two methods of increasing solidity, increase of chord and increase of number of blades, the **former** is the **easier**, the **latter** the **more efficient**. The so-called paddle blades are examples of the former method. But there is a limit to this, first, because the poor aspect ratio makes the blades less efficient.

So, all in all, an increase in the number of blades is the most attractive proposition, and that is why we saw, first, the two-blader (yes, there has been a one-blader! – but only one); then, in turn, three, four, five, and six blades; and we might have gone to eight- and ten-bladers had not jet propulsion come along at the critical time.

After four or, at the most, five blades, it becomes inconvenient to fit all the blades into one hub, and it is, in effect, necessary to have two propellers for each engine. If we are going to have two propellers, we may as well rotate them in opposite directions (Fig. 4K) and so gain other advantages which will become more apparent when we have considered the effects of the propeller on the aeroplane.

Fig. 4K   Contra-rotating propellers
Four sets of these six-bladed contra-rotating propellers were employed to propel the old Shackleton patrol aircraft. The noise inside the fuselage, which had no padding or sound absorption material, made the long duration flights decidedly arduous for the crew.

THE SLIPSTREAM

The propeller produces thrust by forcing the air backwards, and the resultant stream of air which flows over the fuselage, tail units, and other parts of the aeroplane is called the **slipstream**.

The extent of the slipstream may be taken roughly as being that of a cylinder of the same diameter as the propeller. Actually there is a slight contraction of the diameter a short distance behind the propeller.

The velocity of the slipstream is greater than that at which the aeroplane is travelling through the air; the increase in velocity may be as much as 100 per cent, or even more, at the stalling speed of the aeroplane. This means that the velocity of the air flowing over all those parts in the slipstream is twice that of the airflow over the other parts, and so **the drag is four times as great** as corresponding parts outside the slipstream. At higher forward speeds the difference is not as great, being only about 50 per cent at normal speeds, and as little as 10 per cent at high speeds. The extra velocity of the slipstream may be beneficial **in providing more effective control** for rudder and elevators, especially when the aeroplane is travelling slowly through the air, e.g. when taxying, or taking off, or flying near the stalling speed. With jet propulsion, however, it is not advisable for the hot jet to strike the tail plane which, in consequence, is often **set very high** (Fig. 5F).

In addition to increased velocity, the propeller imparts a **rotary motion** to the slipstream in the same direction as its own rotation; so it will strike one side only of such surfaces as the fin, and so may have considerable effects on the directional and lateral balance of the aeroplane. If these effects are compensated for in normal flight – e.g. by offsetting the fin so that it does not lie directly fore and aft – then the balance will be upset when the engine stops and the slipstream ceases to exert its influence.

GYROSCOPIC EFFECT

The rotating mass of the propeller or the compressor in the case of a jet engine may cause a slight **gyroscopic effect**. A rotating body tends to resist any change in its plane of rotation, and if such change does take place there is superimposed a tendency for the plane of rotation to change also in a direction at right angles to that in which it is forced. This can easily be illustrated with an ordinary bicycle wheel, if the wheel, while rapidly rotating, is held on a horizontal shaft and the holder attempts to keep the shaft horizontal while he turns, the shaft will either tilt upwards or downwards according to whether he turns with the opposite or the same sense of rotation as that of the wheel. Thus if the propeller rotates clockwise when viewed from the pilot's cockpit (the usual method of denoting the rotation), the nose will tend to drop on a right-hand turn and the tail to drop on a left-hand turn. It is only in exceptional cases that this effect is really appreciable, although it used to be very marked in the days of rotary engines when the rotating mass was considerable.

SWING ON TAKE-OFF

There is often a tendency for an aeroplane to swing to one side during the take-off run. This must be due to some asymmetric feature of the aircraft, and it is an interesting problem to try to track down the real villain that is causing the swing.

The pilot should be the first suspect. He himself is not symmetrical, he may be right-handed (or left-handed), he probably looks out on one side of the aeroplane and may even sit on one side. Certain it is that some aircraft which have swung violently when the pilot has tried to keep them straight have gone as straight as a die when left to themselves!

The second and main suspect is undoubtedly the propeller. But which of its asymmetric effects is the chief cause of swing in any particular aircraft is not so easy to determine. If the propeller rotates clockwise, the **torque reaction** will be anti-clockwise, the left-hand wheel will be pressed on the ground and the extra friction should **tend to yaw the aircraft to the left**. **But** let us not forget that the torque reaction may be compensated and, in that case, the behaviour of the aeroplane will depend on how it is compensated.

The slipstream – assuming the same clockwise propeller – will itself rotate clockwise and will probably strike the fin and rudder on the left-hand side, again **tending to yaw the aircraft to the left**. But the slipstream too may be compensated.

The **gyroscopic effect** will only come in **when the tail is being raised**. Again the tendency will be to **swing to the left** if the propeller rotates clockwise. Try it with the bicycle wheel.

Apart from the compensating devices already mentioned the tendency to swing can be largely, if not entirely, eliminated by opposite rotating propellers on multi-engined aircraft (Fig. 9A), by contra-rotating propellers on single-engined aircraft and by jet propulsion or rocket propulsion instead of propellers.

Contra-rotating propellers (Fig. 4K) not only give the greater blade area, or solidity, that is required to absorb large power, but they eliminate or very nearly eliminate all the asymmetrical effects of slipstream, propeller torque, and gyroscopic action. It is curious that the average pilot hardly realised the existence of these asymmetrical effects – until he lost them. Pilots who flew behind contra-rotating propellers for the first time reported that the aircraft was easy to handle and nice to fly. This is hardly surprising; what perhaps is surprising is that the previous ill-effects of one-way rotation had been so little noticed. The second propeller straightens the slipstream created by the first and so causes a straight high-speed flow of air over wings and tail; this improves the control and there is little or no resultant torque tending to roll the aircraft in one direction, and therefore no need to counteract such tendency; the gyroscopic effects are also neutralised. All this means that there should be no tendency to swing to one side during take-off, no roll or yaw if the throttle is suddenly opened or closed, no difference in aileron or rudder trim, whether the engine is on or off – in short, the aircraft should be easy to handle and nice to fly.

CAN YOU ANSWER THESE?

Some simple questions about thrust and propellers –

1. What is a ramjet?
2. What is meant by the blade angle of a propeller, and why does this angle decrease from boss to tip?
3. Distinguish between the 'advance per revolution', the 'geometric pitch' and the 'experimental pitch' of a propeller.
4. What is slip?
5. What are the advantages of a variable-pitch propeller?
6. Why is the tip speed an important factor in propeller design?
7. Why is solidity important, and how can it be increased?
8. What methods of propulsion can be used outside the earth's atmosphere?

Turn to Appendix 3 for a few simple numerical examples on thrust.

CHAPTER 5
LEVEL FLIGHT

INTRODUCTION

The flight of an aeroplane may be considered as consisting of various stages. First, **the take-off**, during which the aircraft is transferred from one medium to another; then **the climb**, during which the pilot gains the height at which the level part of the flight will be made; then a period of this **steady flight at a constant height**, interrupted in certain cases by periods of manoeuvres, or aerobatics; **the approach** back towards the earth; and finally **the landing**.

On long distance flights the main portion may consist of **a long slow steady climb**, which is more economical than maintaining the same height as fuel is consumed, and the weight of the aircraft is reduced, and so it is often only during a small portion of each flight that the aeroplane may be considered as travelling in straight and level flight at uniform velocity (Fig. 5A).

THE FOUR FORCES

Now, what are the forces which keep the aeroplane in its state of steady flight? First the **lift**, which will be vertically upwards since the direction of motion is horizontal. This we have created with the express object of keeping the aeroplane in the air by opposing the force of gravity, namely, the **weight**. But we can only produce lift if the aeroplane is moved forward and for this we need the **thrust** provided by the propeller or jets. We also know that the forward motion will be opposed by the **drag**.

The aeroplane, therefore, can be said to be under the influence of four main forces –

Fig. 5A   Long haul, large capacity
This Boeing 747-400 'Jumbo' has 350 seats, and a range of 13 528 km that could take it from Europe to Australia non-stop. With higher density seating, up to 600 passengers could be carried.

1. **The Lift of the Main Planes, *L*, acting vertically upwards through the Centre of Pressure.**
2. **The Weight of the aeroplane, *W*, acting vertically downwards through the Centre of Gravity.**
3. **The Thrust of the engine, *T*, pulling horizontally forwards.**
4. **The Drag, *D*, acting horizontally backwards.**

Just as for certain purposes it is convenient to consider all the weight as acting through one point, called the centre of gravity, or all the lift as acting at the centre of pressure, so we may imagine the resultant of all the drag acting at one point which, for convenience, we will call the centre of drag. Its actual position depends on the relative resistance of different parts of the aeroplane.

CONDITIONS OF EQUILIBRIUM

Now, under what conditions will these four forces balance the aeroplane? That is to say, keep it travelling at a steady height at uniform velocity in a fixed direction, a state of affairs which, in the language of mechanics, is known as **equilibrium**. It is sometimes hard to convince a traveller by air that he may travel at 200 knots and yet be in a state of equilibrium; equilibrium simply means that the existing state of affairs is remaining unchanged; in other words, that the aeroplane is obeying Newton's First Law of Motion.

In order to do this the forces acting on it must be balanced – **the lift must be equal to the weight** (this condition will keep the aeroplane at a constant height); and the **thrust must be equal to the drag** (this condition will keep the aeroplane moving at the same steady velocity).

The idea is often prevalent that the lift must be **greater** than the weight, or, as it is often expressed, the lift must 'overcome' the weight; and when it comes to the question of thrust and drag the author has known students dismiss the idea that the thrust **need only be equal** to the drag as 'contrary to common sense'.

There still remains a third condition for equilibrium. In order to maintain straight and even flight, **we must prevent the aeroplane from rotating**, and this depends not only on the magnitudes of the four forces, but also on the positions at which they act. If the centre of pressure is behind the centre of gravity, the nose will tend to drop and the tail to rise, and vice versa if the centre of pressure is in front of the centre of gravity. But we are also concerned with the lines of action of the thrust and drag, for if the line of thrust is high and the line of drag is low, these two forces also will tend to make the nose drop. Such tendencies could be prevented by the pilot using his controls, but it is the aim of the designer to make an aeroplane which will in the words of the pilot, fly 'hands off'. Therefore he must see that the forces act in the right places.

DIFFICULTIES IN BALANCING THE FOUR FORCES

First, the lift. The lift will act through the centre of pressure, which will depend on the position of the wings; so the designer must be careful to place the planes in the correct position along the fuselage. But the problem is complicated by the fact that a change in the angle of attack means a movement of the lift, and usually in the unstable direction; if the angle of attack is increased the pitching moment **about the centre of gravity** will become more nose-up, and tend to increase the angle even further.

Secondly, the weight. This will act through the centre of gravity, which in turn will depend on the weight and position of every individual part of the aeroplane and the loads that it carries. Here alone is sufficient problem, but again there is a possibility of movement of the centre of gravity during flight caused, for instance, by consumption of fuel, dropping of bombs or movement of passengers. In the Concorde fuel is actually moved from one tank to another to adjust the position of the centre of gravity.

Thirdly, the thrust. Here the problem is easier. The line of thrust is settled by the position of the propeller shaft or centre line of the jet, which in turn depend on the position of the engine or engines. In this matter the designer has little choice, but he has to consider such problems as keeping the propeller clear of the ground and giving the pilot a clear view ahead; new problems arise too when the thrust can be deflected as in certain modern types.

Lastly, the drag. This is, perhaps, the most difficult of all. The total drag is composed of the drag of all the separate parts, and the designer must either estimate the drag of each part separately, and so find the total drag and its line of action, or he must rely on wind tunnel experiments on a model; and even when he has found the line of drag, it too will be liable to change at different angles of attack.

What, then, can he do with these unruly forces? He might try to concentrate them all at one point, but he could never be certain that their positions would not alter and upset the balance. Also there is an advantage to be gained by having the lift slightly behind the weight (Fig. 5.1); this will give these two forces a tendency to pitch the aeroplane

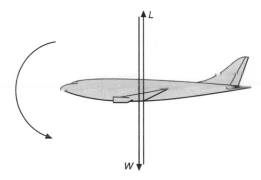

Fig. 5.1   Lift behind weight: nose-down moment

nose downwards, which, in the event of engine failure, will automatically put it in a position ready for a glide, whereas if the lift were in front of the weight there would be a tendency to stall. If, then, he places the wings so that the lift is behind the weight, he must at the same time counteract this nose downwards tendency while the engine is running and the aeroplane is in normal horizontal flight. The obvious way in which to do this is to arrange that the line of drag shall be above the line of thrust so that these two forces will cause a 'tail downwards' tendency, as in Fig. 5.2. In order to do this the thrust must be low, and this causes difficulty because it means either making the propeller of small diameter or bring it close to the ground, both of which are bad features in design. This problem is rather easier to solve in aircraft driven by jets or rockets in which a lower thrust line is possible. Even so, it is not easy to arrange for a high drag line, especially in aircraft with

Fig. 5.2   Drag above thrust: nose-up moment

fixed undercarriages or, if the undercarriage is retractable, when it is in the down position. On a **flying boat** (Fig. 9F), and even on a **floatplane** (Fig. 5B), where **the line of drag is exceptionally low,** and where **the thrust must be even higher** than on a landplane – so that the propeller may clear the waves and spray – the proposition of having the drag above the thrust does, in fact, become virtually impossible. But if we reverse matters and have the thrust line higher than the drag, this will result in **a nose-down pitching moment** which must be counteracted for normal flight. This can be arranged if we reverse the positions of lift and weight, putting the lift in front of the

Fig. 5B   Floatplane
(By courtesy of Cessna Aircraft Company, USA)

weight; then if the engine stops, the thrust will cease to exist, the tail downwards tendency of lift and weight will take charge, and the machine will tend to stall.

THE TAIL PLANE

So much for the problem. What solution can be found? In the first place, where circumstances permit, the forces will be arranged as in Fig. 5.3, which we may call the ideal disposition. Where this is impossible, the arrangement of these four forces alone cannot be considered satisfactory, and we need to look for help from some new quarter altogether. **We find it in the tail plane.**

Fig. 5.3   Forces on an aeroplane in normal flight

At a considerable distance behind the main planes, as the chief lifting aerofoils are usually called, we fit a small plane whose duty it is to provide the upward or downward force necessary to counteract the behaviour of the four main forces. The force on the tail plane need only be a small one, because, owing to its leverage, even a small force will produce a large correcting moment.

'TAIL-LESS' AND 'TAIL-FIRST' AEROPLANES

The reader will probably have realised by now that the existence of this auxiliary plane – the stabiliser, as the Americans rather aptly call it – is a necessity rather than a luxury, because even if the four main forces can be balanced for one particular condition of flight, they are not likely to remain so for long. What then of the so-called **tail-less** type of aeroplane?

This type has had followers from the very early days of flying – and

among birds from prehistoric times – and although the reasons for its
adoption have changed somewhat, a common feature has been a large
degree of sweepback, or even delta-shaped wings, so that although this
type may appear to have no tail, the exact equivalent is found at its wing
tips, the wings being, in fact, swept back so that the tip portion can fulfil
the functions of the tail plane in the orthodox aeroplane. In fact, it is
true to say that **the 'tail-less' type has two tails instead of one!**
(Figs 5C and 5D).

Fig. 5C   Tail-less – old type
(By courtesy of '*Flight*')
The Westland Hill Pterodactyl.

More of a freak is the **tail-first** or **canard** aeroplane. A good example
of this was the original Wright machine which made the first power-
driven flight. Although at various times certain advantages have been
claimed for this type, it has never proved really successful, and there can
be little doubt that the serious drawbacks inherent in such a design have
far outweighed any small advantages. But like so many other old ideas, it
is coming back (Fig. 5E); this time for missiles and very high speed
aircraft (see later chapters). In this case the small auxiliary surface **in
front** of the main planes (which we can hardly call a tail!) fulfils the
duties of the normal tail plane.

Fig. 5D   Tail-less – new type
This popular popular form of powered microlight aircraft has been derived from hang-glider technology.

Fig. 5E   Tail-first
Apart from the canard layout, the Rutan Vari-Eze shows many unusual features such as a pusher propeller, composite construction, and a nosewheel that can be retracted in flight or when parked.

LOADS ON TAIL PLANE

But to return to the normal aeroplane. Where the four main forces can be satisfactorily balanced in themselves, the duty of the tail plane is merely to act as a 'stand-by'. Therefore it will be set at such an angle of attack that **it carries no load in normal flight; at high speeds it must carry a down load** (Fig. 5.4), because at high speed the main aerofoils will be at a small angle of attack, the centre of pressure will move backwards, the wing pitching moment about the centre of gravity will be nose-down, and so the tail must be held down to counteract this tendency; correspondingly, **at low speeds – i.e. at large angles of attack of the main planes – the tail plane must carry an upward load** (Fig. 5.5).

Fig. 5.4   High speed – down load on tail

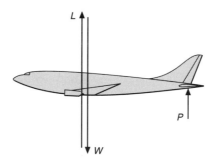

Fig. 5.5   Low speed – up load on tail

Since the tail plane is equally likely to have to carry an upward or a downward load, it is usually of symmetrical camber, and therefore provides no lift when the angle at which it strikes the airflow is 0°. But when the four main forces **cannot** be satisfactorily balanced in themselves, the tail plane may be called upon to provide a more or less permanent balancing force either upwards or downwards. It is then called a **lifting tail**. To provide the lift it may be cambered in the same way as an ordinary aerofoil, or, as the force required is often a downward one, it may even be shaped like an inverted aerofoil.

EFFECTS OF DOWNWASH

In many types of aircraft **the air which strikes the tail plane has already passed over the main planes, and the trailing vortices from these will cause a downwash on to the tail plane** (Fig. 5.6).

Main plane.
Riggers' angle of incidence 4°
Angle of attack 4°

Tail plane.
Riggers' angle of incidence 2°
Angle of attack 0°

Fig. 5.6   Effect of downwash on the tail plane

The angle of this downwash may be at least half the angle of attack on the main planes, so that if the main planes strike the airflow at 4°, the air which strikes the tail plane will be descending at an angle of 2°, so that if the tail plane were given a riggers' angle of incidence of 2°, it would strike the airflow head-on and, if symmetrical, would provide no force upwards or downwards. Again, the angle of downwash will, of course, change with the angle of attack of the main planes, and it is for this reason that the angle at which the tail plane should be set is one of the difficult problems confronting the designer.

As we shall discover later, its setting also affects the stability of the aeroplane, and further difficulties arise from the fact that in a propeller-driven aircraft the tail plane is usually in the slipstream, which is a rotating mass of air and will therefore strike the two sides of the tail plane at different angles. In jet-driven aircraft the tail plane is often set very high (Fig. 5F), to keep it clear of the hot jets, and this in turn may cause trouble since it may be shielded by the main planes at large angles of attack, resulting in what is called a deep stall and general instability, hence the low tail position illustrated in Fig. 5G.

THE ADJUSTABLE TAIL PLANE

Even when the designer has settled the size and angle of the tail plane, the forces which it will experience will be outside the control of the pilot, and although they may automatically compensate for such things as changes in pitching moment on the main planes, they cannot possibly

Fig. 5F   High tail
The tail of a Lockheed C-5 Galaxy. The elevators are split into a number of sections to provide redundancy for reasons of safety.

Fig. 5G   Low tail
The Alpha-Jet has a relatively low tail plane with pronounced anhedral.

deal with changes of weight such as are caused by the dropping of bombs, consumption of fuel, and changes in the disposition of passengers and cargo. For this reason the **adjustable tail plane** was introduced, this being operated by an actuating gear and controlled by a wheel or lever in the pilot's cockpit. This enabled the pilot to **alter the angle of setting of the tail plane during flight** and so cause downward or upward forces upon it at will, thus balancing the aeroplane under all conditions. Although this device was originally intended to cure such vices as nose- or tail-heaviness resulting from changes in the distribution of weight, it was found so useful in compensating for slight differences of balance due to flying or climbing at different angles of attack, gliding with the engine off, and so on, that it was soon employed on nearly all aeroplanes. It was, in fact, one of those many simple devices which made one wonder why 'someone didn't think of it before'.

But it was not long before someone thought of something even better; a device which did the same job as effectively, and was even lighter and simpler than the adjustable tail plane – this was the trimming tab, which will be considered under the heading of control.

Since they also come under the heading of control, we have not yet mentioned the **elevators**, the hinged flaps behind the tail plane which provide the pilot with the means of momentarily adjusting the balance or manoeuvring the aircraft, as distinct from the more permanent effect of the tail plane itself.

SLAB TAIL PLANES

But in spite of the ingenuity and effectiveness of the trimming tab, the variable incidence tail plane is back again – this time often, though not always, in the form of a **slab** tail plane, i.e. one that hasn't even got elevators let alone trimming tabs (Fig. 5H). This again is a return to an old idea, but one that has much greater advantages now than it had in the early days. **The simple moveable tail plane – all in one piece –** has less drag at high speeds than the combination of tail plane, elevator and tabs; it has more rigidity and so is less liable to flutter; and the one device acts in three different capacities – **as an elevator, to provide control – as a stabiliser, to give stability –** and as a very necessary **adjuster of balance** to cover all the conditions of the wide speed range which is a feature of modern aircraft. In some types the fin is also of the all-moving slab kind.

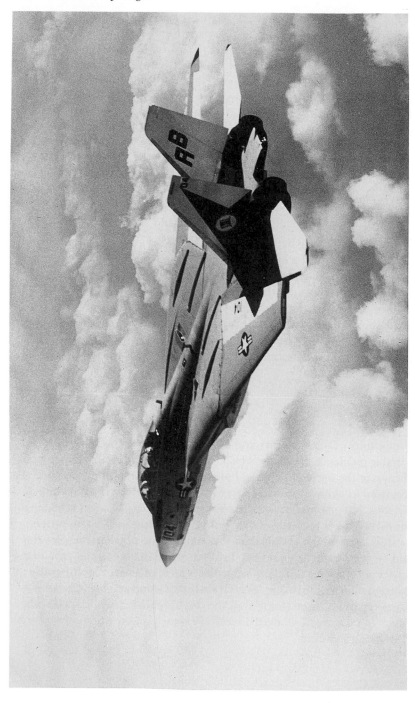

CONDITIONS OF BALANCE CONSIDERED NUMERICALLY

We have so far avoided any numerical consideration of the forces which balance the aeroplane in straight and level flight; in simple cases, however, these present no difficulty.

When there is no load on the tail plane the conditions of balance are these –

1. **Lift = Weight, i.e. L = W.**
2. **Thrust = Drag, i.e. T = D.**
3. **The 'nose-down' pitching moment of L and W must balance the 'tail-down' pitching moment of T and D.**

The two forces, L and W, are two equal and opposite parallel forces, i.e. a couple; their moment is measured by 'one of the forces multiplied by the perpendicular distance between them'. So, if the distance between L and W is x metres, the moment is Lx or Wx newton-metres.

Similarly, T and D form a couple and, if the distance between them is y metres, their moment is Ty or Dy newton-metres.

Therefore the third condition is that –

$$Lx \text{ (or } Wx) = Ty \text{ (or } Dy)$$

To take a numerical example: Suppose the mass of an aeroplane is 2000 kg. The weight will be roughly 20 000 N. $L = W$. So $L =$ 20 000 N.

Now, what will be the value of the thrust and drag? The reader must beware of falling into the ridiculous idea, which is so common among students, that the thrust will be equal to the weight! The statement is sometimes made that the 'Four forces acting on the aeroplane are equal', but nothing could be farther from the truth. $L = W$ and $T = D$, but these equations certainly do not make $T = W$. This point is emphasised for the simple reason that out of over a thousand students to whom the author has put the question, 'How do the think the thrust required to pull the aeroplane along in normal flight compares with the weight?' more than 50 per cent have suggested that the thrust must obviously equal the weight and over half the remainder have insisted that, on the other hand, the thrust must be many times greater than the weight! Such answers show that the student has not really grasped the

Fig. 5H   Slab tail planes (opposite)
(By courtesy of the Grumman Corporation)
Large slab tail planes and twin fins are evident on this photograph of the F-14.

fundamental principles of flight, for is not our object to obtain the maximum of lift with the minimum of drag, or, what amounts to much the same thing, to lift as much weight as possible with the least possible engine power? Have we not seen that the wings can produce, at their most efficient angle, a lift of 20, 25 or even more times as great as their drag?

It is true that there is a great deal of difference between an aeroplane and a wing; for whereas a wing provides us with a large amount of lift and a very much smaller amount of drag, all the other parts help to increase the drag and provide no lift in return. Actually it is not quite true to say that the wings provide all the lift, for by clever design even such parts as the fuselage may be persuaded to help. Efforts to increase the lift may be well worth while, but even total increase of lift from such sources will be small, whereas the addition of the parasite drag of fuselage, tail and undercarriage will form a large item; in old designs it might be as much, or even more than the drag of the wings, but in modern aircraft with 'clean' lines the proportion of parasite drag has been much reduced. None the less it may result in the reduction of the lift/drag ratio of the aeroplane, as distinct from the aerofoil, down to say 10, 12 or 15.

So it will be obvious that **the ideal aeroplane** must be one in which there is no parasite drag, i.e. in the nature of a **'flying wing'**; we should then obtain a lift some forty times the drag. At present we are a long way off this ideal, but there is no reason to doubt that we shall approach much nearer to it.

In our numerical example, let us assume that the lift is ten times the total drag, this being a reasonable figure for an average aeroplane. Then the drag will be 2000 N; so thrust will also be 2000 N. **This means that an aeroplane of weight 20 000 N can be lifted by a thrust of 2000 N; but the lift is not direct, the work is all being done in a forward direction – not upwards.** The aeroplane is in no sense a helicopter in which the thrust is vertical, and in which the thrust must indeed be at least equal to the weight.

To return to the problem –

$$L = W = 20\,000 \, \text{N}$$
$$T = D = 2000 \, \text{N} \text{ (This is merely an approximation for a good type of aeroplane)}$$
$$Lx = Ty$$

So $20\,000\,x = 2000\,y$

and $x = \frac{1}{10}y$

So if T and D are 1 metre apart, L and W must be 1/10 metre apart, i.e.

Fig. 5.7   Balance of forces – no load on tail

100 mm. In other words, **the lines of thrust and drag must be farther apart than the lines of lift and weight in the same proportion as the lift is greater than the drag** (Fig. 5.7).

FINDING THE TAIL LOAD

This conclusion only applies when there is no force on the tail plane. When there is such a force the problem is slightly more complicated, but can still be solved by the principle of moments. Consider a further example –

An aeroplane weighs 10 000 N; the drag in normal horizontal flight is 1250 N; the centre of pressure is 25 mm (0.025 m) behind the centre of gravity, and the line of drag is 150 mm (0.15 m) above the line of thrust. Find what load on the tail plane, which is 6 m behind the centre of gravity, will be required to maintain balance in normal horizontal flight.

Let the lift force on the main planes $= Y$ newtons
Let the force on the tail plane $= P$ newtons (assumed upwards)
Then total lift $= L = Y + P$ (Fig 5.8)
    But $L = W = 10\,000\,\text{N}$
$\therefore Y + P = 10\,000\,\text{N}$ (1)
    Also $T = D = 1250\,\text{N}$ (2)

Take moments about any convenient point; in this case perhaps the most suitable point is 0, the intersection of the weight and thrust lines.

Nose-down moments about 0 are caused by Y and P
W and T will, of course, have no moments about 0,

So total nose-down moments $= 0.025\,Y + 6\,P$
                                (all distances being expressed in metres)

Fig. 5.8   The four forces and the tail load

Tail-down moment about 0 is caused by D only,

So total tail-down moment $= 0.15\,D$

$\therefore 0.025\,Y + 6\,P = 0.15\,D$
i.e. $Y + 240\,P = 6\,D$ \hfill (3)

But from (1), $Y + P = 10\,000$
$\therefore 239\,P = 6\,D - 10\,000$

But from (2), $D = 1250$
$$\therefore 239\,P = 6 \times 1250 - 10\,000$$
$$= 7500 - 10\,000$$
$$= -2500$$
$$\therefore P = -(2500/239) = -10.4\,\text{N}$$

Therefore a small **downward** force of 10.4 N is required on the tail plane, the negative sign in the answer simply indicating that the force which we assumed to be upwards should have been downwards.

The student is advised to work out similar examples which will be found in Appendix 3 at the end of the book.

LEVEL FLIGHT AT DIFFERENT AIR SPEEDS

So far we may seem to have assumed that there is only one condition of level flight; but this is not so. Level flight is possible over the whole speed range of the aeroplane, from the maximum air speed that can be attained down to the minimum air speed at which the aeroplane can be kept in the air, both without losing height. This **speed range** is often very wide on modern aircraft; the maximum speed may be in the region of 1000 knots or even more, and the minimum speed (with flaps lowered) less than 150 knots. Mind you, though level flight is **possible** at any speed within this range, it may be very **inadvisable** to fly unduly fast when considerations of fuel economy are involved, or to fly unduly slowly if an enemy is on your tail! There is nearly always a correct speed to fly according to the circumstances.

RELATION BETWEEN AIR SPEED AND ANGLE OF ATTACK

An aeroplane flying in level flight at different air speeds will be flying at different angles of attack, i.e. at different attitudes to the air. Since the flight is **level**, this means different attitudes to the ground, and so the pilot will be able to notice these attitudes by reference to the horizon (or to the 'artificial horizon' on his instrument panel).

For every air speed – as indicated on the air speed indicator – there is a corresponding angle of attack at which level flight can be maintained (provided the weight of the aeroplane does not change).

Let us examine this important relationship more closely. It all depends on our old friend the lift formula, $L = C_L \cdot \frac{1}{2}\rho V^2 \cdot S$. To maintain level flight, the lift must be equal to the weight. Assuming for the moment that the weight remains constant, then the lift must also remain constant and equal to the weight. The wing area, S, is unalterable. Now, if we look back, or think back, to Chapter 2 we will realise that $\frac{1}{2}\rho V^2$ represents the difference between the pressure on the pitot tube and on the static tube (or static vent), and that this difference represents what is read as air speed on the air speed indicator; in other words, the **indicated air speed**. There is only one other item in the formula, i.e. C_L (the lift coefficient). Therefore if $\frac{1}{2}\rho V^2$, or the indicated air speed, goes up, C_L must be reduced, or the lift will become greater than the weight. Similarly, if $\frac{1}{2}\rho V^2$ goes down, C_L must go up or the lift will become less than the weight. Now C_L depends on the angle of attack of the wings; the greater the angle of attack (up to the stalling angle), the greater the value of C_L. **Therefore for every angle of attack there is a corresponding indicated air speed**.

This is most fortunate, since the pilot will not always have an instrument on which he can read the angle of attack, whereas the air speed indicator gives him an easy reading of air speed. That is why a pilot always talks and thinks in terms of **speed**, landing **speed**, stalling **speed**, best gliding **speed**, climbing **speed**, range or endurance **speed**, and so on. The experimenter on the ground, on the other hand, especially if he does wind tunnel work, is inclined to talk and think in terms of **angle**, stalling **angle**, **angle** of attack for flattest glide, longest range, and so on. **This difference of approach is very natural.** The pilot, after all, has little choice if he does not know the angle of attack

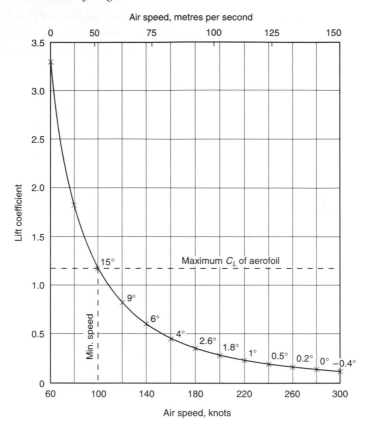

Fig. 5.9   Air speed, lift coefficient and angle of attack
Aeroplane of weight 50 kN (5100 kgf); wing area 25.05 m²; wing loading 1920 N/m²
(196 kgf/m²); aerofoil section characteristics Figs 3.13, 3.15, 3.16 and 3.17.

but does know the speed. To the experimenter on the ground, speed is
rather meaningless; he can alter the angle of attack and still keep the
speed constant – something that the pilot **cannot** do. But, however
natural the difference of outlook, it is unfortunate; and it is undoubtedly
one of the causes of the gap between the two essential partners to
progress, the practical man and the theoretical man.

Let us examine our general statement more critically by working out
some figures. Suppose the mass of an aeroplane is 5100 kg (so that its
weight will be approximately 50 000 newtons and that its wing area is
26.05 m², i.e. a wing loading of about 1920 N/m². Assume that the
aerofoil section has the lift characteristics shown in the lift curve on page
85 (Fig. 3.13). Consider first the ground level condition, the air density
being 1.225 kg/m³.

Whatever the speed, the lift must be equal to the weight, 50 kN; but the lift must also be equal to $C_L \cdot \frac{1}{2}\rho V^2 \cdot S$, so

$$50\,000 = C_L \times \tfrac{1}{2} \times 1.225 \times V^2 \times 26.05$$

from which

$$C_L = 3134/V^2$$

Now insert values of V of 60, 80, 100 and other values up to 300 knots, converting them, of course, to m/s, and work out the corresponding values of C_L; then by referring to Fig. 3.13 read off the angle of attack for each speed. The result will be something like that shown in the following table and in Fig. 5.9. The angles given in the table are approximate since all the values are small, and it is not easy to read off small parts of a degree.

Air speed		V^2	$C_L = 3134/V^2$	Angle of attack
knots	m/s			
60	30.8	949	3.31	–
80	41.2	1697	1.85	–
100	51.5	2652	1.18	15°
120	61.8	3819	0.82	9°
140	72.0	5184	0.60	6°
160	82.4	6790	0.46	4°
180	92.6	8580	0.37	2.6°
200	103.0	10 600	0.30	1.8°
220	113.2	12 800	0.24	1°
240	123.7	15 400	0.20	0.5°
260	134.0	18 000	0.17	0.2°
280	144.0	20 740	0.15	0°
300	154.0	23 700	0.13	−0.4°

Now let us see what this table and graph mean; we shall find it very interesting. In the first place at speeds below about 100 knots, the lift coefficient needed for level flight is greater than the maximum lift coefficient (1.18) provided by the aerofoil, therefore level flight is not possible below this speed. Secondly, as the speed increases to 120, 140, 160 knots, etc., the angle of attack decreases from 15° to 9°, 6°, 4°, etc.; and for each speed there is a corresponding angle of attack. We should notice, in passing, that at comparatively low speeds there is much greater change in angle of attack for each 20 knots increase in air speed than there is at the higher speeds, e.g. the angle of attack at 120 knots is 6° less than at 100 knots, whereas at 280 knots it is only 0.2° less than at 260 knots. This change in proportion is interesting, and is one of the arguments for an **angle of attack indicator, which is sensitive at**

low speeds, which is just where the air speed indicator is most unsatisfactory.

We could, of course, continue the table to speeds higher than 300 knots, and we should find that we needed even smaller lift coefficients, and even more negative angles of attack (though never less than $-1.8°$ since at this angle there would be no lift, whatever the speed). But at this stage we must begin to consider another factor affecting speed range, namely **the power of the engine**. What we have worked out so far is accurate enough, provided we can be sure of obtaining sufficient thrust. It may be that at speeds of 300 knots or above or, for that matter, at 100 knots or below, we shall not be able to maintain level flight for the simple reason that we have not sufficient engine power to overcome the drag. So **the engine power will also determine the speed range**, not only the top speed, but also to some extent the minimum speed.

If the reader thinks that the minimum speed of this aeroplane is rather high, we should perhaps point out, first, that we have not used flaps; secondly, that the aerofoil does not given a very high maximum lift coefficient; and, thirdly, that it has a fairly high wing loading, or ratio of weight to wing area, which, as we shall see later, has an important influence on minimum speed. All we wish to establish now is the relationship between air speed and angle of attack, and this is clearly shown by the table.

EFFECT OF HEIGHT

The table was worked out for ground level conditions. What will be the effect of height on the relationship between air speed and angle of attack? The answer, once it is understood, is simple – but very important.

Whatever the height, the air speed indicator reading is determined by the pressure $\frac{1}{2}\rho V^2$. In this expression, V is the **true** air speed. As has already been explained in Chapter 2, when the air speed indicator reads 200 knots at 3000 m, it simply means that the difference in pressure between pitot and static tubes (i.e. $\frac{1}{2}\rho V^2$) is the same as when the air speed was 200 knots at ground level. Now, it is not only the pitot pressure that depends on $\frac{1}{2}\rho V^2$; so do the lift and the drag. Therefore, at the same value of $\frac{1}{2}\rho V^2$, i.e. **at the same indicated speed**, the lift and drag will be the same as at ground level, other things (such as C_L) being equal. Therefore the table remains equally true **at all heights**, provided the air speed referred to is the **indicated speed**, and not the true speed. Thus the **angle of attack, or the attitude of the aeroplane to the air, is the same in level flight at all heights, provided the indicated air speed remains the same**.

EFFECT OF WEIGHT

The table was worked out for a constant weight of 50 kN. What will be the effect of changes of weight such as must occur in practical flight owing to fuel consumption, etc.? The answer to this is not quite so simple.

Suppose the weight is reduced from 50 kN to 40 kN. At the same indicated air speed, the angle of attack would be the same, and the lift would be the same as previously, i.e. 50 kN. This would be too great. Therefore, in order to reduce the lift, we must adjust the attitude, so that the wings strike the air at a smaller angle of attack, or we must reduce the speed, or both. Whatever we do, we shall get a slightly different relationship between air speed and angle of attack: the reader is advised to work out the figures for a weight of 40 kN. Although the relationship will differ from that for 50 kN weight, it will again remain constant at all heights for the same indicated speeds. To sum up the effect of weight, we can say that **the less the total weight of the aircraft, the less will be the indicated air speed corresponding to a given angle of attack**. A little calculation will show that the indicated air speed for the same angle of attack will be in proportion to the square root of the total weight.

FLYING FOR MAXIMUM RANGE – PROPELLER PROPULSION

Whether in war or peace, we shall often wish to use an aircraft to best advantage for some particular purpose – it may be to fly as fast as possible, or as slowly as possible, or to climb at maximum rate, or to stay in the air as long as possible, or, perhaps, most important of all, to achieve the maximum distance on a given quantity of fuel. Flying for maximum range is one of the outstanding problems of practical flight, but it is also one of the best illustrations of the principles involved. To exploit his engine and aircraft to the utmost in this respect, a pilot must be not only a good flier, but also an intelligent one.

The problem concerns engine, propeller, and aircraft; it also concerns the wind. In this book we are interested chiefly in the aircraft, but we cannot solve this problem, and indeed we can solve few, if any, of the problems of flight, without at least some consideration of the engine and the propeller, or jet, or rocket, or whatever it may be, and how the pilot should use them to get the best out of his aircraft. As for the wind, we shall, as usual in this subject, first consider a condition of still air.

The object in any engine regardless of type is to burn fuel so as to get energy and then to convert this energy into mechanical work. **In order** to get the greatest amount of work from a given amount of fuel, we must, first of all, get the maximum amount of energy out of it, and then we must change it to mechanical work in the most efficient way. Our success or otherwise will clearly depend to some extent on the use of the best fuel for the purpose, and on the skill of the engine designer. But the pilot, too, must play his part. To get the most heat from the fuel, it must be properly burned; this means that the mixture of air to fuel must be correct. In a piston engine, what is usually called 'weak mixture' is, in fact, not so very weak, but approximately the correct mixture to burn the fuel properly. If we use a richer mixture, some of the fuel will not be properly burned, and we shall get less energy from the same amount of fuel: we may get other advantages, but we shall not get economy. Both the manifold pressure and the revolutions per minute will affect the efficiency of the engine in its capacity of converting energy to work. The problem of the best combination of boost and rpm, though interesting, is outside the scope of this book and at this stage, too, the principles of the reciprocating and turbine engine begin to differ, while the rocket or ramjet has not got any rpm! For the reciprocating engine we can sum up the engine and propeller problem by saying that, generally speaking, the pilot will be using them to best advantage if he uses weak mixture, the highest boost permissible in weak mixture, and the lowest rpm consistent with the charging of the electrical generator and the avoidance of detonation. All this has assumed that he has control over such factors, and that the engine is supercharged and that the propeller has controllable pitch. Without such complications, the pilot's job will, of course, be easier; but the chances are that, whereas a poor pilot may get better results, the good pilot will get worse – far, far worse.

Before we leave the question of the engine and propeller, let us look at a problem which affects **all** engines in which fuel is burnt to give mechanical energy – not just piston engines.

The problem is that we cannot convert all the energy produced by burning the fuel into mechanical work, however hard we try. What is more, although in a sense we are always trying to do this (and then call the engine inefficient because we do not succeed!) we know quite well why we cannot and never will do it – it is just contrary to the laws of thermodynamics, the laws that govern the conversion of energy into mechanical work. All we can get, even in the best engines and in the hands of the best pilots, is something like 30 per cent of this figure. From each litre of fuel we can expect to get about 31 780 000 joules of thermal energy and hence only $0.3 \times 31\,780\,000$ joules i.e. about 9 500 000 joules or newton metres of mechanical energy.

This is what the engine should give to the propeller; and we may lose

20 per cent of it due to the inefficiency of the propeller, and so the aircraft will only get about 80 per cent of 9 500 000, i.e. **about 7 600 000 joules, or newton metres**.

It still seems a large figure – it is a large figure – but, as we shall see, it will not take the aircraft very far. However at this stage we are not so much concerned with the numerical value, as with the unit, and to think of it in the form of the **newton metre**. We have found that a litre of fuel, if used in the best possible way (we have said nothing about how quickly or slowly we use it) will give to the aircraft so many newton metres. **Suppose, then, that we want to move the aircraft the maximum number of metres, we must pull it with the minimum number of newtons, i.e. with minimum force. That simple principle is the essence of flying for range.**

Let us examine it more closely. It means that we must fly so that the propeller gives the least thrust with which level flight is possible. Least thrust means least drag, because drag and thrust will be equal.

FLYING WITH MINIMUM DRAG

Now, on first thoughts, we might think that flying with minimum drag meant presenting the aeroplane to the air in such an attitude that it would be most streamlined; in other words, in the attitude that would give least drag if a model of the aircraft were tested in a wind tunnel. But if we think again, we shall soon realise that such an idea is erroneous. This 'streamlined attitude' would mean high speed, and the high speed would more than make up for the effects of presenting the aeroplane to the air at a good attitude; in a sense, of course, it is the streamlined attitude that enables us to get the high speed and the high speed, in turn, causes drag. We are spending too much effort in trying to go fast.

On the other hand, we must not imagine, as we well might, that we will be flying with least drag if we fly at the minimum speed of level flight. This would mean a large angle of attack, 15° or more, and the induced drag particularly would be very high – we would be spending too much effort in keeping up in the air.

There must be some compromise between these two extremes – it would not be an aeroplane if there was not a compromise in it somewhere. Perhaps, too, it would not be an aeroplane if the solution were not rather obvious – **once it has been pointed out to us! Since the lift must always equal the weight**, which we have assumed to be constant at 50 kN, **the drag will be least when the lift/drag ratio is greatest**. Now, the curve of lift/drag ratio given on page 89 (Fig. 3.16) refers to the aerofoil only. The values of this ratio will be less when applied to the whole aeroplane, since the lift will be little, if any greater, than that of the wing alone, whereas the drag will be considerably more, perhaps twice as much. Furthermore, the change of the ratio at different angles of attack, in other words, the shape of the curve, will not be quite the same for the whole aeroplane. None the less, there will be a maximum value of, say 12 to 1 at about the same angle of attack that gave the best value for the wing, i.e. at 3° or 4°, and the curve will fall off on each side of the maximum, so that the lift/drag ratio will be less, i.e. the drag will be greater, whether we fly at a smaller or a greater angle of attack than 4°; in other words, at a greater or less speed than that corresponding to 4°, which our table showed to be 160 knots.

A typical lift/drag curve for a complete aeroplane is shown in Fig. 6.3, and these are the sort of figures we shall get from it –

Air speed		Angle of attack	L/D ratio		Total drag newtons
knots	m/s		Wing	Aeroplane	
100	51.5	15°	10.7	6.0	8330
120	61.8	9°	17.2	10.6	4720
140	72.0	6°	20.6	11.8	4240
160	82.4	4°	22.7	12.0	4170
180	92.6	2.6°	23.8	10.7	4670
200	103.0	1.7°	22.8	8.5	5880
220	113.2	1°	20.8	7.2	6940
240	123.7	0.5°	18.8	6.0	8330
260	134.0	0.2°	16.4	5.2	9615
280	144.0	0°	13.9	4.5	11 110
300	154.0	−0.4°	12.8	3.8	13 160

This table is very instructive, and shows quite clearly the effect of different air speeds in level flight on the total drag that will be experienced. It shows, too, that the least total drag is at the best lift/drag ratio, which in this case is at 4° angle of attack, which, in turn, is at about 160 knots air speed (Fig. 5.10).

The angle of attack that gives the best lift/drag ratio will be the same whatever the height and whatever the weight; it is simply a question of presenting the aeroplane to the air at the best attitude, and has nothing to do with the density of the air, or the loads that are carried inside the aeroplane, or even the method of propulsion.

This means that **the indicated air speed**, which is what the pilot must go by, **will be the same, whatever the height, but will increase slightly for increased loads**. The same indicated air speed means the same drag at any height, and therefore **the same range**.

On the other hand, the higher speed which must be used for increased weights means greater drag, because, looking at it very simply, even the same lift/drag ratio means a greater drag if the lift is greater. **So added weight** means added drag – in proportion – and therefore **less range** – also in proportion.

Let us go back to the newton metres, the 7 600 000 joules that we hope to get from 1 litre of fuel. How far can we fly on this? At 100 knots we shall be able to go 7 600 000 divided by 8330, i.e. about 912 metres; at 120 knots 1610 m; at 140 knots 1792 m; at 160 knots 1822 m; at 180 knots 1627 m; at 200 knots 1292 m; and at 220, 240, 260 and 280 knots, 1095, 912, 790, 684 metres respectively, and at 300 knots only 577 metres. These will apply at whatever height we fly. If the load is 60 kN instead of 50 kN each distance must be divided by 60/50, i.e. by 1.20; if the load is less than 50 kN each distance will be correspondingly greater.

Now to sum up this interesting argument: **in order to obtain the maximum range, we must fly at a given angle of attack, i.e. at a given indicated air speed, we may fly at any height**, and we **should carry the minimum load**; but **if we must carry extra load, we must increase the air speed.**

That is the whole thing in a nutshell from the aeroplane's point of view. Unfortunately, there are considerations of engine and propeller efficiency, and of wind, which may make it advisable to depart to some extent from these simple rules, and there are essential differences between jet and propeller propulsion in these respects. We cannot enter into these problems in detail, but a brief survey of the practical effects is given in the next paragraphs.

Another way of thinking of the significance of flying with minimum drag is to divide the total drag into induced drag – which decreases in proportion to the square of the speed – and all-the-remainder of the drag – which increases in proportion to the square of the speed. This idea is well illustrated in Fig. 5.10 and in the numerical examples (Appendix 3).

EFFECTS OF HEIGHT – PROPELLER PROPULSION

So far as the aeroplane is concerned, we will get the same range and we should fly at the same indicated speed, whatever the height. Now, **although the drag is the same** at the same indicated speed at all heights – the **power is not**. This may sound strange, but it is a very important fact. If it were not so, if we needed the same power to fly at the same indicated speed at all heights, then the advantage would always be to fly high, the higher the better, because for the same power the

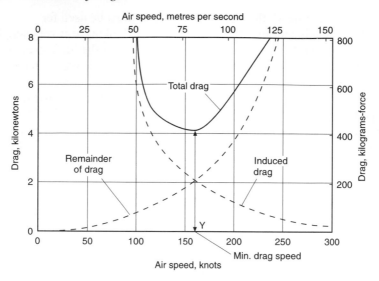

Fig. 5.10　Flying for maximum range: how the total drag is made up
Induced drag decreases with square of speed. Remainder of drag increases with
square of speed. Total drag is the sum of the two; and is a minimum when they are
equal.

higher we went, the greater would be our true speed. However, it can
hardly be considered a proof that an idea is incorrect simply because it
would be very nice if it were correct. The real explanation is quite
simple. Power is the **rate of doing work**. Our fuel gives us so many
newton metres, **however long we take to use it**. But if we want the
work done quickly, if we want to pull with a certain **thrust** through a
certain **distance** in a certain **time**, then the power will depend on the
thrust **and** the distance **and** the time, in other words, on the **thrust** and
the **velocity**. But which velocity, indicated or true? Perhaps it is easier
to answer that if we put the question as, which **distance**? Well, there is
only one distance, the actual distance moved, the **true** distance. So it is
the **true air speed that settles the power. The higher we go, the
greater is the true air speed for the same indicated speed and
therefore the greater the power required**, although the thrust and
the drag remain the same.

　Now a reciprocating engine can be designed to work most efficiently
at some considerable height above sea-level, if it is supercharged. If we
use it at sea-level, and if we fly at the best speed for range, the thrust
will be a minimum, that is what we want, but, **owing to the lower
speed, little power will be required from the engine**. That may
sound satisfactory, but actually it is not economical; the engine must be

throttled, the venturi tube in its carburettor is partially closed, the
engine is held in check and does not run at its designed power, and,
what is more important, does not give of its best efficiency; we can say
almost literally that it does not give best value for money. In some cases
this effect is so marked that it actually pays us, if we **must** fly at sea-
level, to fly considerably faster than our best speed and use more power,
thereby using the aeroplane less efficiently but the engine more
efficiently. But to obtain maximum range, both aircraft and engine
should be used to the best advantage, and this can easily be done if we
choose a greater height such that when we fly at the correct indicated
speed from the point of view of the aeroplane, the engine is also working
most efficiently, that is to say, the throttle valve is fully open, but we can
still fly with a weak mixture. At this height, which may be, say, 15 000 ft
(4570 m), we shall get the best out of both aeroplane and engine, and so
will obtain maximum range.

Here the reader may be wondering what governs the operating height
the designer chooses. This may be selected for terrain clearance, cruise
above likely adverse weather conditions or the engine may be sized for
take-off performance and the cruising altitude follows as a by-product to
give full throttle cruise.

What happens at greater heights? At the same indicated speed we
shall need more and more power; but if the throttle is fully open, we
cannot get more power without using a richer mixture. Therefore we
must either reduce speed and use the aircraft uneconomically, or we
must enrich the mixture and use the engine uneconomically.

**Thus there is a best height at which to fly, but the height is
determined by the engine efficiency** (and to some extent by
propeller efficiency) **and not by the aircraft, which would be
equally good at all heights.** The best height is not usually very
critical, nor is there generally any great loss in range by flying below that
height. It may well be that considerations of wind, such as are explained
in the next paragraph, make it advisable to do so.

RANGE FLYING – EFFECTS OF WIND

If the flight is to be made from A towards B and back to A, then **wind
of any strength from any direction will adversely affect the
radius of action.** This fact, which at first sounds rather strange, but
which is well known to all students of navigation, can easily be verified
by working out a few simple examples, taking at first a head and tail
wind, and then various cross-winds. But the wind usually changes in
direction and increases in velocity with height, and so a skilful pilot can

sometimes pick his height to best advantage and so gain more by getting the best, or the least bad, effect from the wind than he may lose by flying at a height that is slightly uneconomical from other points of view. It may pay him, too, to modify his air speed slightly according to the strength of the wind, but these are really problems of navigation rather than of the mechanics of flight.

FLYING FOR ENDURANCE – PROPELLER PROPULSION

We may sometimes want to stay in the air for the longest possible time on a given quantity of fuel. This is **not** the same consideration as flying for maximum range. **To get maximum endurance, we must use the least possible fuel in a given time, that is to say, we must use minimum power.** But, as already explained, power means drag × velocity, the velocity being true air speed. Let us look back to the table on page 178 showing total drag against air speed; multiply the drag by the air speed and see what happens.

This table shows that, although **the drag is least at about 160 knots, the power is least at about 125 knots** (see also Fig. 5.11). The explanation is quite simple; by flying slightly slower, we gain more (from the power point of view) by the reduced speed than we lose by the increased drag. **Therefore the speed for best endurance is less than the speed for best range** and, since we are now concerned with **true** speed, the **lower the height, the better.**

The endurance speed is apt to be uncomfortably low for accurate flying; even the best range speed is not always easy and, as neither is very critical, the pilot is often recommended to fly at **a somewhat higher speed.**

The reader who would like to consider endurance flying a little

Air speed		Drag newtons	Drag × Air speed	Power kW
knots	m/s			
100	51.5	8330	428 995	429
120	61.8	4720	291 696	292
140	72.0	4240	305 280	305
160	82.4	4170	343 608	344
180	92.6	4670	432 442	432
200	103.0	5880	605 640	606
220	113.2	6940	785 608	786
240	123.7	8330	1 030 421	1030
260	134.0	9615	1 288 410	1288
280	144.0	11 110	1 599 840	1600
300	154.0	13 160	2 026 640	2027

Fig. 5.11   Flying for maximum endurance
Maximum endurance is at speed of minimum power (X). Maximum range
(minimum drag) is at speed *B* since AB/OB = Power/Speed = (Drag ×
Speed)/Speed = Drag, and AB/OB is least when OA is a tangent to the power
curve.

further should look back to page 95. Here he will find some of the
desirable qualities of a good aerofoil. Among these was a high maximum
value of $C_L^{\frac{3}{2}}/C_D$ – the quality which means minimum power, i.e.
maximum endurance. There we were considering only the aerofoil, and
the aeroplane is not quite the same thing as regards values, but the idea
is the same. So **for endurance flying we must present the
aeroplane to the air at the angle of attack that gives the best
value of** $C_L^{\frac{3}{2}}/C_D$ **(for the aeroplane), and this will be a greater angle
of attack and so a lower speed, than for range.**

FLYING FOR RANGE – JET PROPULSION

In trying to get maximum range or endurance out of any aircraft we are,
in effect, simply trying to get maximum value for money, the value
being the range or endurance and the money being the fuel used. We
shall only get the maximum overall efficiency if in turn we get the
maximum efficiency **at each stage** of the conversion of the fuel into
useful work done. The three main stages are **the engine**, the **system of
propulsion**, and the **aeroplane**.

This applies to **every type of aircraft** – it is necessary to emphasise this point because there seems to be a growing tendency to think that jet or rocket propulsion involves completely new principles. This is not so – the principles are exactly the same, the only difference lies in the degree of importance of the various efficiencies.

From the point of view of an aeroplane, flying for maximum range means flying with minimum drag. It is in that condition that the aeroplane is most efficient no matter by what means it is driven. But if, when we fly with minimum drag, either the propulsive system, or the engine, or both, are hopelessly inefficient – then, rather obviously, it will pay us to make some compromise, probably by flying rather faster than the minimum drag speed.

From the point of view of an aeroplane, as an aeroplane, we shall obtain the same range at whatever height we fly, provided we fly in the attitude of minimum drag. But if at some heights the propulsive system, or the engine, or both, are more efficient than at other heights – then, rather obviously, it will pay us to fly at those heights so as to get the maximum overall value out of the engine-propulsion-aeroplane system.

Now an aeroplane is an aeroplane whether it is driven by propeller or jet and, as an aeroplane, the same rules for range flying will apply. But when the efficiencies of the propulsion system and the engine are included the overall effects are rather different. In the propeller-driven aeroplane we do not go far wrong if we obey the aeroplane rules although, even so, it usually pays us to fly rather faster than the minimum drag speed because, by so doing, engine and propeller efficiency is improved – and flying is more comfortable. It also definitely pays us to fly at a certain height because at that height the engine-propeller combination is more efficient. But in the **main it is the aeroplane efficiency that decides the issue. Not so with the jet aircraft.**

There are two important reasons for the difference –

1. **Whereas the thrust of a propeller falls off as forward speed increases, the thrust of a jet is nearly constant at all speeds (at the same rpm).**

2. **Whereas the fuel consumption in a reciprocating engine is approximately proportional to the power developed, the fuel consumption in jet propulsion is approximately proportional to the thrust.**

Both of these are really connected with the fact that the efficiency of the jet propulsion system increases with speed, and this increase in efficiency is so important that it is absolutely necessary to take it into account, as well as the efficiency of the aeroplane. When we do so we find that **we shall get greater range if we fly a great deal faster than the minimum drag speed.** The drag will be slightly greater – not much, because we are on the low portion of the curve (Fig. 5.12) – the thrust, being equal to the drag, will also of course be slightly greater, and so will the fuel consumption in litres per hour. The speed, on the other hand, will be considerably greater and so we shall get more miles per hour. Everything, in fact, depends on getting the **maximum of speed compared with thrust**, or speed compared with drag. In short, we **must fly at minimum drag/speed** which as the figure shows, **will always occur at a higher speed than that giving minimum drag** (Fig. 5.12). **So to get maximum range jet aircraft must fly faster than propeller-driven aircraft** – the difference being due to the different relationship between efficiency and speed in the two systems.

As a matter of interest let us go back to our table of figures and work out for each speed the value of drag/speed –

Air-speed	Drag/Speed	Air-speed	Drag/Speed
100	8330/100 = 83.3	220	6940/220 = 31.5
120	4720/120 = 39.3	240	8330/240 = 34.7
140	4240/140 = 30.3	260	9615/260 = 37.0
160	4170/160 = 26.1	280	11 110/280 = 39.7
180	4670/180 = 25.9	300	13 160/300 = 43.9
200	5880/200 = 29.4		

Since, in this instance, we are only concerned with the air speed at which minimum drag/speed occurs, there is no need to convert the knots to m/s.

Note that the minimum value for drag/speed is at about 175 knots, so the range speed for this aircraft, if driven by jets, is 175 instead of 160 knots; but, what is more important, *note the shape of the drag/speed curve* (Fig. 5.13); whereas the other curves rise fairly steeply above the minimum value the drag/speed curve hardly rises at all between 170 and 280 knots, so with jet propulsion we can get good range anywhere between these speeds.

At what height shall we fly? That is an easy one to answer. We know that it makes no difference as far as the efficiency of the aeroplane is concerned – **but it makes all the difference to the efficiency of jet propulsion.** The aircraft will be in the same attitude, and we shall get the same drag and the same thrust, if we fly at the same indicated speed at altitude – **but the true speed will be greater.** Now it is the true speed that largely settles the overall efficiency so at **40 000 ft (12 200 m),** where the true speed is doubled, **the efficiency will be greatly**

Fig. 5.12   Flying for range: jet propulsion
The tangent OP to the curve of total drag gives the air speed Q at which drag/speed is a minimum.

increased, and, provided the fuel consumption remains proportional to thrust, **the range will be similarly increased. So to get range on jet aircraft – fly high.**

Since modern flights in jet aircraft may take place at heights such as 40 000 or 50 000 ft (12 200 m or 15 200 m) quite a large proportion of the flights may be spent in climbing and descending, and in order to obtain maximum range rather different speeds may be required say for climbing and for the level portion of the flight. The best speed, for instance, for a cruising climb may be 1.3 × speed for minimum drag, and for level flight 1.2 × speed for minimum drag, i.e. for the aeroplane which we have considered, 208 and 192 knots respectively.

FLYING FOR ENDURANCE – JET PROPULSION

If the argument has been followed so far, there will be no difficulty in understanding the problem of maximum endurance for jet-driven aircraft. Since fuel consumption is roughly proportional to thrust, we shall get **maximum endurance by flying with minimum thrust,** i.e. **with minimum drag**. So the endurance speed of a jet aircraft corresponds closely to the range speed of a propeller-driven aircraft, and from the comfort point of view, this makes the jet aircraft easier to fly in the condition of maximum endurance.

Since the thrust, and hence the consumption, should be the same at the same indicated speed at any height, it should not matter at what height we fly for endurance.

SUMMARY

The table below and Fig. 5.13 summarise the difference between jet and propeller-driven aircraft so far as range and endurance are concerned. They must be considered as a first approximation only – they take into account the aeroplane efficiency for the propeller-driven type (neglecting propeller and engine efficiency), and both aeroplane and propulsive efficiency for the jet-driven type (neglecting engine efficiency). All this means is that the more important factors have been

	Propeller	*Jet*
Speed for maximum range	Minimum drag (160)	Minimum drag/speed (175 up to 280)
Height for maximum range	Unimportant	High
Speed for maximum endurance	Minimum power (125)	Minimum drag (160)
Height for maximum endurance	Low	Unimportant

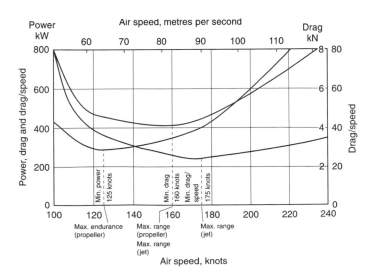

Fig. 5.13   Power, drag and drag/speed curves

taken into account, and the less important factors have been neglected. It is **not** the whole story, and should not be considered as such.

The figures in brackets are the speeds in knots for the particular aircraft that has been considered in this chapter.

Before leaving this important subject it should be made clear that flying control regulations, made in the interests of safety, may sometimes make it necessary to fly at flight levels which do not exactly correspond to the best interests of either aircraft or engines.

CAN YOU ANSWER THESE?

1. What are the four most important forces which act upon an aeroplane during flight?
2. What are the conditions of equilibrium of these four forces?
3. Are these forces likely to alter in value, and to move their line of action during flight?
4. Explain how it is that an aeroplane can fly level at a wide range of air speeds.
5. Is the relationship between air speed and angle of attack the same at height as at sea-level?
6. What is the effect of weight on the relationship between air speed and angle of attack?
7. On a propeller-driven aircraft –
   (*a*) Why will we get less range if we fly too high?
   (*b*) At what height should we fly for best endurance?
   (*c*) Why is the air speed for best endurance different from the air speed for best range?
8. On a jet-driven aircraft –
   (*a*) Under what conditions should we fly for maximum range?
   (*b*) At what height should we fly for maximum range?
   (*c*) At what speed and height should we fly for maximum endurance?

In Appendix 3 you will find some simple numerical examples on the problems of level flight.

GLIDING AND LANDING

GLIDING

Let us next consider the flight of an aeroplane while **gliding** under the influence of the force of gravity and without the use of the engine.

Of the four forces, we are now deprived of the thrust, and therefore when the aeroplane is travelling in a steady glide it must be kept in a state of equilibrium by the lift, drag, and weight only. This means that **the total force – that is to say, the resultant of the lift and drag – must be exactly equal and opposite to the weight** (see Fig. 6.1). But the

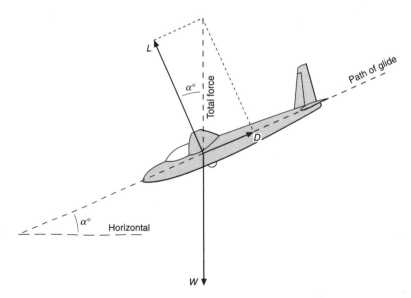

Fig. 6.1   Forces acting on an aeroplane during a glide

lift is now at right angles to the path of the glide, while the drag acts directly backwards parallel to the gliding path.

GLIDING ANGLE

By a process of simple geometry, it is easy to see that the angle formed between the lift and the total force is the same as the angle α between the path of the glide and the horizontal, which is called the **gliding angle**. Therefore

$D/L = \tan \alpha$.

This means that the less the value of D/L – i.e. **the greater the value of L/D – the flatter will be the gliding angle**.

From this simple fact we can very easily come to some important conclusions; for instance –

1. **The tangent of the gliding angle is directly dependent on the L/D**, which is really the 'efficiency' of the design of the aeroplane, and therefore the more 'efficient' the aeroplane, the farther it will glide, or, expressing it the other way round, the **measurement of the angle of glide will give a simple estimate of the efficiency of the aeroplane**.

The word 'efficiency' is apt to have a rather vague meaning, and we are using it here in a particular sense – we are not concerned with efficiency as regards engine power, nor with the merits of its internal structure from the standpoint of weight for strength. We are concerned only with the success or otherwise of the designer in obtaining the maximum amount of lift with the minimum of drag, or what might be called the 'aerodynamic' merit of the aeroplane. For instance, our conclusion shows that any improvement which reduces the drag will result in a flatter gliding angle.

It will be noticed that this is the same criterion as for maximum range, so that an aeroplane that has a flat gliding angle should also be efficient at flying for range, neglecting the influence of the propulsion efficiency.

2. **If an aeroplane is to glide as far as possible, the angle of attack during the glide must be such that the lift/drag is a maximum.**

The aeroplane is so constructed that the riggers' angle of incidence is a small angle of, say, 2° or 3°. This particular angle is chosen because it is the most suitable for level flight. As was explained when considering the characteristics of aerofoils, the modern tendency is to make this angle rather less than the angle of maximum L/D (because we are out for speed), but, even so, it will be within a degree or so of that angle, so it is true to say that the angle of attack during a flat glide will be very nearly the same as that during straight and level flight, and almost exactly the same as when flying for maximum range with piston engines.

The pilot finds it fairly easy to maintain ordinary horizontal flight at the most efficient angle because his fuselage is then in a more or less horizontal position. When gliding however, the task is not always so easy. Sophisticated modern aircraft may be fitted with an angle of attack indicator, but on older and simpler types this is not normally the case. Fortunately, as in level flight, there is a direct connection between the air speed and the angle of attack, and therefore the air speed can be found which gives the best gliding angle, and this acts as a guide to the pilot. The fact remains, however, that it requires considerable skill, instinct, or whatever one likes to call it, on the part of a pilot to glide at the flattest possible angle. This is the type of skill which is especially needed by the pilot of a motorless glider or sailplane.

It should be noted that, although there is a relationship between air speed and angle of attack on the glide just as there is in level flight, the relationship is not exactly the same, and the speed that gives the flattest gliding angle is usually rather less than the speed that gives maximum range. The difference, however, is small and the principle is the same.

3. **If the pilot attempts to glide at an angle of attack either greater or less than that which gives the best L/D, then in each case the path of descent will be steeper.**

Perhaps this conclusion may be considered redundant because it is simply another way of expressing the preceding one. It is purposely repeated in this form because there seems to be such a strong natural instinct on the part of pilots to think that if the aeroplane is put in a more horizontal attitude it will glide farther. Even if one has never flown it is not difficult to imagine the feelings of a pilot whose engine has failed, and who is trying to reach a certain field in which to make a forced landing. It gradually dawns on him that in the way in which he is gliding he will not reach that field. What, then, could be more natural than that he should pull up the nose of his aeroplane in his efforts to reach it? What happens? In answer to this question the student often says that he will stall the aeroplane. Not necessarily. He should in the first place have been gliding nowhere near the stalling angle, but at an angle of attack of only about 3° or 4°, so that he has many degrees through which to increase the angle before stalling. But what will most certainly happen is that the increase in angle will decrease the value of L/D and so **increase the gliding angle**, and although the aeroplane will lie flatter to the horizontal, it will glide towards the earth at a steeper angle and will not reach even so far as it would otherwise have done. The air speed during such a glide will be less than that which gives the best gliding angle.

Suppose, on the other hand, that, when a pilot is gliding at the angle of attack which gives him the greatest value of L/D, he puts the nose of the aeroplane down, this will decrease the angle of attack which, as before, will decrease the value of L/D and therefore **increase the steepness of the gliding path**, the air speed this time being greater than that which gives the best gliding angle.

It is not easy to visualise the angle of attack during a glide, and the reader, like the pilot, must be careful not to be confused between the direction in which the aeroplane is pointing and the direction in which it is travelling. It is hoped that the figures may help to make this important point clear (Figs 6.2 and 6.3).

Fig. 6.2   How the angle of attack affects the gliding angle
(*a*) Slow glide at slope of 1 in 8 (7°). Angle of attack 13°. Speed 115 knots.
(*b*) Fast glide at slope of 1 in 8 (7°). Angle of attack $1\frac{1}{2}$°. Speed 210 knots.
(*c*) Flattest glide at slope of 1 in 12 (5°). Angle of attack 4°. Speed 155 knots.
Note. In the diagram the gliding angles, and the differences between them, have been exaggerated so as to bring out the principles.

In the previous chapter we discovered that the ratio of lift to drag of complete aeroplanes may be in the neighbourhood of 8, 10 or 12 to 1. These values correspond to gliding angles of which the tangents are 1/8, 1/10 and 1/12, i.e. approximately 7°, 6° and 5° respectively. Thus, neglecting the effect of wind, a pilot will usually be in error on the right side if he assumes that he can glide a kilometre for every 200 metres of height, i.e. if he reckons on a gliding angle of which the tangent is 1/5.

REAL AND APPARENT ANGLES OF GLIDE

Let us remember once again that gliding must be considered **as relative to the air**. To an observer on the ground an aeroplane gliding

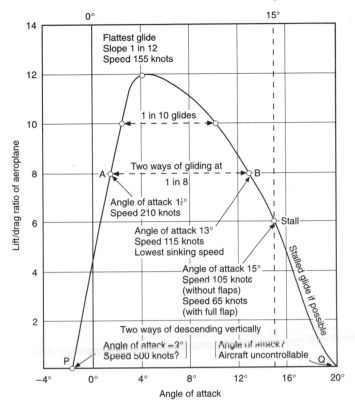

Fig. 6.3 Lift/drag curve and gliding angles of aeroplane

into the wind may appear to remain still or, in some cases, even to ascend. In such instances there must be a wind blowing which has both a horizontal and an upward velocity, and to an observer travelling on this wind in a balloon the aeroplane would appear to be travelling forwards and descending. When viewed from the ground an aeroplane gliding against the wind will appear to glide more steeply, and will, **in fact**, glide more steeply relative to the ground (Fig. 6.4); and when

Fig. 6.4 Effect of wind on angle of glide relative to the earth

gliding with the wind it will glide less steeply than the real angle
measured relative to the air – the angle as it would appear to an observer
in a free balloon.

EFFECT OF WEIGHT ON GLIDING

It is commonly thought that heavy aeroplanes should glide more steeply
that light aeroplanes, but a moment's reflection will make one realise
that this is not so, since the gliding angle depends on the ratio of lift to
drag, which is quite independent of the weight. Probably another
moment's reflection will reveal the simple fact that modern aeroplanes,
which are heavier than older types, have a much flatter angle. **No,
neither in principle nor in fact does weight have an appreciable
influence on the gliding angle, but what it does affect is the air
speed during the glide.**

Look back for a moment at Fig. 6.1. Imagine an increase in the line
representing the weight; there will need to be a corresponding increase
in the total force, and a greater lift and a greater drag. But the
proportions will all remain exactly the same, the same lift/drag ratio, the
same gliding angle. But the greater lift, and greater drag, can only be got
by greater speed. If we now think back to flying for range, it will be
remembered that the condition was the same: greater weight meant
greater speed. But there is an interesting and important difference in
this case. In flying for range, greater speed meant greater drag, greater
thrust, and so less range. In gliding without engine power, greater speed
means greater drag, but now the 'thrust' is provided by the component
of the weight which acts along the gliding path and this, of course, is
automatically greater because the weight is greater. **So greater weight
does not affect the gliding angle and does not affect the range,** on
a pure glide – **but it does affect the speed.**

ENDURANCE ON THE GLIDE

The conclusion of the previous paragraph might perhaps lead one to ask
whether, in that case, there is any need for a sailplane to be built of light
construction. The answer is definitely – **Yes.** A sailplane (Fig. 6A) must
have a flat gliding angle if it is to get any distance, any range from its
starting point; but, even more important, it must have a **low rate of
vertical descent or sinking speed**; it must be able to stay a long time
in the air and be able to take advantage of every breath of rising air,
however slight. Sailplane pilots do sometimes add ballast so as to
increase the flight speed as this can be useful under certain

Fig. 6A   Gliding
(By courtesy of Slingsby Sailplanes Ltd)
The Skylark 4 with large aspect ratio and good value of lift/drag; needing spoilers to increase gliding angle when necessary

circumstances. However, a description of such advanced sailplane techniques is best left to books devoted specifically to that subject. It is easy to see that **the rate of vertical descent** depends **both** on the **angle of glide** and on the **air speed during the glide**. Therefore to get a low rate of descent we need a good lift/drag ratio, i.e. good aerodynamic design, and a low air speed, i.e. low weight.

Actually we shall get a lower rate of descent **by reducing speed below that which gives the flattest glide**; this is because we gain more by the lower air speed than we lose by the steeper glide. **Thus there is an 'endurance' speed for gliding just as for level flight and, as before, it is lower than the range speed, and corresponds to the speed for minimum power requirement.**

DISADVANTAGES OF FLAT GLIDING ANGLE

It should not be thought that a flat gliding angle is always an advantage; when approaching a small airfield near the edge of which are high obstacles, it is advisable to reach the ground as soon as possible after passing over such obstacles. In these circumstances a flat gliding angle is a definite disadvantage, and even if the aeroplane is dived steeply it will pick up speed and will tend to float across the airfield before touching the ground.

The gliding angle can be steepened **by reducing the ratio of lift to drag**; this can be done by decreasing the angle of attack (resulting in too high a speed), or by increasing the angle of attack (resulting in an air speed which may be too low for safety), or by using an **air brake** (Fig. 6.5). The last is by far the most satisfactory means, and the air brake may take the form of some kind of flap, such as was described in the chapter on aerofoils; but the modern tendency is **to use the lift types of flap when lift is required**, and **separate air brakes or spoilers when drag** is **required**.

(a) Spoiler on top of wing

(b) Spoiler below wing

(c) Split flap

(d) Double flap

(e) Spoiler round fuselage

(f) Tail parachute

(g) Also reversible pitch propellers and jet engine thrust reversers

Fig. 6.5   Air brakes

LANDING

The art of landing an aeroplane consists of bringing it in contact with the ground **at the lowest possible vertical velocity** and, at the same time, **somewhere near the lowest possible horizontal velocity relative to the ground**. It is true that in certain circumstances a fast landing may be permissible, and that some modern aircraft are **flown onto the ground** in a definitely unstalled condition, but the general rule applies to the landings of many slower and lighter types, and especially to forced landings in which everything usually depends on the minimum horizontal velocity being achieved.

The reader will have noticed that it is the horizontal velocity **relative to the ground** which must be reasonably low. Now, the first step in this direction is to land against the wind and so reduce the ground speed. This, however, is entirely up to the pilot; in our present problem we are only concerned with a low air speed. Given this low air speed, the pilot can, by landing into the wind obtain a low ground speed. In the case of

Fig 6B   Deck landing
(By courtesy of the former British Aircraft Corporation, Preston)
The Jaguar, designed by Breguet and BAC, landing on a deck.

landing on the decks of ships (Fig. 6B), if the ship herself steams into
the wind, the ground speed will be still further reduced. Supposing, for
instance, the minimum air speed of an aeroplane is 80 knots, the wind
speed is 20 knots, and the ship is steaming at 30 knots into the wind,
then the 'ground' speed of the aeroplane when landing will be only 30
knots; while if the wind speed had been 50 knots, the 'ground' speed
would have been reduced to nil – a perfectly possible state of affairs.

In an early chapter it was mentioned that the wind speed is apt to be
irregular near the ground, and it is when landing that such irregularity
may be important. If the wind speed suddenly decreases, the aircraft,
owing to its inertia, will tend to continue at the same ground speed and
will therefore lose air speed, and, if already flying near the critical speed,
may stall. Similarly, if the wind speed suddenly increases, the aircraft
will temporarily gain air speed and will 'balloon' upwards, making it
difficult to make contact with the ground at the right moment. Such
instances may occur in changeable and gusty winds, in up-currents
caused by heating of parts of the earth's surface, in cases of turbulence
caused by the wind flowing over obstructions such as hills and hangars,
and due to wind gradient. Of these, wind gradient is probably the most

important, and the most easily allowed for. An aeroplane, when landing against a high wind, will encounter a decreasing wind speed as it descends through the last few feet and will be in danger of stalling unless it has speed in hand to compensate for the air speed lost. If landing up a slope or towards a hangar, one may suddenly run into air which is blanketed by the obstruction, or a head wind may even become a following wind blowing up the hill or towards the hangar. In a really high wind, and when flying a small light aircraft, these conditions may be dangerous, and the obvious moral is to allow for them by approaching to land at a higher speed than usual.

The vertical velocity of landing can be reduced to practically nothing provided the forward velocity is sufficient to keep the aeroplane in horizontal flight – that is to say, provided the lift of the wings is sufficient to balance the weight of the aeroplane.

We have already seen that there is a definite relationship between the indicated air speed and the angle of attack. Fig. 6.6 illustrates the attitudes of an aeroplane at various speeds and the corresponding angles of attack required to maintain level flight: (*a*) shows the attitude of maximum speed; (*b*) that of normal cruising flight; (*c*) that for an ordinary landing; and (*d*) the attitude when fitted with flaps and slots and flying as slowly as possible.

Now, since lift must equal weight, and must also equal $C_L . \frac{1}{2}\rho V^2 . S$, it is **quite obvious that if V is to be as small as possible, C_L must be as large as possible**. The pilot may never have heard of a lift coefficient, and he may be none the worse a pilot for that; but, consciously or unconsciously, he **will increase C_L by increasing the angle of attack until he decides** (it matters not whether his decision is based on scientific knowledge, instinct or bitter experience!) **that any further increase in the angle of attack will decrease rather than increase the lift**; in other words, until he has come near to that **stalling angle** which we considered so fully when dealing with aerofoils. At this angle (about 15° to 20° in the case of an ordinary aerofoil), C_L is at its maximum, and therefore V is a minimum.

If the pilot, through lack of any of the three qualities mentioned above, exceeds this angle, then both C_L and V will decrease; therefore $C_L . \frac{1}{2}\rho V^2 . S$ can no longer equal the weight and the aeroplane will commence to drop. For 20 m, 50 m, or more, the vertical component of velocity will increase and the nose of the aeroplane will drop, therefore the pilot must beware that, when he does this experiment of flying as slowly as possible, he is either very near the ground or at a considerable height above it. In fact, slow landings should not be practised between 1 and 500 m from the earth's surface, and the whole skill of the pilot is exercised in approaching the ground in such a manner that he has

Fig. 6.6   Attitudes for level flight

reached the correct condition of affairs just as he skims the surface of the runway, provided, of course, that he has sufficient clear run in front in which to pull up after landing.

STALLING SPEED

Much of what has been said applies not only to level flight, but to stalls when gliding, climbing or turning; for instance, when banking on a turn the lift on the wings must be greater than the weight, and therefore the stalling speed is higher than when landing. Also at height the air density ρ will be less, and this means that in order to keep $C_L \cdot \frac{1}{2}\rho V^2 \cdot S$ equal to the weight, the stalling speed V will be greater than at ground level. This fact is not, however, of great importance, for two reasons – first because at a considerable height an accidental stall does not matter, as there is plenty of time to recover; and, secondly, because, although the stalling speed is in reality greater, the **air speed indicator**, which is in itself worked by the effect of the air density, **will record the same speed when the aeroplane stalls** as it did at ground level; in other words, the **indicated stalling speeds** will remain the same at all heights.

But on **high airfields**, such as are found in mountainous countries, **the true landing speed of an aeroplane will be appreciably higher** than on sea-level airfields; and in **tropical countries** the air density is decreased owing to the high temperatures, and **the true landing speed is consequently increased**. The taking-off speed, and the run required, are also increased in both these instances, and this is perhaps an even more important consideration.

When stalling intentionally the aeroplane is pulled into a steeply climbing attitude and the air speed allowed to drop to practically nil until the nose suddenly drops or, as frequently happens, one wing drops and the aeroplane commences to dive or spin. It is simply a case of 'the same, only more so'.

Before leaving the subject of stalling it might be as well to mention that there has always been difficulty in deciding upon an exact definition of stalling or stalling speed. The stall occurs because the smooth airflow over the wing becomes separated – **but this is a gradual process**. At quite small angles of attack there is some turbulence near the trailing edge; as the angle increases, the turbulence spreads forward. What is even more important is that it also spreads spanwise, usually from tip to root on highly-tapered wings, and from root to tip on rectangular wings. **If we define the stall as being the break up of the airflow, when did it occur?** There may be buffeting of the tail plane or main planes, but this too may be slight and unimportant, or it may be violent. As a result of the change from smooth to turbulent airflow the curve of lift coefficient reaches a maximum and then starts to fall. We defined the stalling angle in Chapter 3 as the **angle at which the lift coefficient is a maximum. But how does the pilot know that it is at its maximum value? All the pilot knows is that if he tried to fly below a certain speed he gets into difficulties.** How great the difficulties depends on the type of aircraft, and the extent to which the pilot can overcome them depends on a lot of things but particularly on his own skill. **Some pilots can fly at lower speeds than others – so which is the stalling speed?**

In fact, there are different definitions of stalling according to the point of view of the person who wishes to define it – the pilot looks at it one way, the aerodynamicist another, and so on. What is important is that each should realise that it is his own definition, and that all these things do not necessarily occur at the same time.

POSSIBILITIES OF LOWER MINIMUM SPEEDS

What are the possibilities of reducing this minimum velocity of flight?

In all forms of transport, with the exception of flying, the maximum speed attainable is the chief consideration; in fact, it is really the **only** consideration because the minimum speed may be reduced to nothing or even reversed. But in the exceptional case of flight it is equally important to obtain a low minimum speed as it is to obtain a high maximum speed. This low speed is of such importance that it is apt to be exaggerated at the expense of the maximum speed. Whatever we say about obtaining a low landing speed, **we must never forget that the chief advantage of flight over other means of transport depends on the high speed obtainable.** But, provided we bear this in mind, everything must be done to reduce the landing (and taking off) speed, because only in this way can flying be made a popular and safe means of transport. The minimum speed of most light aeroplanes is as much as 50 or 60 knots, and of some more than 100 knots; imagine a motor car which could not travel at less than 185 km/h!

HIGH LIFT AEROFOILS

What, then, has been done, and what can yet be done to decrease minimum speed? If V is to be small, C_L must be as large as possible. In other words, we must have a larger lift coefficient. So the aerofoils which give the largest maximum lift coefficient will give the lowest minimum speeds. Unfortunately, however, these aerofoils are usually those with a large drag, and so they seriously affect the high speed end of the range. Therefore we must turn to some device by which the shape of the aerofoil can be altered during flight, and so we naturally think of flaps and slots.

In an earlier chapter (Fig. 3.32) we noticed the effect of various kinds of flaps and slots on maximum lift and speed range. The idea of variable camber is an old one, but it is only in recent years, when maximum speeds have increased so much, that the problem has become really urgent and these devices have come into their own. In this respect necessity has proved to be the mother of invention, and many and varied have been the devices which have been tried. It is not easy to compare the respective merits of all these types of slots and flaps, or combinations of slot and flap, because so many conflicting qualities are required. If a low speed was our **only** aim, the problem would be comparatively simple, the device giving the highest maximum lift coefficient being the most suitable. But what we really need is a **low minimum speed and a high maximum speed, i.e. a good speed range**. This condition means that the device must be such that it can be altered, or will alter automatically, from the position giving maximum lift (e.g. slot open or flap down), to the position of minimum drag (e.g. slot closed or flap neutral). Even that is not the end of our requirements for, having landed as slowly as possible, we must pull up quickly after landing. The former (slow landing) needs high lift, the latter (quick pull-up) needs much drag, lift being of no consequence at all. For a quick pull-up we really need a definite **air brake** which will assist the wheel brakes. Notice, however, that **an air brake cannot reduce actual landing speed**, it can only improve the pull-up **after** landing. Once we are on the ground we want to get rid of lift as quickly as possible to achieve minimum wheel brake effectiveness; hence the use of lift dumpers after touchdown. These are usually devices which disrupt the flow over the top of the wing, increasing drag, and decreasing lift.

Yet another aspect of the problem, so far as landing is concerned, is the question of **attitude**, and in this respect some of the otherwise most effective types of slots and flaps are at a disadvantage, for they attain their maximum lift coefficient at a greater angle of attack than the ordinary aerofoil; this means that in order to make full use of them the angle of attack when landing may need to be 25°, or even more. But when an aeroplane with a tail-wheel type of undercarriage rests with its main and tail wheels on the ground, the angle of inclination of the wings is only about 15°. With a nose-wheel type of undercarriage the problem is if anything, worse – as the reader will no doubt realise – for, in order to land at an angle of 25°, we are faced with four possibilities all of which have serious drawbacks –

1. **To allow the tail to touch the ground before the main wheels** (Fig. 6.7a). This is hardly a practical proposition.

2. **To have a much higher undercarriage** (Fig. 6.7b). This will cause extra drag and generally do more harm than good.

3. **To provide the main planes with a variable incidence gear**

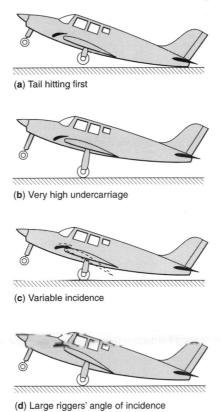

(a) Tail hitting first

(b) Very high undercarriage

(c) Variable incidence

(d) Large riggers' angle of incidence

Fig. 6.7   Difficulties of landing at large angles of attack

similar to that which is sometimes used for tail planes (Fig. 6.7c). This involves considerable mechanical difficulties.

4. **To set the wings at a much greater angle to the fuselage** (Fig. 6.7d). This means that in normal flight the rear portion of the fuselage sticks up into the air at an angle which not only looks ridiculous, but which is inefficient from the point of view of drag. It gives the appearance of a 'broken back', but has sometimes been used for aircraft designed for deck landing since it not only gives a low landing speed, but a quick pull-up after landing.

The only real answer lies in the design of flaps and slots which must be such that the effective camber of the wing can be altered so as to give maximum lift **and still maintain a reasonable attitude for landing**. Improvements in design on these lines have resulted in a real, and by no means negligible, reduction in landing speed (or perhaps, more correctly, has halted further increases in landing speed) but at the expense of ever more sophistication – and ever more complication.

WING LOADING

We assumed at an earlier stage, that the area of the wings was bound to remain constant, but inventors have from time to time investigated the problem of providing wings with **variable area**.

Since $W = C_L \cdot \frac{1}{2}\rho V^2 \cdot S$

$W/S = C_L \cdot \frac{1}{2}\rho V^2$

The fraction W/S, weight divided by wing area, is called the **wing loading** of the aeroplane. An increase in the wing area should reduce the value of W/S, and so also reduce the minimum velocity at which level flight is possible.

The objections to variable area are chiefly mechanical; the operating gear means extra weight and so, since W will increase, it by no means follows that W/S will actually decrease if the wing area is increased; also, once again, the extra sophistication means more complication, more levers for the pilot to fiddle with, more chances of something going wrong. Some flap systems do, however, give an increase in wing area during landing as they extend beyond the trailing edge of the wing.

Apart from the question of altering the wing area during flight, the equation $W/S = C_L \cdot \frac{1}{2}\rho V^2$ shows us that, **other things being equal, the aeroplane with a low wing loading will have a lower minimum speed than one with a high wing loading**. The so-called 'light aeroplane' may have a high wing loading, and therefore a high landing speed; in other words, it is not a question of weight, but of weight compared with wing area, that settles the minimum speed. The wing loading of a sailplane may be less than $100\,\mathrm{N/m^2}$, of a training aeroplane 300 to $1000\,\mathrm{N/m^2}$, of a fighter, bomber or airliner anything from 1500 up to 2000, 3000 or more $\mathrm{N/m^2}$. **The modern tendency is to increase wing loading by reducing wing area and thus raising the maximum speed, and then using flaps to keep down the landing speed.** The student is advised to work out the wing loading of existing aeroplanes and to compare the figures obtained with their landing speeds; in making this comparison, however, he must be careful to notice the above phrase **'other things being equal'**, because the maximum lift coefficient of the aerofoil used also affects the result. An old example of high wing loading, very high for that time, was the 1977 $\mathrm{N/m^2}$ of the S.6b Schneider Trophy racing seaplane; the corresponding figure for fighters like the 'Spitfire' and 'Hurricane' at the beginning of the Second World War was $1187\,\text{N}/\text{m}^2$, and for the German fighter, Messerschmitt 109, $1522\,\text{N}/\text{m}^2$. Modern figures may be considerably higher than these; the Concorde is about $4800\,\text{N}/\text{m}^2$, and this is not by any means the highest.

METHOD OF FINDING MINIMUM LANDING SPEED

It is easy to work out simple problems on minimum or landing speeds by using the now familiar formula –

$$\text{Weight} = \text{Lift} = C_L \cdot \tfrac{1}{2}\rho V^2 \cdot S$$

If we denote the maximum value of the lift coefficient by C_L max, and the landing speed by V_L, then our formula becomes

$$W = C_L \cdot \text{max } \tfrac{1}{2}\rho V_L^2 \cdot S$$

Consider this problem –

Find the wing area required for an aeroplane of mass 1500 kg, if the minimum landing speed is to be 35 knots (65 km/h) and the maximum value of the lift coefficient for the aerofoil used is 1.2 (assume the air density to be $1.225\,\text{kg}/\text{m}^3$).

> *Data:* $W = 14715\,\text{N}$
> $\rho = 1.225\,\text{kg}/\text{m}^3$
> $V_L = 65\,\text{km/h} = 18\,\text{m/s}$
> $C_L\,\text{max} = 1.2$
> $S = ?$
> So $14715 = 1.2 \times \tfrac{1}{2} \times 1.225 \times 18 \times 18 \times S$
> $\therefore S = \mathbf{62\,m^2}$ **approx**

This is rather a large wing area for an aeroplane of this weight, and it is doubtful whether the structure involved would not make the total weight greater than 14 715 N, in which case, of course, the landing speed would be above 35 knots.

Suppose we could use a flapped wing with a maximum lift coefficient of 1.8 instead of 1.2, then, neglecting any small increase in weight, the necessary wing area to produce the same landing speed would be –

$$(1.2/1.8) \times 62 = \mathbf{41\,m^2}\textbf{ approx}$$

It would certainly be much easier to design a wing structure of this size so as to conform to a total weight of 14 715 N and, further, the reduced wing area would enable a much greater maximum speed to be obtained.

As another problem, let us compare the minimum landing speeds of the following –

(*a*) A sailplane of wing loading $100 \, \text{N/m}^2$.
(*b*) A training machine of wing loading $400 \, \text{N/m}^2$.
(*c*) A fighter of wing loading $1500 \, \text{N/m}^2$.
(*d*) The S.6b of wing loading about $2000 \, \text{N/m}^2$.
(*e*) An airliner of wing loading $3000 \, \text{N/m}^2$.

Supposing other things to be equal, e.g. taking ρ as $1.225 \, \text{kg/m}^3$, and assuming each machine is fitted with an aerofoil section having a maximum lift coefficient of 1.12, then –

(*a*) Wing loading $= W/S = 100 = C_L \max \frac{1}{2}\rho V_L{}^2$
$$= 1.12 \times \tfrac{1}{2} \times 1.225 \times V_L{}^2$$

$\therefore V_L{}^2 = (100 \times 2)/(1.12 \times 1.225) = 146$
$\therefore V_L = \textbf{12 m/s, i.e. 23 knots or 43 km/h}$

Similarly for

(*b*) Landing speed $= 47$ knots (87 km/h)
(*c*) 91 knots (169 km/h)
(*d*) 106 knots (195 km/h)
(*e*) ? knots (? km/h). Reader, work it out.

Such is the type of problem which confronts the designer of an aeroplane in the very early stages, when, by a process of calculations, he has to decide such important items as the wing area, the type of aerofoil, and the landing speed. It will now be obvious that in order to settle these he must know the weight – the weight of an aeroplane which he has not yet commenced to design! Here comes the first great guess; but it is a guess based on experience, and often proves remarkably accurate. A decision as to landing speed and as to the type of aerofoil will then decide the wing area, on which the whole lay-out of the aeroplane depends, so it will be seen how important is this question of landing speed and its influence on the whole design of the finished aeroplane.

LANDING SPEEDS AND THE FUTURE

So far as landing speed is concerned, we are reaching an interesting stage in the history of aviation. Wing loadings are still going up; they went up slowly but surely for the first thirty years of flight, and rather less slowly, but more surely, during the Second World War – and since. There is at the moment no sign of any halt in this progress – for progress it certainly is. **We must assume, therefore, that wing loadings will go still farther.**

During this time, slots and flaps, and then better flaps, have been invented, and the maximum value of C_L has gone up from just over 1 to about 3, or even 4, for a good modern aerofoil section with slotted flaps and slots extending along 60 per cent of the wing span. When the maximum C_L was 1.22 (RAF 15), a wing loading of 500 N/m² was considered high; but with a C_L max of 3, even 5000 N/m² has already been exceeded. Now a C_L max of 1.22 and W/S of 500 gives a landing speed of about 50 knots (93 km/h) whereas a C_L max of 3 and W/S of 5000 gives a landing speed of about 102 knots (188 km/h).

So we have accepted a considerably higher landing speed – **but how long can this go on?** The increase in wing loading has already had a greater effect than the increase in maximum lift coefficient, but so far we have discovered better and better flaps. Now, however, it would seem that there is not much hope of any further great improvement in flaps – so what of the future? The first thing we must do is clear enough – flaps, and slots too, **must extend along the whole span of the wing, perhaps also under the fuselage**; this has already been done in some types of aircraft, sometimes by arranging that the ailerons act also as flaps, or more hopefully by **dispensing with the ailerons altogether** and adopting an alternative form of lateral control, such as spoilers – this will be discussed later.

Such methods might give us another 40 per cent increase in maximum lift coefficient and a landing speed of about 86 knots (158 km/h) for a wing loading of 5000 N/m². But it is only a temporary reprieve. What next? Are we to accept higher and higher landing speeds? Surely there must be a limit somewhere? Are we to call a halt to increase in wing loading? We may try, but it is doubtful whether we shall succeed. Or may we discover some altogether new device for increasing C_L max? Who knows?

A value of 6.5 for C_L max has been achieved in wind tunnel tests on an 'Augmentor' wing by De Havilland, Canada, and another interesting example of efforts in this direction is **the 'Custer' (USA) channel wing** in which the high-speed flow induced by the engines and pusher propellers flows over the upper surface of those parts of the wings which are curved to house the engines (Fig. 6.8), and even over the ailerons which are just outboard of the curved portions. A considerable increase in C_L max is claimed for this patented device, but it is reported that it has taken 45 years to develop!

Fig. 6.8   The idea of the 'Custer' channel wing
The engines – with pusher propellers – are suspended in the channels; the ailerons are just outboard of the channels.

STOL AND VTOL

The search for means of reducing landing speeds, and take-off speeds, has resulted in the introduction of two new words, or rather sets of initials, into the aeronautical vocabulary – STOL and VTOL. They are clumsy expressions, but at least they have the merit of saying more or less what they mean, that is, if they are expressed in full as **Steep** (or **Short**, as you wish) **Take Off** and **Landing**, and **Vertical Take Off** and **Landing**.

THE GYROPLANE

Perhaps the first really practical achievement in this direction was the Cierva Autogiro in 1923 (Fig. 6C). Although often described as such, **the Autogiro was not a helicopter, but a gyroplane**, and a gyroplane differs from a helicopter in one vital particular in that, whereas in a helicopter the wings or blades are rotated by the power of the main engine, **in a gyroplane the rotating wings are not driven** except by the action of the air upon them, and this in turn is caused by the forward or downward air speed of the aircraft. **Thus forward speed is necessary in a gyroplane, just as it is in a conventional aeroplane** and, as in the latter, it is provided by the thrust of an ordinary engine and propeller. What then was the secret of the success of the original Autogiro in obtaining such remarkably low landing speeds?

Space does not permit a full consideration of the principles underlying this very interesting flying machine, and it must suffice to say that the secret lies in the fact that by a suitable inclination backwards of the axis the wings rotate automatically in such a way that even when the forward speed of the aircraft is far lower than the stalling speed of a conventional aeroplane the rotating wings are still striking the air at a considerable velocity, and can thus provide sufficient lift to keep the aircraft in the air. In this way the forward speed can be reduced to 5 or

10 knots which, in a slight head wind, means a ground speed of practically nil.

When the Cierva Autogiro first appeared it succeeded in achieving, in one fell swoop, **a landing speed such as the experimenters of all nations had considered to be a dream of the distant future.** Moreover later developments resulted, in 1935, in a 'jump start', or true vertical take-off, achieved by rotating the wings on the ground (by engine power), and then suddenly increasing the pitch, or angle of attack of the blades.

The gyroplane, sometimes called an autogyro (the spelling Autogiro was a trade name), still has its advocates, and still exists in various forms, including very light and simple single-seaters (Fig. 6C).

THE HELICOPTER

In a true helicopter in normal flight, the upward thrust of the revolving blades must be equal to the weight; forward motion is produced by inclining the effective axis of the rotor forward which normally entails tilting the nose of the helicopter down. The actual means by which this is achieved is rather complicated and involves altering the cyclic variations in rotor blade incidence. Owing to the reaction of the torque of the lifting blades **the whole aircraft will try to rotate in the opposite direction**, this resulting in a tendency to yaw which corresponds to the rolling tendency due to the propeller on a fixed-wing aircraft. The yawing tendency can be counteracted either **by an auxiliary propeller** or **by a jet reaction system** at the tail; this can also be used to provide directional control and to save the necessity of having a rudder (Fig. 6D).

The development of successful helicopters has involved the solution of many other practical problems. The blade going into the wind (the wind produced by the motion of the aircraft) gets more lift, and drag, than the blade going down wind. This happens whenever the machine moves in any direction, forwards, backwards or sideways; but it becomes an even more serious problem when moving at high speeds because the tip portions of the blade going into the wind meet compressibility problems before the aircraft itself is moving anywhere near the speed of sound; this fact has so far limited the speed of helicopters to something like 200 knots (370 km/h). The helicopter has other failings too, vibration is apt to be excessive, as is the noise when compared with aircraft of similar power.

Attempts to solve these and other problems have resulted in wings that must not only rotate, but also have variable incidence, hinges to

Fig. 6C   STOL
Top: The autogyro after almost disappearing as an historical curiosity has made a comeback in the form of a popular ultralight.
Bottom: The de Havilland Canada Dash-7, original mainstay of the London City (STOL) Airport. The four propellers provide a high-energy airflow over much of the wing.
(Photo by courtesy of Terry Shwetz, de Havilland Canada.)

Fig. 6D   VTOL – helicopters
Top: The tiny but popular Robinson R–22, with piston engine and rubber belt
drive. Flying it feels like trying to ride a monocycle at first.
Bottom: At the other extreme, the massive Russian Mil Mi-26 TM capable of lifting
a load of 196 000 newtons (20 tonnes).

give variable dihedral (resulting in a kind of flapping motion of the blades), drag hinges to allow the blades to bend backwards, and tilting axes of rotation. Then there are the auxiliary propellers at the tail, or jets, some have had jets too at the wing tips to rotate the blades, some have had more than one set of rotating wings (like contra-rotating propellers), and there have been many other ingenious devices, all helping to some extent to solve the problems, but all contributing to the complication and weight of an already complicated type of aircraft.

One development of considerable interest in the rotating-wing type of aircraft – all of which come under the general description of **rotorcraft** – was the Fairey Rotodyne, another trade name. Although it did not prove a commercial success, this experimental type was **both gyroplane and helicopter**, and to some extent **a conventional aeroplane too**. For take-off and landing it was a helicopter, with all the advantages of VTOL; it remained a helicopter, the wings being rotated by jets at the wing-tips, up to forward speeds of about 80 knots; between 80 and 120 knots it was a combination of conventional aeroplane, twin-engine turboprop, and helicopter; above about 120 knots it was aeroplane plus gyroplane, the wings now auto-rotating, and in this condition it was economical, could achieve 170 knots, and could glide in the event of power failure.

Modern types of helicopter can do extraordinary things; not only can they take off and land vertically, they can hover at will, fly sideways or backwards or forwards, land in confined spaces including the heart of cities, drop and pick up mail, passengers and even parts of buildings, and they have proved invaluable in rescue operations, both over land and sea – including the retrieving of space-craft and their crews. All these things they can do better than conventional aircraft – but still they cannot fly at high speeds.

There has not yet been a real break-through as regards maximum speed, but **its weight-lifting capacity**, one of its great assets, **has increased even more rapidly than that of its fixed-wing counterpart**. Moreover at the present time great attention is being given to the solution of the other problems, including that of maximum speed.

Recent developments have included the use of so-called **'rigid' rotors** (Fig. 8E on p. 272) which means dispensing with both the drag and flapping hinges, but making the blades themselves flexible enough to bend upwards and backwards under the lift and drag forces – not exactly rigidity! This has reduced complication to some extent, and has resulted in lighter rotor heads, of titanium, and lighter blades, of reinforced plastics. Another development has been the shrouded tail-rotor, enclosed in the fin, reducing danger and giving less drag, the fin itself also having an unsymmetrical camber so as to counteract some of the torque in forward flight.

One difficulty is to keep down the disc loading, which corresponds to the wing loading of ordinary aircraft. This means some form of 'compound' aircraft, with a small auxiliary fixed wing taking up to 80 per cent of the lift, and with propeller propulsion for forward flight (the propeller being situated behind the fin with its shrouded tail rotor), the main rotor revolving at a lower speed than in a normal helicopter; this should allow a forward speed up to 270 knots. Another possibility is the **turbogyro**, in which two gas turbines are used, one giving power to the main rotor for hovering flight (with gearing to the tail rotor), while both give jet propulsion for forward flight, the main rotor becoming, in effect, an autogyro. Or again the **rotojet**, in which, as the name implies, jets are used to drive the blades directly (at about 40 per cent of the radius rather than at the tips), and again an auto-rotating rotor, auxiliary wing and jet propulsion for forward flight.

But the only hope for speeds approaching anywhere near those of fixed-wing aircraft seems to lie in using twin side-by-side rotors, which can be turned through 90° to become propellers for forward flight, or in retracting the rotors by telescoping the blades, and using ordinary propellers or jets for forward flight.

MORE STOL AND VTOL

In a conventional aeroplane with light wing loading we can of course go a long way, towards STOL at least, by the use of slots, and flaps, and slotted flaps, and double slotted flaps, and, as has already been explained, by extending these all along the wing span (Fig. 6C). But, as has also been explained, light wing loading is quite contrary to the modern trend, and inevitably limits high speed.

What then can be done? Well, there are various possibilities, for instance –

(*a*) **The whole aircraft can be tilted** (Fig. 6E) and can take off in a steeply inclined or even vertical position; then, having gained sufficient height, flatten out into normal flight. This system has the great advantage that the aeroplane need not have movable wings or engines, but it does involve most complicated launching arrangements. As in nearly all systems, great thrust is needed at no forward speed, and in the event of engine failure shortly after take-off, there is little or no chance of making a safe landing.

(*b*) **The engines and wings can be tilted** for take-off and landing,

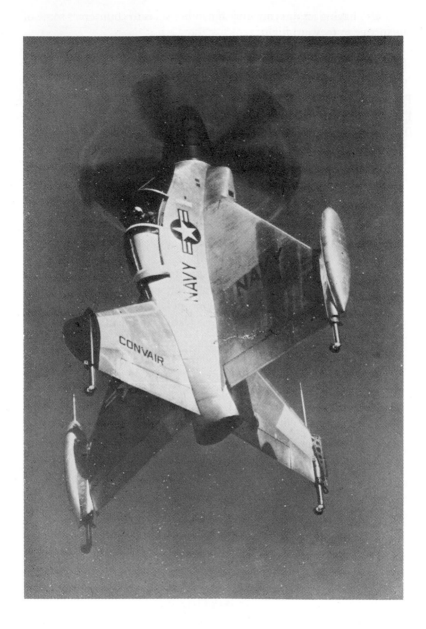

Fig. 6E   VTOL
(By courtesy of the General Dynamics Corporation, USA)
Pogo aircraft built for US Navy; after vertical take-off it could level off and fly
horizontally; for landing the pilot pointed the nose straight up, and backed down tail
first onto small wheels at trailing edges of wings and fins.

the body of the aircraft remaining horizontal. This introduces the considerable complication of movable wings, but has advantages from the point of view of changing from vertical flight to horizontal, and vice versa, and it should be safer in the event of engine failure.

(*c*) **The engines or propellers alone can be tilted** (Fig. 6F). The difficulty in this system is the location of the engines so that the propellers, slipstream, or jet efflux can clear parts of the aeroplane for all positions of the engines.

(*d*) **Slipstream or jet efflux can be deflected** (Fig. 6G). From the structural point of view this is a much more feasible proposition, especially as applied to jets.

(*e*) **There can be two completely different sets of engines**, one for vertical flight and one for horizontal flight. The great disadvantage in this system is that in vertical flight the horizontal engines are useless,

Fig. 6F   Convertiplane
(By courtesy of Bell Helicopter Textron)
The Boeing-Bell V-22 Osprey. Capable of VTOL and VSTOL. The engines can be rotated to provide conventional forward flight after take-off.

Fig. 6G   V/STOL
The Harrier is capable of vertical take-off, very short take-off, and conventional
take-off using its vectored-thrust Pegasus turbofan engine; it can also hover and fly
sideways or backwards. Control at low speed is by reaction jet nozzles situated at the
wing-tips and either end of the fuselage.
(Photo courtesy of Nigel Cogger)

but have to be lifted; and in horizontal flight the vertical engines are
useless, but have to be carried.

All these devices have been tried in one form or another; and so far, at
least, the last two, (*d*) and (*e*), seem the most promising, and we have
already seen VTOL aircraft that can fly across the Atlantic at reasonable
speed – far and away above that of any helicopter – and, perhaps even
more significant, looking like aeroplanes rather than 'flying bedsteads'.
There are still many problems – of safety, of control and, not least, of
noise (so much noise that some people think that we should be
concentrating on QTOL – **Quiet Take Off** and **Landing** – rather than
V/STOL).

HOVERCRAFT

If helicopters and gyroplanes are considered as aeroplanes, what of the
'hovercraft'? Somehow I feel that this interesting machine hardly comes
under the heading of an aeroplane. The way in which it differs from all
other types of machines that fly is in **the limitation of the height at
which it can fly, and this limitation is not one that will**

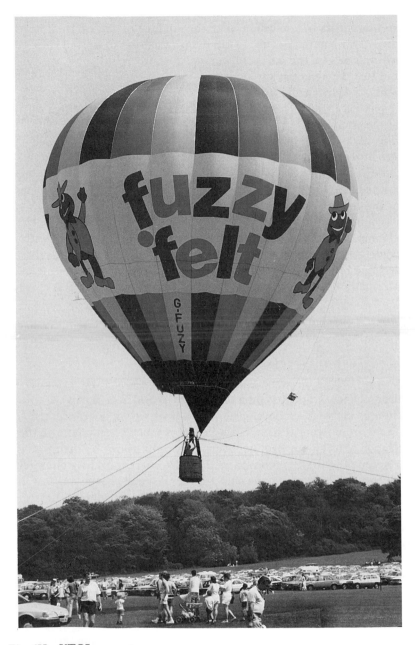

Fig. 6H   VTOL – another way
After a century of neglect, the hot-air balloon has made a major comeback. This one
is tethered and is being used for pilot training. The wicker basket provides good
impact absorption on landing.

disappear with progress, but one of principle. The hovercraft **depends for its lift on being near the surface of the earth**, it floats or rides, or whatever you like to call it, on a **cushion** of air, but **this cushion rests on the earth**. Of course, if we are to be honest, we must admit that the weight of any aeroplane, even when flying at considerable heights, is distributed over the earth's surface; and also that all aircraft do experience, and do take advantage of this same 'cushion' effect when near the ground.

The cushion of air on which the hovercraft floats may be provided either by jet engines or by propellers; and the efflux is often from a ring round the periphery of the craft, the airflow being directed downwards or turned in towards the centre of the craft. So long as the craft remains near the ground, or water, and particularly if there is a kind of curtain or skirt round the rim to prevent the air from escaping outwards, the cushioning effect is such that the machine is lifted off the ground with considerably less thrust than would be required to lift it to any height. Once clear of the ground, there is no great difficulty in propelling it horizontally – so long, of course, as the ground is horizontal. It is clearly more suitable for transport over the sea – at least so long as the sea is reasonably horizontal!

THE COMPLETE APPROACH AND LANDING

During the few years preceding the Second World War, a new technique in flying was developed. In this book we are not concerned with the art of flying as such, but we are very much concerned with the alteration of technique **because it was brought about by progress in the science of flight accompanied by the corresponding changes in aeroplane design**. To the outward eye the main change was that the monoplane, rather suddenly, took precedence over the biplane. Less obvious, perhaps but more important, was the **increase in wing loadings** – which incidentally was itself the reason behind the ascendancy of the monoplane. With increase in wing loading came higher landing speed, higher landing speed meant that flaps – once a luxury – became a necessity, and flaps, in their turn, were largely responsible for the new technique, especially in so far as it affects the approach and landing.

A pilot, when approaching a landing ground, may find that he is either undershooting or overshooting. If he is undershooting there is little that he can do (assuming, of course, that he is already gliding at the best angle) except use his engine power to flatten his glide. In the old days such a method was considered bad flying; if the engine was functioning satisfactorily, it showed lack of judgment; if the engine was out of action – that is to say in the case of a forced landing – it could not be done. **Nowadays the engine-assisted approach,** as it is called, **is a standard method of approaching to land.** It might almost be said that on the most modern machines the glide approach is only used as practice for a forced landing.

In the anxiety to avoid undershooting there is a natural tendency to overshoot, especially since it would seem to be easy to lose any unnecessary height. In practice it is not quite so easy. In older types of aircraft the following methods were available as means of losing height in the event of overshooting –

(*a*) Sideslipping.
(*b*) Prolonging the glide by S-turns.
(*c*) Putting the nose down and gliding fast.
(*d*) Holding the nose up and gliding slowly.

The objections to the last two have already been explained; the first two methods, on the other hand, were successfully employed for very many years.

Then came the modern type of aircraft – its **superior streamlining gave it a very flat gliding angle,** so flat that even a slight degree of overshooting caused it to float much too far before landing. But that was not all – **it did not like sideslipping** (reasons for this will be given later), and with its very flat angle of glide the **S-turn did not result in sufficient loss of height**.

Necessity may be the mother of invention, but in this little bit of aviation history the invention existed before the necessity arose; flaps had been in use for many years, but they had not really been fully applied to their modern purposes. These purposes can best be described by considering the process of approaching and landing a light private aircraft by the old-fashioned glide approach technique. This typically consists of five separate phases (Fig. 6.9) – **the glide, the flattening-out, the float or hold-off, the landing,** and **the pull-up.** In each and all of these flaps have their part to play. Let us consider them in turn.

First, then, let it be understood that **the last 150 m or so of the glide should be straight, without any slipping or turning to one side or the other.** This can only be done if the pilot has means of controlling the gliding angle relative to the earth without unduly raising or lowering the air speed. Flaps can give him the means to do this, at any rate over a limited range of gliding angles. As the flaps are lowered both lift and drag are increased. The **increase in lift** tends to **flatten the gliding angle** and to make it possible to **glide at a slower air speed** without approaching dangerously near to the stall. The **increase**

Fig. 6.9　The approach and landing
Air speeds shown are approximate only, but indicate at what parts of the approach and landing there is a loss of air speed.

in drag tends to **steepen the gliding angle**, and gliding attitude, of the aeroplane **for the same air speed**. The net effect depends on whether lift or drag has the greater proportional increase, i.e. on whether the L/D ratio is raised or lowered, and that in turn depends on how much the flaps are lowered and on the type of flap. This aspect of flaps has already been discussed. But, as an illustration of how our ideas change – in an earlier edition of this book, the next sentence read: '**All that need be added is that the split flap is the type most used.**' What must be added now is '**that the split flap is hardly ever used!**' The real interest of the change is not in the bare fact, **but in the reasons behind it**. The split flap was popular because it was simple, because it gave a reasonable amount of both lift and drag, and **to some extent** a choice between them – when lowered by 60° or more the L/D ratio was definitely reduced and the angle of glide steepened. But now we have **more flexibility in getting lift when we need it from the greatly improved design and variety of lift flap, and drag when we need it with air brakes** (Fig. 6I).

So much then for the actual glide – we can sum it up by saying that flaps give us at least some control over the gliding angle.

Next comes the process of **flattening out**; this involves a change of direction, and so an acceleration, and force, towards the centre of the curved path. This force must be provided by the wings which must therefore have more speed and more angle, so in effect the stalling speed is higher. Now **the steeper the original glide, the greater the change in flight path involved, and so the more speed must there**

Fig. 6I   Ready for landing
The huge Antonov AN-124 with Fowler flaps and leading edge slats extended, and
undercarriage lowered, showing the ten pairs of main wheels.

be in hand for flattening out. All this is very annoying – it means that
the more steeply we glide, the faster must we glide; just what we were
trying to avoid. The solution is to use engine power, as we shall consider
later.

After flattening out, we must lose any excess speed – this may be
called **the float or hold-off.** In this the drag of flaps or air brakes play
their part, as do the wings themselves as they are brought to the angle
for actual landing.

After the float comes **the landing** (Fig. 6J). This, in a sense, is
momentary only but the landing speed is of the utmost importance
because it settles both the gliding speed and the distance to pull up after
landing. The problem of landing speed has already been fully discussed,
but it must be emphasised – because it is so often misunderstood – that
drag, whether caused by flaps or anything else, **cannot reduce
landing speed; that is entirely a question of lift.**

After the landing, **the pull-up.** This at least is easy to understand;
what we need for a quick pull-up is drag – wheel brakes and air
brakes – the more the better, provided the aircraft can stand it and does
not tip on its nose. In addition to actual air brakes, some types of flap,
when fully lowered, give good braking effect, and so do the wings at
their angle of 16° or so with a tail-wheel type of undercarriage. The lack

Fig. 6J   Landing
The BAe 146 about to touch down, Fowler flaps extended and rear air-brake doors open.

of air drag during the landing run is one of the few disadvantages of the nose-wheel type – an effective substitute, sometimes used, is a tail parachute (Fig. 6K), and, perhaps most effective of all, reversible pitch propellers or reversed thrusts of jets. **The air-braking effect is greatest at the beginning of the landing run, the wheel-braking later when it can more safely be used.** The problem of brakes is a straightforward one of mechanics, but apart from the question of coefficient of friction between wheels and ground, and the serious danger of 'aquaplaning' when there is water on the runway, there are some aspects of brakes which are peculiar to aircraft. For instance, the centre of gravity is high above the wheels, though not so much with jets as with propellers; it is also, with a tail-wheel type of undercarriage, only a short distance behind the wheels, and so, if the brakes are applied violently there is an immediate tendency to go over on to the nose. Another difficulty is that if the aeroplane starts to swing, the centre of gravity, being behind the wheels, will cause the swing to increase. This may be checked by the differential action of the brakes, but it is interesting to note that the **tricycle or nose-wheel undercarriage** (Figs 4D, 5G, 7A, 11C and others) can remove the cause of this and other troubles. When this type of undercarriage is used, as it is almost universally on modern aircraft, the centre of gravity is in front of the main wheels and there is no tendency to swing, and at the same time the aircraft is prevented from going on to its nose by the front wheel. The

Fig. 6K   Pulling up after landing
A MiG-29 using an arrester parachute to reduce the landing run.

effect on braking, and consequent shorter pull-up has to be seen – or, better still, tried – to be believed. And there are other advantages too.

So much for the glide approach and landing; for forced landings it is the only way. For approaching over high obstacles at the edge of an airfield it may also come in useful. **In all other circumstances, with modern types of aircraft,** the engine-assisted approach **is better,** so let us sum up the reasons as to why it is preferred –

1. By slight adjustments of the throttle **the path of glide can be flattened or steepened at will.**

2. **The gliding path is flatter,** so there is less change of path in flattening out, less acceleration, less extra lift required, less increase in stalling speed, and thus less excess speed is needed, and the glide can safely be made more slowly.

3. In propeller-driven aircraft the extra speed of the slipstream over elevators and rudder **makes these controls more effective** – their effectiveness enables us to counteract wind gradient and turbulence effects near the ground.

4. Since there is less excess speed to be lost, **the float is reduced.**

5. **An engine, already running, will respond more readily** to

the throttle if it is found necessary to make another circuit before landing.

6. For all the reasons already given **less judgment is required** – the whole process is easier.

Well, all that sounds pretty convincing, and not only is it perfectly sound reasoning but it is amply confirmed by experience.

As was mentioned in Chapter 3 many modern types of aircraft are literally **'flown onto the ground'** at speeds well above the stalling speed; this is where real **lift dumpers** come into their own – these are larger than normal air brakes or spoilers, and **cannot be operated unless the undercarriage is locked down, and the aircraft is bearing on the wheels** and **the engines throttled back** – they **'kill' the lift, keep the aircraft on the ground** and **make the wheel brakes more effective**. They are essentially for use **after landing**.

EFFECT OF FLAPS ON TRIM

The lowering or raising of flaps affects the airflow not only over the lower surface of the wing, but also over the upper surface – probably the more important effect – and in front of the wing and behind the wing (Fig. 6.10). The airflow in turn affects the pressure distribution and the forces and moments on the wing and on the tail plane. It is hardly surprising, therefore, that the trim may be affected, but it may seem

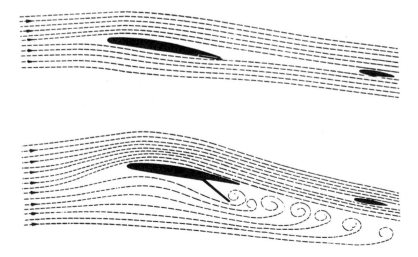

Fig. 6.10   Flow over wing and tail – flaps up and flaps down

curious that the lowering of the flap sometimes tends towards nose-heaviness, sometimes tail-heaviness.

Consider the top surface of the wing. When the flap is lowered, the air flows faster over the top, especially near the leading edge. There will be greater suction here and the chances are that **the centre of pressure on the top surface will move forward**, thus tending towards **tail-heaviness**.

The downwash behind the wing will be large; and if the tail plane is so situated as to receive the full benefit of this downwash, there will be a **downward force on the tail plane**, tending towards **tail-heaviness**.

In a low-wing aircraft the **low position of the drag** on the flap, especially when fully lowered, will tend towards **nose-heaviness**. On a high wing aircraft the drag, being high, may tend towards **tail-heaviness**.

The net effect on the pitching moment **depends entirely on the type of flap or slots used, on how much they are lowered**, and **on the situation of the tail plane**. Slotted flaps, and flaps that move backwards so increasing the rear portion of the wing area, will nearly always cause a nose-down moment which sometimes has to be counteracted by leading edge slots and flaps.

Sometimes, too, the change of trim is in one direction for the first part of the lowering of the flap, usually tail-heavy; and in the other direction, nose-heavy, when full flap is lowered. In some aircraft the effects, whether by design or good luck, so cancel each other that there is little or no change of trim, and no one is more pleased than the pilot.

It should be noted that the technique of landing a modern high performance aircraft, be it military or civilian, is nothing like the straightforward seat-of-the-pants procedure described earlier for the case of a light private aircraft. For such modern aircraft, the speed, height, angle of descent and a host of other factors such as the flap and power setting must be correct within very close limits as the aircraft crosses the airfield threshold. If not, the pilot must abort the approach and try again. Landing a modern airliner manually requires a great deal of skill and concentration, and the majority of landings are nowadays made under automatic control with the pilot merely keeping a watchful eye, and being ready to take over at any instant in the event of a system malfunction. In order to maintain their skill (and their licence) pilots are, however, required to make a certain proportion of landings under manual control. It would be virtually impossible to land machines such as the Space Shuttle without some form of computer assistance.

CAN YOU ANSWER THESE?

1. If, when an aeroplane is gliding at its minimum angle of glide, the pilot attempts to glide farther by holding the nose of the aeroplane up, what will be the result, and why?
2. Discuss the effect of flaps on the gliding angle.
3. How does the load carried in an aeroplane affect the gliding angle and gliding speed?
4. Does the flattest glide give the longest time in the air? If not, why not?
5. Does (*a*) the stalling speed, (*b*) the stalling angle, change with height?
6. What are the advantages of the engine-assisted approach?
7. Why may the lowering of flaps affect the trim of an aeroplane?
8. You are flying an aeroplane well out to sea when the engine fails; there is a good airfield just on the coast and it is touch and go whether you can reach it; you have disposable load on board, luggage, bombs, and fuel; should you jettison your load, and if so when, and what should be your tactics in an endeavour to reach the airfield? (*Note.* There is more in this question than one might at first think, e.g. your tactics should be different for different wind conditions, so consider conditions of no wind, head winds, and tail winds; if you have surplus speed, what can you do with it?; can you reduce your drag?; should you use flaps?; at what speed should you fly?; should you jettison your load and, if so, when?)

Numerical examples on gliding and landing will be found in Appendix 3.

CHAPTER 7

PERFORMANCE

It may seem rather illogical that we should first consider level flight, then gliding and landing, and now the take-off, climb and general performance of the aeroplane. But there is method in our madness. Level flight is, as it were, the standard condition of flight with which all other manoeuvres are compared; whether we are learning to fly, or merely learning to understand the principles of flight, it is right and proper that we should first learn the problems of level flight. Gliding, too, involves simple fundamental principles, in some ways more elementary than those of level flight, but not so closely linked with the principles of other conditions of flight. Landing we have used to illustrate the principles of flight at low speeds. All these have followed quite naturally one on the other, but the take-off is a problem on its own, having little connection with anything else, and the climb involves very little in the way of new principles.

TAKING-OFF

The pilot needs skill and practice before he can be sure of making a good take-off, one of the main problems being to keep the aircraft on a straight and narrow path. This difficulty applies mainly to propeller-driven aircraft, and has already been discussed in Chapter 4. In general, it may be said that **the object during the take-off is to obtain sufficient lift to support the weight with the least possible run along the ground.** In order to obtain this result the angle of attack is kept small during the first part of the run so as to reduce the drag; then, when the speed has reached the minimum speed of flight, if the tail is lowered and the wings brought to about 15° angle of attack, the aeroplane will be capable of flight. Although by this method the aeroplane probably leaves the ground with the least possible run, it is apt to be dangerous because, once having left the ground, any attempt to climb by further increase of angle will result in stalling and dropping back on to the ground. Therefore **it is necessary to allow the speed to increase beyond the stalling speed before 'pulling-off'**, and sometimes the aeroplane is allowed to continue to run in the tail-up position until it takes off of its own accord (Fig. 7A).

Fig. 7A   Taking-off
The Myasischev M-55 high-altitude aircraft for surveillance and geophysics research.

The process of taking-off is largely influenced by such things as the runway surface, and although of extreme interest, the subject is too practical (if considered from the practical point of view) and too highbrow (if considered from the theoretical point of view) to be within the scope of this book. In order to reduce the length of run, and increase the angle of climb after leaving the ground – so as to clear obstacles on the outskirts of the airfield – the take-off will, when possible, be made against the wind. Other aids to taking-off are slots, flaps or any other devices which increase the lift without unduly increasing the drag, and, essential in propeller-driven high-speed aircraft, the variable-pitch propeller.

The question as to whether or not flaps should be used for taking-off depends upon whether the increased lift of the flap, with the resulting decrease in taking-off speed, makes up for the lower acceleration caused by the increased drag of the flap. But the problem is a little more complicated than that, because while we wish to avoid drag throughout all the take-off run, we only really need the extra lift at the end, when we are ready to take off. No doubt we could get off most quickly **by a sudden application of flap at this stage**, but such a method would certainly be **dangerous**. The lift type of flap helps the take-off considerably, other types may have some beneficial effect if used at a moderate angle, and in practice some degree of flap is nearly always used for take-off in modern high-speed types of aircraft if only because it reduces the otherwise very high take-off speed with consequent wear of tyres.

Some interesting problems arise in connection with the take-off. Modern undercarriages may tuck away nicely during flight, but when lowered they are less streamlined than a fixed undercarriage and their drag may hamper the take-off quite considerably; the lower undercarriage that can be used with jets is a great advantage in this respect. Again, just as landing speeds go up with high wing loading, so do take-off speeds, and the length of run needed to attain such speeds is liable to become excessive. The idea of catapulting is an old one; it has been used with some success, but its application has never been very wide, and it raises many new problems of its own. The assistance of rockets gives much the same effect as catapulting and has great advantages in that it does not require bulky apparatus on the airfield and so it can be used away from the main base; another point in favour of rockets is that they may have alternative uses, as for instance, for special accelerations when required during flight. Refuelling in the air does not sound like a form of assisted take-off, but it does present possibilities in that an aircraft can be taken off lightly loaded as regards fuel. Perhaps the most interesting experiment in this direction has been the pick-a-back method in which a small very highly loaded aircraft was released from its large mother craft during flight.

Finally, although we have considered STOL and VTOL in their effect on landing, we must not forget that they are at least as important for take-off, and QTOL is even more difficult to achieve.

CLIMBING

During level flight the power of the engine must produce, via the propeller, jet or rocket, a thrust equal to the drag of the aeroplane at that particular speed of flight. If now the engine has some reserve of power in hand, and if the throttle is further opened, **either** –

(*a*) The pilot can put the nose down slightly, and **maintain level flight at an increased speed and decreased angle of attack,** or

(*b*) **The aeroplane will commence to climb** (Fig. 7B).

A consideration of the forces which act upon an aeroplane during a climb is interesting, but slightly more complicated than the other cases which we have considered.

Assuming that the path actually travelled by the aeroplane is in the

Fig. 7B Climbing
An Airbus A-330 using flaps to demonstrate its steep climb capability.

same direction as the thrust, then the forces will be as shown in Fig. 7.1. If *a* is the angle of climb, and if we resolve the forces parallel and at right angles to the direction of flight, we obtain two equations –

(1) $T = D + W \sin a$
(2) $L = W \cos a$

Translated into non-mathematical language, the first of these equations tells us that during a climb the thrust needed is greater than the drag and increases with the steepness of the climb. This is what we would expect. If a vertical climb were possible, *a* would be 90° and therefore sin *a* would be 1, so the first equation would become $T = D + W$, which is obviously true because in such an extreme case the thrust would have the opposition of both the weight and the drag. Similarly if $a = 0°$ (i.e. if there is no climb), sin *a* = 0. Therefore $W \sin a = 0$. Therefore $T = D$, the condition which we have already established for straight and level flight.

The second equation tells us that the lift is less than the weight, which is rather interesting because one often hears it said that an aeroplane climbs when the lift is greater than the weight! One must admit, however, that the misunderstanding is largely due to the rather curious definition which we have assigned to the word 'lift'. Let us

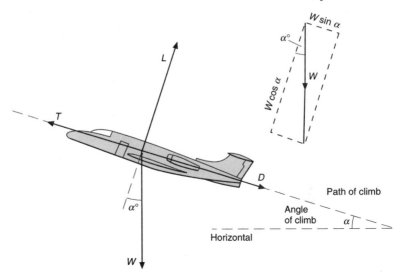

Fig. 7.1 Forces acting on an aeroplane during a climb

consider the second equation under extreme conditions. If the climb were vertical, cos 90° = 0. Therefore $L = 0$. So that in a vertical climb we have no lift. This simply means that all the real lift is provided by the thrust, the wings doing nothing to help. If, on the other hand, $a = 0°$, cos $a = 1$, and therefore $L = W$, which we already know to be the condition of straight and level flight.

POWER CURVES – PROPELLER PROPULSION

A more interesting and more practical way of approaching the climbing problem is by means of what are called **performance curves**. By estimating the **power which will be available** from the engine and the **power which is required** for level flight at various speeds, we can arrive at many interesting deductions. It is largely by this method that forecasts are made of the probable performance of an aeroplane, and it is remarkable how accurate these forecasts usually prove to be.

The procedure for jet and rocket systems of propulsion is rather different because, as already mentioned, we must think in terms of thrust rather than power. They will therefore be dealt with separately, and the following discussion relates primarily to piston-engined aircraft.

The deduction of the curve which gives the power output of the engine is outside the scope of this book, as it depends on a knowledge of the characteristics of the piston (or gas turbine) engine for a propeller-driven aircraft. From this curve must be subtracted the power which is lost through the inefficiency of the propeller (the efficiency of a good propeller at reasonable speed, but falling off on both sides depending on rpm, is about 80 per cent). The resulting curve (Fig. 7.2) shows the power which is **available** at various forward speeds of the aeroplane.

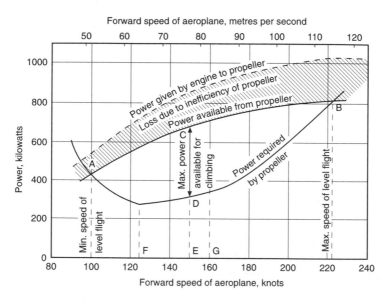

Fig. 7.2   **Power available and power required**

The power which will be **required** is found by estimating the drag. For this purpose the wing drag and the parasite drag are usually found separately, the former from the characteristics of the aerofoils and the latter by estimating the drag of all the various parts and summing them up. Another method of finding the total drag is by measuring the drag of a complete model in a wind tunnel and scaling up to full size. After the total drag has been found at any speed, the power is obtained by multiplying the drag by the speed, as in flying for endurance in Chapter 5, e.g. if the total drag is 4170 N at 82.4 m/s –

Power required $= 4170 \times 82.4 = $ **344 kW**

And in a similar way the power required is found at other speeds, the lower curve in Fig. 7.2 illustrating a typical result. The reader may be puzzled as to why the power required increases so rapidly at low speeds; the explanation is that in order to maintain level flights at these low speeds, a very large angle of attack is required, and this results in an increase of drag in spite of the reduction in speed. If the argument sounds familiar, it is simply because we are returning to the same argument as when discussing range and endurance. The figures we have just quoted are taken from that argument, and the curve of power required in Fig. 7.2 is based on the aeroplane of Chapter 5. This follow-up of the same aeroplane will make the power curves more interesting and instructive.

It should be noted that there will be no fundamental difference in the shape of the power required curve for jet and propeller-driven aircraft. It is in the power available that the difference lies.

MAXIMUM AND MINIMUM SPEEDS OF HORIZONTAL FLIGHT

From the combination of the two curves (Fig. 7.2) some interesting deductions can be made. **Wherever the power available curve is above the power required curve, level flight is possible,** whereas both to the left and right of the two intersections it becomes impossible for the rather obvious reason that we would require more power than we have available! Therefore the intersection **A shows the least possible speed** (51 m/s, as we had discovered before for other reasons), and the intersection **B the greatest possible speed** (115 m/s), at which level flight can be maintained. Between the points A and B the difference between the power available and the power required at any particular speed, i.e. the distance between the two curves, represents the amount of **extra power** which can be used for climbing purposes at that speed, and **where the distance between the two curves is greatest,** i.e. at **CD, the rate of climb will be a maximum,** while the corresponding point **E shows that the best speed for climbing is 77 m/s.** From the weight of the aeroplane 50 kN, and the extra power CD (680 − 320, i.e. 360 kW) available for climbing, we can deduce the vertical rate of climb, for if this is x m/s, then the work done per second in lifting 50 kN is 50 000 x watts,

So 50 000 x = 360 000
and x = 360 000/50 000
= 7.2 m/s

This represents the best rate of climb for this particular aeroplane, **but it will only be attained if the pilot maintains the right speed of 77 m/s.** As in gliding, there is a natural tendency to try to get a better climb by holding the nose up higher but, as will be seen from the curves, if the speed is reduced to 62 m/s only about 250 kN will be available for climbing, and this will reduce the rate of climb to 250 000/50 000, i.e. 5 m/s. Similarly, at speeds above 77 m/s the rate of climb will decrease, although it will be noticed that between certain speeds **the curves run roughly parallel to each other** and **there is very little change in the rate of climb between 72 and 88 m/s**; obviously at 51 m/s and again at 115 m/s, the rate of climb is reduced to nil, while below 51 and above 115 m/s the aeroplane will lose height.

As a matter of interest, the speeds for maximum endurance (F), 64 m/s, and maximum range (G), 82 m/s, have also been marked. As was explained in Chapter 5, these are the best speeds from the point of view of the aeroplane, but they may have to be modified to suit engine conditions. Note that the speed for maximum endurance (F) could be deduced from the curve, since it is the lowest point on the curve, i.e. the point of minimum power required for level flight. The speed for maximum range (G), however, must be obtained from the table on page 178, which showed the drag at various speeds.

EFFECT OF CHANGES OF ENGINE POWER

We have so far assumed that for a certain forward speed of the aeroplane the power available is a fixed quantity. This, of course, is not so, since the power of the engine can be varied considerably by manipulating the engine controls. If the curve shown in Fig. 7.2 represents the power available at some reasonably economical conditions and in weak mixture, then we shall be able to get more power by using rich mixture, and the absolute maximum power by opening the throttle to the maximum permissible boost and using the maximum permissible rpm – with fixed-pitch propellers this will simply be a case of full throttle. From this we shall get a curve of emergency full power (Fig. 7.3). It will be noticed that **the minimum speed of level flight is now slightly lower** – very slightly, so slightly as to be unimportant. **The maximum speed is**, as we might expect, **higher** – perhaps not so much higher as we might expect (118 instead of 115 m/s). **The most important change is in the rate of climb**: 460 kW surplus power is now available for climbing, and the rate of climb is 9.2 m/s instead of 7.2 m/s.

Except in special circumstances, it is inadvisable to fly with the engine 'flat out', and, even so, full power must be used only for a limited time or there will be a risk of damage to the engine. The effects of decreasing the power are also shown in Fig. 7.3. From the point of view of the aeroplane, it makes no difference whether the power is decreased by reducing boost, or lowering the rpm, or both; but for fuel economy it is generally advisable to lower the rpm. It will be noticed that as the power is reduced, the minimum speed of level flight becomes slightly greater, the maximum speed becomes considerably less, and the possible rate of climb decreases at all speeds.

Fig. 7.3   Effects of engine power

All this is what we might expect, with the possible exception of the fact, which pilots often do not realise, that **the lowest speeds can be obtained with the engine running at full throttle.** However, this flight condition cannot easily be sustained in practice because a small, inadvertent, decrease in speed would mean an increase in required power and a simultaneous decrease in available power. The speed reduction would then 'run away'; a condition called speed instability.

Eventually, as the engine is throttled down, we reach a state of affairs at which there is only one possible speed of flight. This is the speed at which least engine power will be used, and at which we shall therefore obtain maximum endurance. It is rather puzzling to find that this speed (72 m/s) is different from the speed (64 m/s) at the lowest point of the power required curve. This is because the engine and propeller efficiency is slightly better at 72 than at 64 m/s.

EFFECT OF ALTITUDE ON POWER CURVES

We have not yet exhausted the information which can be obtained from these magic performance curves, for if we can estimate the corresponding curves for various heights above sea-level we shall be able to see how performance is affected at different altitudes. There is much to be said, and much has been said, on the subject of whether it is preferable to fly high or to fly low when travelling from one place to another. It is one of those many interesting problems about flight to which no direct answer can be given, chiefly because there are so many conflicting considerations which have to be taken into account. Some of them, such as the question of temperature, wind and the quantity of oxygen in the air, have already been mentioned when dealing with the atmosphere, but the most important problem is that of performance.

How will the performance be affected as the altitude of flight is increased? At first one is tempted to think that since the density of the air is decreased, resistances will be less and therefore speeds will be higher. Unfortunately, however, the decrease in the air density has more far-reaching effects which may be summarised as follows –

1. Decrease in lift.
2. Decrease in drag.
3. Decrease in propeller thrust.
4. Decrease in the weight of air supplied to the engine, and hence a falling-off in the power of the engine.
5. Decrease in the air supply, and therefore in the amount of oxygen, to the occupants.

Some of these can be compensated for, and others interact on each other to such an extent that the problem becomes very complicated. For instance, since the lift must be kept equal to the weight, any loss in lift due to decrease in air density must be made up for by an increase in the angle of attack or increase in air speed, or both; and these, in turn, will increase the drag and thus balance, or probably overbalance, the decrease in drag, which is directly due to the change in the air density. The loss in the propeller thrust can to some extent be compensated for by the use of a propeller having variable pitch, the air supply to the engine can be augmented by supercharging, and the aircraft can be pressurised.

Whatever attempts are made to mitigate the difficulties of flight at high altitudes, in propeller-driven aircraft the general tendency remains for the **power available to decrease** and the **power required to increase with the altitude** (Fig. 7.4). This will cause the curves to close in towards each other, resulting in a gradual **increase in the minimum speed** and a **decrease in the maximum speed**, while the distance between the curves, and therefore the **rate of climb, will also become less**. Any pilot will confirm that this is what actually happens in practice, although, as previously mentioned, he may be somewhat deceived by the fact that the air speed indicator is also affected by the change in density and consequently reads lower than the true air speed. This is really what accounts for the curve of power required moving over to the right as the altitude increases; if the curves were plotted against indicated speed, the curves for 3000 m and 6000 m would simply

Fig. 7.4   Effects of altitude

be displaced upwards, compared with that for sea-level. The difference between true and indicated speed also accounts for another apparent discrepancy in that the curves as plotted (against **true** speed) suggest that the air speed to give the best rate of climb **increases with height**. This is so, but the **indicated** speed for best rate of climb **falls with height**.

For certain purposes **good performance at high altitudes may be of such importance that it becomes worth while to design the engine, propeller and aeroplane to give their best efficiencies at some specified height,** such as 10 000 metres. It may then happen that performance at sea-level is **inferior** to that at the height for which the machine was designed, and this is a feature of many modern aircraft. Even so, above a certain critical height, 6000, 9000 m or whatever it may be, performance will inevitably fall off and so the performance curves will be very similar, except that the highest curve of power available will correspond to the critical height. In such aircraft it may well be that the advantages of flying high outweigh the disadvantages.

CEILING

This process of improving performance at altitude cannot be continued indefinitely and we shall eventually reach such a height that **there is only one possible speed for level flight** and the **rate climb is nil**. This is called the **ceiling**.

It requires extreme patience and time to reach such a ceiling, and, owing to the hopeless performance of the aeroplane when flying at this height, it is of little use for practical purposes, and therefore the idea of a **service ceiling** is introduced, this being defined as that height at which the rate of climb becomes less than 0.5 m/s, or some other specified rate.

There is a story which dates from the early days of so-called 'light aeroplanes', which relates how, in a competition in which marks were awarded for 'speed range', competitors were required to fly between two points, first as fast as possible, and secondly as slowly as possible; one competitor succeeded in flying faster when he was flying as slow as possible than when he was flying as fast as possible! Such an occurrence is not so strange as it may at first appear to be; it simply means that this machine, although flying but a few feet off the ground, was practically at its ceiling. As a matter of fact, such was the case with many of those first under-powered light aeroplanes, some of which succeeded in winning large prizes, although they often had considerable difficulty in leaving the ground.

EFFECT OF WEIGHT ON PERFORMANCE

It is sometimes important to be able to calculate what will be the effect on performance of increasing the total weight of an aeroplane by carrying extra load. Here again the performance curves will help us.

If the weight is increased, the lift will also have to be increased. So we must either fly at a larger angle of attack or, if we keep the same angle of attack, at a higher speed. This speed can easily be calculated thus –

Let old weight $= W$, new weight $= W_1$.
Let V be the old speed, and V_1 be the new speed at the same angle of attack.

Since angle of attack is the same, C_L will be the same.

$$\therefore W = C_L \cdot \tfrac{1}{2}\rho V^2 \cdot S$$
$$\text{and } W_1 = C_L \cdot \tfrac{1}{2}\rho V_1^2 \cdot S$$
$$\text{so } V_1/V = \sqrt{(W_1/W)}$$

Such problems always become more interesting if we consider actual figures; so suppose that we wish to carry an extra load of 10 000 N on our aeroplane which already weighs 50 000 N; then –

$$V_1/V = \sqrt{(60\,000/50\,000)} = \sqrt{1.2} = 1.095$$

Since the angle of attack is the same, the lift/drag ratio remains constant, and the corresponding drag, D, will be

$$D_1 = D\frac{L_1}{L} = D \times \frac{60\,000}{50\,000} = 1.2D$$

The corresponding power, P, is thus:

$$P_1 = D_1V_1 = 1.2 \times 1.095DV = 1.2P$$

So, for example, if we take the point on the power-required curve (Fig. 7.4) marked A, the corresponding point A, will be at

$$V_1 = 1.095V, \qquad P_1 = 1.2P$$

i.e. $V_1 = 1.095 \times 60 = 65.7\,\text{m/s}$
and $P_1 = 1.2 \times 310 = 372\,\text{kW}$

In a similar way, for each angle of attack, new speed and new power can be calculated, and thus a new curve of power required can be drawn for the new weight of the aeroplane.

It is interesting to note that the net effects of the additional weight are exactly the same as the effects of an increase of altitude, i.e. –

1. **Slight reduction in maximum speed.**
2. **Large reduction in rate of climb.**
3. **Increase in minimum speed.**

In short, the curve of power required is again displaced upwards. It will be noticed that there is too a **slight increase in the best speed to use for climbing.** (This must not be confused with rate of climb.)

In spite of the similarity in effect of increase of weight and increase of altitude it should be noted that the increase of weight does not affect the reading of the air speed indicator, and so the results apply equally well whether we consider true or indicated air speed.

INFLUENCE OF JET PROPULSION ON PERFORMANCE

We have so far considered the performance of an aircraft mainly from the point of view of the reciprocating–engine propeller-driven type. The substitution of a gas turbine driving a propeller, or turboprop, will make very little difference to these considerations but, as with range and endurance, there will be **certain differences for the pure jet-driven type, or turbojet.** The differences all turn, as before, on the increase of propulsive efficiency with speed in the jet-driven aircraft (Fig. 7C).

As we have seen in Chapter 5 this increase in efficiency results in the **thrust remaining approximately constant at all speeds, and**

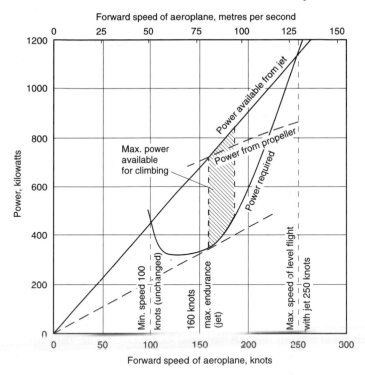

Forward speed of aeroplane, metres per second

Fig. 7.5 Power available and power required – jet propulsion

therefore the thrust power available, i.e. thrust × speed, will rise
in proportion to the speed. So the curve of power available rises
much more steeply than with the propeller-driven aircraft; it is, in fact,
practically a straight line through the origin (Fig. 7.5). It will be evident
from the figure that as a result of this difference **there is a tendency in
the jet-driven type for all the speeds of best performance to be
higher.** Assuming the same power required curve as before (and since
this only concerned the **aeroplane**, there is no reason why it should be
different), the minimum power available at which it is possible to fly will
be represented by the straight line which is tangential to the curve. With
increase of rpm the straight line will swing upwards. The maximum
performance of the aircraft will, of course, depend on **how much**

Fig. 7C   Cruise performance (opposite)
(By courtesy of the Boeing Company)
Two models of the Boeing 747. The nearer aircraft is a 747-400 which cruises at
around 965 km/h (600 mph) at 9150 m, and has a cruise range of 13 528 km with 412
passengers.

power is available at maximum rpm. If, as in the figure, we assume that the same power is available at 82 m/s for both jet and propeller, it will be seen that the **maximum speed of level flight is 129 (instead of 115 m/s)**, but the minimum speed is practically unchanged at 51 m/s in this particular case; **the speed corresponding to best rate of climb** is again not very critical, but the **average value is about 88 (instead of 77 m/s); the speed for maximum endurance** has already been discovered in Chapter 5 as **being 82 (instead of 64 m/s)** this corresponds to the point where the tangent from the origin touches the curve, since this represents the least slope to the line of power against speed, but power is thrust × speed, so the minimum value of power/speed is minimum value of (thrust × speed)/speed, i.e. minimum thrust, i.e. minimum drag, which in turn means maximum endurance for jet propulsion. Finally **the speed for maximum range with jet propulsion** (see Chapter 5) **is 90 m/s instead of 82 m/s.** Figure 7.6 (p. 243) gives a good idea of the speed range of the aircraft we have been considering if equipped with either propeller or jet propulsion. It will be seen from this figure that the stalling speed may be slightly lower for the propeller-driven aircraft. This is because part of the wing is beneficially influenced by the propeller slipstream; a feature which is not present in a conventional jet aircraft. If the 'flaps up' minimum flying speed is near wing stall, the 'power required' curve may be similarly influenced. However in Figs 7.5 and 7.6 it is assumed that this is not the case.

Other features in which the performance of jet-driven types differs from propeller-driven types may be summed up under the following headings –

1. **Jet aircraft**, though uneconomical at low speeds, **become comparatively more economical at high speeds**.

2. **In a jet aircraft the true air speed attainable in level flight remains approximately constant at all altitudes** (the indicated speed therefore decreases), whereas in a propeller-driven aircraft the true air speed varies considerably, usually increasing and then decreasing, but depending especially on the methods of supercharging. In both types, of course, the air speed attainable at altitude may be influenced by compressibility effects (see later chapters).

CAN YOU ANSWER THESE?

1. 'When an aeroplane is climbing, the lift is less than the weight.'
   Explain why this statement is not so inconsistent as it sounds.

Fig. 7.6   Effect of power plant on speed range

2. What is the effect of altitude on the maximum and minimum speeds of an aeroplane?
3. Distinguish between 'ceiling' and 'service ceiling'.

4. In attempting to climb to the ceiling, should the air speed be kept constant during the climb?
5. If the load carried by an aeroplane is increased, what will be the effects on performance?

Numerical examples on Performance will be found in Appendix 3.

CHAPTER 8

MANOEUVRES

In a sense, any motion of an aeroplane may be considered as a manoeuvre. In no other form of transport is there such freedom of movement. An aeroplane may be said to have six degrees of freedom which are best described in relation to its three axes, defined as follows –

The **longitudinal axis** (Fig. 8.1) is a straight line running fore and aft through the centre of gravity and is horizontal when the aeroplane is in 'rigging position'.

The aeroplane may travel backwards or forwards along this axis. Backward motion – such as a tail-slide – is one of the most rare of all manoeuvres, but forward movement along this axis is the most common of all, and is the main feature of **straight and level flight**.

Any rotary motion about this axis is called **rolling**.

The **normal axis** (Fig. 8.1) is a straight line through the centre of gravity, and is vertical when the aeroplane is in rigging position. It is therefore at right angles to the longitudinal axis as defined above.

The aeroplane may travel upwards or downwards along this axis, as in **climbing** or **descending**, but in fact such movement is not very common, the climb or descent being obtained chiefly by the inclination of the longitudinal axis to the horizontal, followed by a straightforward movement along that axis.

Rotary motion of the aeroplane about the normal axis is called **yawing**.

The **lateral axis** (Fig. 8.1) is a straight line through the centre of gravity at right angles to both the longitudinal and the normal axes. It is horizontal when the aeroplane is in rigging position and parallel to the line joining the wing tips.

The aeroplane may travel to right or left along the lateral axis; such motion is called **sideslipping or skidding**.

Rotary motion of the aeroplane about the lateral axis is called **pitching**.

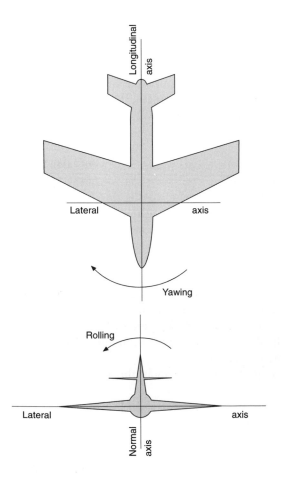

Fig. 8.1   The three axes

These axes must be considered as moving with the aeroplane and always remaining **fixed relative to the aeroplane**, e.g. the lateral axis will remain parallel to the line joining the wing tips in whatever attitude the aeroplane may be, or, to take another example, during a vertical nose-dive the longitudinal axis will be vertical and the lateral and normal axes horizontal.

So the manoeuvres of an aeroplane are made up of one or more, or even of all the following –

1. Movement forwards or backwards.
2. Movement up or down.
3. Movement sideways, to right or left.
4. Rolling.
5. Yawing.
6. Pitching.

Some of these motions, or combinations of motion, are gentle in that they involve only a state of equilibrium. These have already been covered under the headings of level flight, gliding, climbing, and so on. In this chapter we shall deal with the more thrilling manoeuvres, those that involve changes of direction, or of speed, or of both – in other words, **accelerations**. In such manoeuvres the aeroplane is no longer in equilibrium. There is more thrill for the pilot; more interest, but more complication, in thinking out the problems on the ground (Figs 8A and 8E).

Fig. 8A Manoeuvres
A clipped-wing Spitfire in mock combat manoeuvres with a Chance Vought Corsair.

ACCELERATIONS

Now the **accelerations** of an aeroplane **along its line of flight are comparatively unimportant**. They are probably greatest during the take-off, or, in the negative sense, during the pull-up after landing. But the **accelerations due to change in direction of flight are of tremendous importance**.

As we have already discovered, when a body is compelled to move on a curved path, it is necessary to supply a force towards the centre, this force being directly proportional to the acceleration required. Such a force is called the centripetal force. The body will cause a reaction, that is to say an outward force, on whatever makes it travel on a curved path. This reaction is called by some people the centrifugal force.

If an aeroplane is travelling at a velocity of V metres per second on the circumference of a circle of radius r metres, then the acceleration towards the centre of the circle is V^2/r metres per second per second.

Therefore the centripetal (or centrifugal) force is $m \times$ acceleration, where m is the mass of the aeroplane in kilograms,

$= mV^2/r$ newtons

In practice aeroplanes very rarely travel for any length of time on the arc of a **circle**; but that does not alter the principle, since any **small** arc of a curve is, for all practical purposes, an arc of **some** circle with **some** radius, so all it means is that the centre and the radius of the circle keep changing as the aeroplane manoeuvres.

The acceleration being V^2/r shows that the two factors which decide the acceleration, and therefore the necessary force, are velocity and radius, the velocity being squared having the greatest effect. **Thus curves at high speed, tight turns at small radius, need large forces towards the centre of the curve.**

We can easily work out the acceleration, V^2/r. For instance, for an aeroplane travelling at 160 knots (82 metres per second) on a radius of 200 metres, the acceleration is $(82 \times 82)/200 = 34 \, \text{m/s}^2$ approx, which is a little less than $4g$.

The force required to produce an acceleration of $4g$ is $m \times 4g$ or $W/g \times 4g$, i.e. $4W$, so if the aeroplane weighs 2000 newtons the necessary force is 8000 newtons.

Now imagine, for a moment, that the aeroplane is describing this circle in a vertical plane, a loop – but of an even curvature such as never occurs in practice! Then **the acceleration will be $4g$ all round the curve**, so the force **to produce the acceleration** will be $4W$ newtons all round the curve, **but** at the bottom of the curve there will **also be the weight to be lifted**, so the total lift on the aeroplane wings will be

5*W* newtons; at the top of the loop, on the other hand, the weight will help to produce the centripetal force, so the lift on the wings will be 3*W* newtons – downward. At the sides of the loop it will be 4*W* newtons – in a horizontal direction. So it will be seen that while the acceleration is 4*g* all round the curve, the force varies between 3*W* and 5*W* newtons.

This has been emphasised again because many pilots have grown into the habit of talking in terms of *g*'s, and, as so often happens when a scientific word or symbol comes into popular use, the meaning of *g* is often – more often than not – misunderstood and misapplied. You will often hear people (unfortunately pilots are the worst offenders) saying that *g* stands for 'gravity' – in a sense, of course, it does; but what they mean by 'gravity' is 'the force due to gravity', and this it most certainly does **not** stand for; *g* is an acceleration, and it is measured in m/s^2. Those who know a little more may justify their loose talk by saying that the force is **proportional** to the acceleration. That is quite correct, but there is all the difference in the world between proportionality and equality. Also, in the example of the loop given above, notice that the proportion is only true if we consider **only** the total force to produce the acceleration. The acceleration is constant all round the circle but the force produced by the wings is not. It would, of course, be very convenient if we could make the *g*'s and the *W*'s correspond, even in a vertical plane, that is to say if we could call the acceleration at the bottom of the curve 5*g* and at the top 3*g*. This is the same as imagining the original **weight as being equivalent to an initial upward acceleration of** *g*, superimposing, as it were, an upward acceleration of *g* on the whole motion. This is quite legitimate, **provided we realise what we are doing** (there's the rub). It simplifies our thoughts and calculations, and the expert does it quite happily. But how many others do it, too, and how little do they know what they are doing!

PULLING OUT OF A DIVE

In the light of what we have considered, let us take first the manoeuvre of pulling out of a nose-dive. Clearly this is a case of the aeroplane following a curved path at high velocity, with all consequent accelerations and forces. On the assumption that the aeroplane was diving at 60° to the horizontal – a very steep dive – and was travelling at 300 knots (154 m/s) before pulling out, Fig. 8.2 gives some idea of the accelerations involved and their effects on pilot and aircraft if the pull-out is effected with various losses of height. Exact figures cannot be arrived at without making all kinds of doubtful assumptions, but the principle illustrated is true enough.

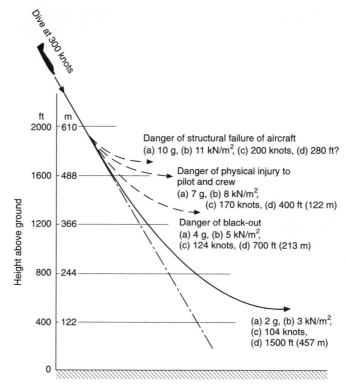

Fig. 8.2   Pulling out of a dive
The figures and symbols represent: (*a*) acceleration, (*b*) wing loading, (*c*) stalling
speed, (*d*) loss of height. Stalling speed in level flight = 60 knots (31 m/s); wing
loading in level flight = 1 kN/m². *Note:* The wing loadings given in figure are those
which occur at the lower part of the manoeuvre, i.e. when the weight is added to the
accelerating force.

The effects of the increase of *g*'s are as follows: Throughout the whole
range **the wing loading is going up; the equivalent weight of the
aeroplane, of the pilot, of every part of aeroplane and pilot, is
increasing; the stalling speed is getting higher and higher**. At
about 4*g* or 5*g* (we are ourselves offending by using *g* in its loose sense –
it is really the loading that matters) the aeroplane is sitting on the air
with four or five times its normal weight and the air is reacting on the
structure in the same proportion, the pilot is sitting on his seat with four
or five times his normal weight, his head feels heavy on his neck, but,
most important of all, his heart, a pressure pump, is having difficulty in
pumping the now heavy blood-stream to his head, and the lack of blood-
pressure causes the sight to diminish and eventually the pilot 'greys out'

or even 'blacks out' altogether. This, in itself, is not serious, since he does not lose all his senses and usually completes the manoeuvre satisfactorily and recovers as soon as the g's are reduced. But if he goes up to 7 or $8g$ it is possible that more serious physical injury may result. For fighting purposes it is important that pilots should be able to stand as many g's as possible so that they can manoeuvre quickly, and experiments have been tried with this end in view. Perhaps we ought to take a lesson from the birds and fly with our bodies horizontal – even that has been tried, and is the standard method for lift-off and re-entry in space-craft. When we reach $9g$ or $10g$ – **if we can reach $9g$ or $10g$** – the loading on the aeroplane has become such that **the structure itself is in a critical condition** and may begin to show signs of breaking up. The designer has made it just a **little** stronger than the pilot, but there would be no point in making it **much** stronger.

All this time the stalling speed has been going up, in proportion to the square root of the wing loading (see page 204). The figure shows the increase of stalling speeds assuming a stalling speed in level flight of 60 knots (31 m/s). Stalling at these high speeds is a very real danger; the aircraft may be near the ground, and when it stalls it will lose height rapidly and may drop a wing and start a spin.

The moral of all this is that the pilot should allow plenty of height to pull out and should do so gently, while the designer must make the aircraft strong enough, and all who work on aeroplanes should understand the problem – that is why we have devoted so much space to it. We have also served our purpose because what has been said applies to nearly all the manoeuvres of an aircraft.

THE LOAD FACTOR

In order to allow for the extra loads likely to be encountered during aerobatics, every part of an aeroplane is given a **load factor**, which varies according to conditions, being usually between 4 and 8. This means that the various parts are made from 4 to 8 times stronger than they need be for straight and level flight.

TURNING

In an ordinary turn (Fig. 8B) the inward centripetal force is provided by the aeroplane **banking** (like a car on a racing track) so that the total lift on the wings, in addition to lifting the aeroplane, can supply a component towards the centre of the turn (Fig. 8.3).

Suppose an aeroplane of weight W newtons to be travelling at a

Fig. 8B   Turning
A Mustang making a tight turn near the ground.

velocity of V metres per second on the circumference of a circle of radius r metres, then the acceleration towards the centre of the circle is V^2/r metres per second per second.

Therefore the force required towards the centre

$= WV^2/gr$ newtons.

If the wings of the aeroplane are banked at an angle of θ to the horizontal, and if this angle is such that the aeroplane has no tendency to slip either inwards or outwards, then the lift L newtons will act at right angles to the wings, and **it must provide a vertical component, equal to W newtons, to balance the weight, and an inward component, of WV^2/gr newtons, to provide the acceleration towards the centre.**

This being so, it will be seen that –

$\tan \theta = (WV^2/gr) \div W = V^2/gr$

This simple formula shows that **there is a correct angle of bank, θ,**

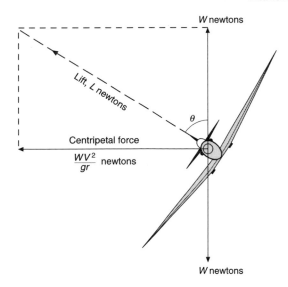

Fig. 8.3   Forces acting on an aeroplane during a turn

for any turn of radius *r* metres at a velocity of *V* m/s, and that this angle of bank is quite independent of the weight of the aeroplane.

Consider a numerical example –

Find the correct angle of bank for an aeroplane travelling on a circle of radius 120 m at a velocity of 53 m/s (take the value of *g* as 9.81 m/s²).

$$V = 53 \text{ m/s}$$
$$r = 120 \text{ m}$$
$$\tan \theta = V^2/gr = (53 \times 53)/(9.81 \times 120) = 2.38$$
$$\therefore \theta = 67° \text{ approx}$$

What would be the effect if the velocity were doubled, i.e. 106 m/s?

$$\tan \theta \text{ would be } 4 \times 2.38 = 9.52$$
$$\therefore \theta = 83° \text{ approx}$$

What would be the effect if the velocity were 53 m/s as in the first example, but the radius was doubled to 240 m instead of 120 m?

$$\tan \theta \text{ would be } 2.38/2 = 1.19$$
$$\therefore \theta = 49° \text{ approx}$$

Thus we see that an increase in velocity needs an increase in the angle of bank, whereas if the radius of the turn is increased the angle of bank may be reduced, all of which is what we might expect from experiences of cornering by other means of transport. Figures 8.4 and 8.5 show the

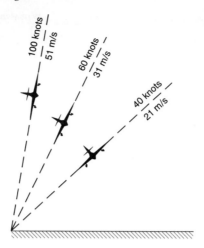

Fig. 8.4   Correct angles of bank
Radius of turn 50 metres

Fig. 8.5   Correct angles of bank
Air speed 60 knots (31 m/s)

correct angle of bank for varying speeds and radii; notice again **how the speed has more effect on the angle than does the radius of turn**.

LOADS DURING A TURN

It will be clear from the figures that the lift on the wings during the turn is greater than during straight flight; it is also very noticeable that the lift increases considerably with the angle of bank. This means that all the lift bracing of the aeroplane, such as the wing covering and spars, will have to carry loads considerably greater than those of straight flight.

Mathematically, $W/L = \cos\theta$, or $L = W/\cos\theta$

i.e. at 60° angle of bank, lift = $2W$, stalling speed, 85 knots (44 m/s)
at 70° angle of bank, lift = $3W$, stalling speed, 104 knots (53 m/s)
at 75° angle of bank, lift = $4W$, stalling speed, 120 knots (62 m/s)
at 84° angle of bank, lift = $10W$, stalling speed, 190 knots (98 m/s)

These figures mean that at these angles of bank, which are given to the nearest degree, the loads on the wing structure are 2, 3, 4, and 10 times respectively the loads of normal flight. This is simply our old friend g again, but in this instance it is certainly better to talk in terms of load than of g because the accelerations, and the corresponding loads, are in a horizontal plane while the initial weight is vertical; it is no longer a question of adding by simple arithmetic.

Whatever the angle of bank, the lift on the wings must be provided by $C_L \cdot \frac{1}{2}\rho V^2 \cdot S$. It follows, therefore, that the value of $C_L \cdot \frac{1}{2}\rho V^2 \cdot S$ must be greater during a turn than during normal flight, and this must be achieved either by increasing the velocity or increasing the value of C_L. Thus it follows that the stalling speed, which means the speed at the maximum value of C_L, must go up in a turn; as before it will go up in proportion of the square root of the wing loading, and the stalling speeds corresponding to the various angles of bank are shown in the table assuming, as for the pull-out of a dive, a stalling speed in level flight of 60 knots (31 m/s). These are all fairly steep banks; for banks up to 45° or so the loads are not serious, there is no danger of blacking out, and the increase of stalling speed is quite small – even so, it needs watching if one is already flying or gliding anywhere near the normal stalling speed, and suddenly decides to turn. At steep angles of bank we have to contend not only with the considerable increase of stalling speeds but with all the same problems as arose with the pull-out, i.e. blacking out, injury to pilot and crew, and the possibility of structural failure in the aircraft. It may seem curious that the angle of bank should be the deciding factor, but it must be remembered that the angle of bank (provided it is the **correct** angle of bank) is itself dependent on the velocity and radius of the turn, and these are the factors that really matter. In the history of fighting aircraft the ability to out-turn an opponent has probably counted more than any other feature, and from this point of view the question of steeply banked turns is one of paramount importance. An aspect of this question which must not be forgotten is that of engine power; **steep turns can only be accomplished if the engine is powerful enough to keep the aeroplane travelling at high speed and at large angles of attack, perhaps even at the stalling angle.** The normal duties of the engine are to propel the aeroplane at high speed at small angles of attack, or low speed at large angles of attack, **but not both at the same time.** The need for extra power in steeply banked tight turns has resulted in a technique in which the pilot embarking on such a manoeuvre suddenly applies all the power available.

CORRECT AND INCORRECT ANGLES OF BANK

We have so far assumed that the aeroplane is banked at **the correct angle** for the given turn. Fortunately the pilot has several means of telling whether the bank is correct or not (Fig. 8.6), and since the methods help us to understand the mechanics of the turn, it may be as well to mention them here.

A good indicator is **the wind itself, or a vane, like a weather cock, mounted in some exposed position.** In normal flight and in a correct bank the wind will come from straight ahead (neglecting any local effects from the slipstream); if the bank is too much, the aeroplane will sideslip inwards and the aeroplane, and pilot if he is in an open cockpit, will feel the wind coming from the inside of the turn, whereas if the bank is too small, the wind will come from the outside of the turn, due to an outward skid on the part of the aeroplane.

Another indication is a plumb-bob hung in the cockpit out of contact with the wind. In normal flight this will, of course, hang vertically; **during a correct bank it will not hang vertically, but in exactly the same position relative to the aeroplane as it did in normal flight,** i.e. it will bank with the aeroplane. If over-banked the plumb-line will be inclined inwards; if under-banked, outwards from the above position. This plumb-bob idea, in the form of a pendulum, forms the basis of the sideslip indicator which is provided by the top pointer of the so-called **turn and bank indicator.** The pointer is geared so as to move in such a way that the pilot must move the control column **away**

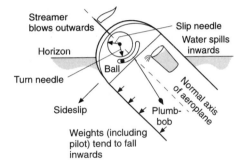

Angle of bank too large

Correct angle of bank

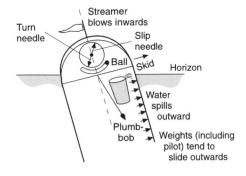

Angle of bank too small

Fig. 8.6 Effects of correct and incorrect angles of bank

from the direction of the pointer, this being the instinctive reaction. Sometimes a curved transparent tube containing a metal ball is used, and again the control column must be moved **away from** the indication given on the instrument. It is interesting to note that in early aeroplanes the slip indicator was, in effect, a spirit level, the tube being curved the opposite way and with a bubble (in liquid) instead of the ball; the pilot was then told to 'follow the bubble' – **not** the instinctive reaction. The figure shows how a tumbler full of water would not spill even when tilted at 80° in a correct bank; if the bank were too small it would spill outwards over the top lip of the tumbler!

Lastly, during a correct bank the pilot will sit on his seat without any feeling of sliding either inwards or outwards; in fact, he will be sitting tighter on his seat than ever, his effective weight being magnified in the same proportions as the lift so that if he weighs 800 N in normal flight he will feel that he weighs 8000 N when banking at 84°! If he over-banks he will tend to slide inwards, but outwards if the bank is insufficient.

OTHER PROBLEMS OF TURNING

In order to get into a turn the pilot puts on bank by means of the ailerons, but once the turn has commenced **the outer wing will be travelling faster than the inner wing and will therefore obtain more lift**, so he may find that not only is it necessary to take off the aileron control but actually to apply opposite aileron by moving the control column against the direction of bank – this is called **holding off bank**.

An interesting point is that this effect is different in turns on a glide and on a climb. On a **gliding turn** the whole aircraft will move the same distance downwards during one complete turn, but the inner wing, because it is turning on a smaller radius, will have descended on a steeper spiral than the outer wing; therefore the air will have come up to meet it at a steeper angle, in other words **the inner wing will have a larger angle of attack** and so obtain more lift than the outer wing. The extra lift obtained in this way may compensate, or more than compensate, the lift obtained by the outer wing due to increase in velocity. **Thus in a gliding turn there may be little or no need to hold off bank.**

In a **climbing turn**, on the other hand, the inner wing still describes a steeper spiral, but this time it is an **upward** spiral, so the air comes down to meet the inner wing more than the outer wing, thus **reducing the angle of attack on the inner wing**. So, in this case, the outer wing has more lift **both** because of velocity **and** because of increased angle, and **there is even more necessity for holding off bank than during a normal turn**.

Another interesting way of looking at the problem of gliding and climbing turns is to analyse the motion of an aircraft around its three axes during such turns. In a flat turn, i.e. a level turn without any bank, the aircraft is yawing only. In a banked level turn, the aircraft is yawing and pitching – in the extreme of a vertically banked turn it would be pitching only. But in a gliding or climbing turn the aircraft is pitching, yawing **and rolling**. In a **gliding** turn it is rolling **inwards**, as it were; in a **climbing** turn, **outwards**. The inward roll of the gliding turn causes the extra angle of attack on the inner wing, the outward roll of the climbing turn on the outer wing. Many people find it difficult to believe this. If the reader is in such difficulty he may be convinced by one of two methods; which will suit him best will depend upon his temperament. The mathematically-minded may like to analyse the motion in terms of the following (Fig. 8.7) –

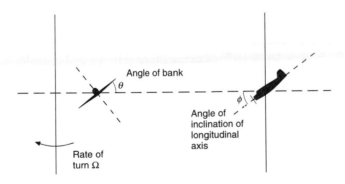

Fig. 8.7   Gliding turns

The rate of turn of the complete aeroplane (about the vertical), Ω.
The angle of bank of the aeroplane, θ.
The angle of pitch of the aeroplane, ϕ.
A little thought will reveal the fact that the

Rate of yaw $= \Omega \,.\, \cos\phi \,.\, \cos\theta$.
Rate of pitch $= \Omega \,.\, \cos\phi \,.\, \sin\theta$.
Rate of roll $= \Omega \,.\, \sin\phi$.

Translating this back into English, and taking one of the extreme examples, when $\theta = 0$, i.e. no bank, and $\phi = 0$, i.e. no pitch, $\cos\theta$ and $\cos\phi$ will be 1, $\sin\theta$ and $\sin\phi$ will be 0.

∴ rate of yaw = Ω = rate of turn of complete aeroplane.

Rate of pitch and rate of roll are zero. All of which we had previously decided for the flat turn.

The reader (mathematically-minded) may like to work out for himself the other extremes such as the vertical bank ($\theta = 90°$) or vertical pitch ($\phi = 90°$), or better still the more real cases with reasonable values of θ and ϕ.

Notice that the rate of roll depends entirely on the angle of pitch, i.e. the inclination of the longitudinal axis to the vertical – if this is zero, there is no rate of roll even though the aircraft may be descending or climbing.

What about the reader who does not like mathematics? Get hold of a model aeroplane, or, failing this, a waste-paper basket and spend a few minutes making it do upward and downward spirals; some people are convinced by doing gliding and climbing turns with their hand and wrist – and their friends may be amused in watching!

At large angles of bank there is less difference in velocity, and in angle, between inner and outer wings, and so the question of holding off bank becomes less important; but much more difficult problems arise to take its place.

First, though, let us go back to the other extreme and consider what is called a '**flat turn**', i.e. one that is all yaw and without any bank at all.

Very slight turns of this kind have sometimes been useful when approaching a target for bombing purposes, but otherwise they are in the nature of 'crazy flying', in other words, incorrect flying, and good pilots always try to keep their sideslip indicator in the central position. Actually flat turns are rather difficult to execute for several reasons. First, the extra velocity of the other wing tends to bank the aeroplane automatically; secondly, the lateral stability (explained later) acts in such a way as to try to prevent the outward skid by banking the aeroplane; thirdly, the side area is often insufficient to provide enough inward force to cause a turn except on a very large radius; fourthly, the directional stability (also explained later) opposes the action of the rudder and tends to put the nose of the aircraft back so that it will continue on a straight path. Taking these four reasons together, it will be realised that an aeroplane has a strong objection to a flat turn!

Modern aircraft have a **small side surface** and if this is coupled with **good directional stability**, for the last two reasons particularly, a flat turn becomes virtually impossible. So much is this so that it is very little use applying rudder to start a turn, the correct technique being to put on bank only.

CONTROLS ON STEEP BANKS

The turning of an aeroplane is also interesting from the control point of view because as the bank becomes steeper **the rudder gradually takes the place of the elevators,** and vice versa. This idea, however, needs treating with a certain amount of caution because, in a vertical bank for instance, the rudder is nothing like so powerful in raising or lowering the nose as are the elevators in normal horizontal flight. Incidentally, the reader may have realised that a vertical bank, without sideslip, is theoretically impossible, since in such a bank the lift will be horizontal and will provide no contribution towards lifting the weight. If it is claimed that such a bank can, in practice, be executed, the explanation must be that a slight upward inclination of the fuselage together with the propeller thrust provides sufficient lift.

This only applies to a **continuous** vertical bank in which no height is to be lost; it is perfectly possible, both theoretically and practically, to execute a turn in which, for a few moments, the bank is vertical, or even over the vertical. In the latter case the manoeuvre is really a combination of a loop and a turn.

Generally speaking, the radius of turn can be reduced as the angle of bank is increased, but even with a vertical bank there is a limit to the smallness of the radius because, quite apart from the question of side-slipping, the lift on the wings (represented by $C_L . \frac{1}{2}\rho V^2 . S$) must provide all the force towards the centre, i.e. $m . V^2/r$ or WV^2/gr.

Thus $\quad WV^2/gr = C_L . \frac{1}{2}\rho V^2 . S$

or $\quad r = 2W/(C_L . \rho S . g)$

Now, in straight and level flight the stalling speed (V_s) is given by the equation

$$W = L = C_L \max . \tfrac{1}{2}\rho V_s^2 . S$$

If we substitute this value of W into our formula for the radius we get

$$r = (2 . C_L \max . \tfrac{1}{2}\rho V_s^2 . S)/(C_L . \rho S . g)$$

i.e. $\quad r = (V_s^2/g) \times (C_L \max/C_L)$

This shows that the radius of turn will be least when C_L is equal to C_L max, i.e. when the angle of attack is the stalling angle, and radius of turn $= V_s^2/g$. It is rather interesting to note that the minimum radius of turn is quite independent of the **actual** speed during the vertical banks; it is settled only by the stalling speed of the particular aeroplane. Thus, to turn at minimum radius, one must fly at the stalling angle, but any speed may be employed provided the engine power is sufficient to maintain it. In actual practice, the engine power is the deciding factor in settling the minimum radius of turn whether in a vertical bank or any other bank, and it must be admitted that it is not usually possible to turn on such a small radius as the above formula would indicate.

This formula applies to some extent to all steep turns and shows that the aeroplane with the lower stalling speed can make a tighter turn than one with a higher stalling speed. (We are referring, as explained above, to the stalling speed in straight and level flight.) But in order to take advantage of this we must be able to stand the *g*'s involved in the steep banks, and we must have engine power sufficient to maintain turns at such angles of bank.

AEROBATICS

The usual aerobatics are loops, spins, rolls, sideslips, and nose-dives, to which may be added upside-down flight, the inverted spin, and the inverted loop. The manoeuvres may also be combined in various ways, e.g. a half loop followed by a half roll, or a half roll followed by the second half of a loop.

There are many reasons why aerobatics should be performed in those types of aircraft which are suitable for them. They provide excellent training for accuracy and precision in manoeuvre, and give a feeling of complete mastery of the aircraft, which is invaluable in all combat flying. They may also be used for exhibition purposes, but modern aircraft are so fast and the radius on which they can turn or manoeuvre is so large that, in many ways, they provide less of a spectacle than older types. Not least, aerobatics increase the joy and sensations of flight to the pilot himself – not quite so much to other occupants of the aircraft!

The movements of the aeroplane during these aerobatics are so complicated that they baffle any attempt to reduce them to the terms of simple mechanics and, indeed, to more advanced theoretical considerations, unless assumptions are made which are not true to the facts.

Figs 8.8 and 8.9 show the approximate path travelled by a slow type of aircraft during a **loop** and the corresponding 'accelerometer' diagram which shows how the force on the wings varies during the manoeuvre. From this it will be seen that, as in many manoeuvres, the greatest loads occur at the moment of entry. Notice also that even at the top of the loop the load is very little less than normal – that is to say, that the pilot is sitting firmly on his seat in the upward direction, the loads will still be in the same direction **relative to the aircraft** as in normal flight, and

Fig. 8.8 A loop

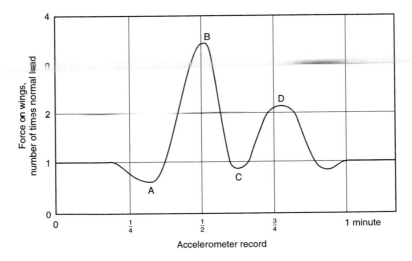

Fig. 8.9 Accelerometer diagram for a loop

our plumb bob will be hanging upwards! Only in a bad loop will the
loads at the top become negative, causing the loads on the aircraft
structure to be reversed and the pilot to rely on his straps to prevent him
from falling out.

Simple theoretical problems can be worked out on such assumptions
as that a loop is in the form of a circle or that the velocity remains
constant during the loop, but the error in these assumptions is so great
that very little practical information can be obtained by attempting to

solve such problems, and it is much better to rely on the results of practical experiments.

A **Spin** (Fig. 8.10) is an interesting manoeuvre, if only for the reason that at one time there stood to its discredit a large proportion of all aeroplane accidents that had ever occurred. It differs from other manoeuvres in the fact that the wings are 'stalled', – i.e. are beyond the critical angle of attack, and this accounts for the lack of control which the pilot experiences over the movements of the aeroplane while spinning; it is, in fact, a form of 'auto-rotation' (Fig. 8.11), which means that there is a natural tendency for the aeroplane to rotate of its own accord. This tendency will be explained a little more fully when dealing with the subject of control at low speeds in the next chapter. In a spin the aeroplane follows a steep spiral path, but the attitude while spinning

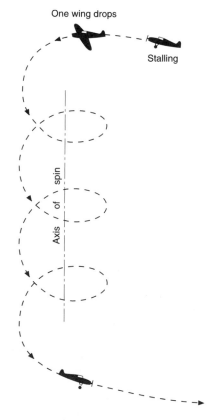

Fig. 8.10   A spin
During spin: large angle of attack; forward speed low; downward speed low; aeroplane stalled; loads above normal, but do not vary very much.

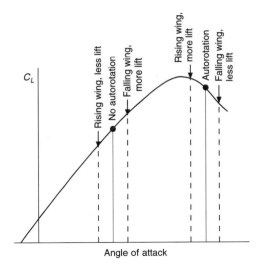

C_L

Rising wing, less lift

No autorotation

Falling wing, more lift

Rising wing, more lift

Autorotation

Falling wing, less lift

Angle of attack

Fig. 8.11 Cause of autorotation

may vary from the almost horizontal position of the 'flat' spin to the almost vertical position of the 'spinning nose-dive'. In other words the spin, like a gliding turn or steep spiral is composed of varying degrees of yaw, pitch and roll. A flat spin is chiefly yaw, a spinning nose-dive chiefly roll. The amount of pitch depends on how much the wings are banked from the horizontal. In general, the air speed during a spin is comparatively low, and the rate of descent is also low. Any device, such as slots, which tend to prevent stalling, will also tend to minimise the danger of the accidental spin and may even make it impossible to carry out deliberately. The area and disposition of the fin, rudder, and tail plane exert considerable influence on the susceptibility of the aeroplane to spinning.

Many of the terrors of a spin were banished once it was known just what it was. We then realised that in order to get out of a spin we must get it out of the stalled state by putting the nose down, and we must stop it rotating by applying 'opposite rudder'. In practice, the latter is usually done first, because it is found that the elevators are not really effective until the rotation is stopped. The farther back the centre of gravity, and the more masses that are distributed along the length of the fuselage, the flatter and faster does the spin tend to become and the more difficult is it to recover. This flattening of the spin is due to the centrifugal forces that act on the masses at the various parts of the aircraft (Fig. 8.12). A spin is no longer a useful combat manoeuvre, nor is it really a pleasant form of aerobatics, but since it is liable to occur accidentally, pilots are taught how to recover from it.

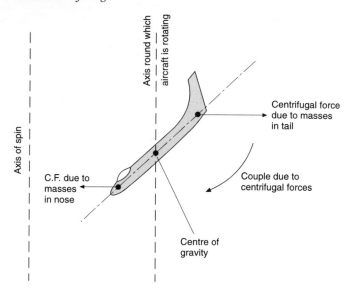

Fig. 8.12   Tendency of a spin to flatten owing to centrifugal force

During a **roll** (Fig. 8.13) the aeroplane rotates laterally through 360°, but the actual path is in the nature of a horizontal corkscrew, there being varying degrees of pitch and yaw. In the so-called slow roll the loads in the 180° position are reversed, as in inverted flight, whereas in the other extreme, the barrel roll, which is a cross between a roll and a loop, the loads are never reversed.

Fig. 8.13   A roll

In a **sideslip** (Fig. 8C) there will be considerable wind pressure on all the side surfaces of the aeroplane, notably the fuselage, the fin and the rudder, while if the planes have a dihedral angle the pressure on the wings will tend to bring the machine on to an even keel. The sideslip is a useful manoeuvre for losing height or for compensating a sideways drift just prior to landing, but, as already mentioned, modern types of aircraft do not take very kindly to sideslipping. The small side area means that they drop very quickly if the sideslip is at all steep, and the directional

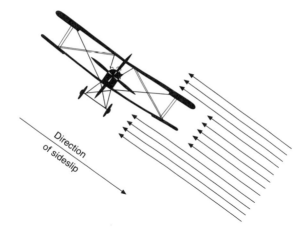

Direction of sideslip

Fig. 8C   A sideslip in an old-type biplane
This is one of the few remaining figures from the first edition of this book; it seems appropriate that an old type of manoeuvre should be illustrated on an old type of aircraft, the SE5A of the First World War.

stability is so strong that it may be impossible to hold the nose of the machine up (by means of the rudder), and the dropping of the nose causes even more increase of speed.

A **nose-dive** is really an exaggerated form of gliding; the gliding angle may be as great as 90° – i.e. vertical descent – although such a steep dive is rarely performed in practice. If an aeroplane is dived vertically it will eventually reach a steady velocity called the **terminal velocity**. In such a dive the weight is entirely balanced by the drag, while the lift has disappeared. The angle of attack is very small or even negative, there is a large positive pressure near the leading edge on the top surface of the aerofoil, tending to turn the aeroplane on to its back, and this is balanced by a considerable 'down' load on the tail plane (Fig. 8.14). In such extreme conditions the terms used are apt to be misleading; for instance, the 'down' load referred to is horizontal, while the lift, if any such exists, will also be horizontal. The terminal velocity of modern aeroplanes is very high, and it makes little difference whether the engine is running or not. They lose so much height in attaining the terminal velocity that, in practice, it is doubtful whether it can ever be reached. As was only to be expected, the problems which accompany the attainment of a speed near to the speed of sound first made themselves felt in connection with the nose-dive, especially at high altitudes. At that time these compressibility effects were a special feature of nose-diving, but there was one consoling feature – as one got nearer the earth, terminal velocities were lower (owing to the greater density), and the

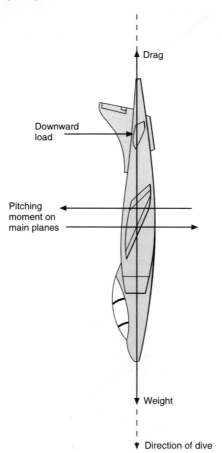

Fig. 8.14   Nose-dive

speed of sound was higher (owing to the higher temperature). So if one got into trouble high up, there was always a chance of getting out of it lower down. But nowadays, when more and more aircraft can exceed the speed of sound in level flight, and when compressibility troubles are no longer associated with fears of the unknown, these ideas are out of date, and the whole subject of flight at and above the speed of sound will be considered in Chapters 11 and 12.

The nose-dive, and the pulling-out of a nose-dive, are two entirely different problems, and the latter has already been fully dealt with.

INVERTED MANOEUVRES

Real upside–down flight (Fig. 8.15) is not so often attempted as is commonly supposed, and should be distinguished from a **glide** in the inverted position, which does not involve problems affecting the engine. If height is to be maintained during inverted flight, the engine must, of course, continue to run and this necessitates precautions being taken to ensure a supply of fuel and, with a piston engine, the proper functioning of the carburettor. The aerofoil will be inverted, and therefore, unless of the symmetrical type, will certainly be inefficient; while in order to produce an angle of attack, the fuselage will have to be in a very much 'tail-down' attitude. The stability will be affected, although some aircraft have been more stable when upside-down than the right way up, and considerable difficulty has been experienced in restoring them to normal flight. In spite of all the disabilities involved, some aeroplanes are capable of maintaining height in the inverted position (Fig. 8D).

The **inverted spin** is in most of its characteristics similar to the normal spin; in fact, in some instances pilots report that the motion is more steady and therefore more comfortable. As in inverted flight, however, the loads on the aeroplane structure are reversed and the pilot must rely on his straps to hold him in the machine.

The **inverted loop**, or 'double bunt', in which the pilot is on the outside of the loop (Fig. 8.16), is a manoeuvre of extreme difficulty and danger. The difficulty arises from the fact that whereas in the normal loop the climb to the top of the loop is completed while there is speed and power in hand and engines and aerofoils are functioning in the normal fashion, in the inverted loop the climb to the top is required during the second portion of the loop, when the aerofoils are in the inefficient inverted position. The danger is incurred because of the large reversed loads and also because of the physiological effects of the pilot's

Fig. 8.15   **Upside-down flight**

Fig. 8D   Inverted flight
(By courtesy of '*Flight*')
And to illustrate upside-down flight, the Hurricane of the Second World War; note
the tail-down attitude usually associated with inverted flight.

blood being forced into his head. It was a long time before this
manoeuvre was successfully accomplished, and once it had been, so
many foolhardy pilots began to attempt it – often with fatal results –
that it had to be forbidden, except under very strict precautions and
regulations.

'BUMPY' WEATHER

In addition to the loads incurred during definite aerobatics, all aircraft
are required to face the effects of **unsteady weather conditions**.
Accelerometer records show that these may be quite considerable, and
they must certainly be reckoned with when designing commercial
aircraft. Where aeroplanes are, in any case, required to perform
aerobatics, they will probably be amply strong enough to withstand any
loads due to adverse weather. The conditions which are likely to inflict
the most severe loads consist of strong gusty winds, hot sun,

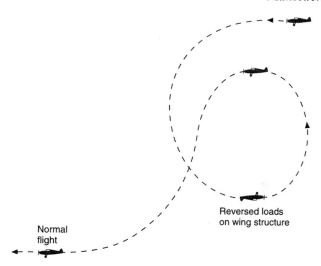

Reversed loads
on wing structure

Normal
flight

Fig. 8.16   The inverted loop

intermittent clouds, especially thunder clouds in which there is often considerable turbulence, and uneven ground conditions; a combination of all these factors will, almost certainly, spell a 'rough passage' that will rival any crossing of the Atlantic by sea. The turbulence and up-currents that may be encountered in severe thunderstorms and in cumulo–nimbus clouds can sometimes be such as to tax the strength of the aeroplane and the flying skill of the pilot.

MANOEUVRABILITY

Before leaving the subject of manoeuvres we ought to mention that the inertia of an aeroplane – or, to be more correct, **the moment of inertia of the various parts** – will largely determine the ease or otherwise of handling the machine during manoeuvres. Without entering into the mathematical meaning of moment of inertia, we can say that, in effect, it means the natural resistance of the machine to any form of rotation about its centre of gravity. Any heavy masses which are a long distance away from a particular axis of rotation will make it more difficult to cause any rapid movement around the axis; thus masses such as engines far out on the wings result in a resistance to rolling about the longitudinal axis; and a long fuselage with large masses well forward or back will mean a resistance to pitching and yawing.

In this question of quick manoeuvrability the biplane had a decided advantage over the monoplane, since it was in every way more compact, the masses concentrated nearer to the centre of gravity. Another advantage from the point of view of manoeuvrability was its greater rigidity, which prevented dangerous distortion of the structure taking place during manoeuvres. It was largely for these reasons that the biplane was for so long the more popular type for military purposes, where quick manoeuvrability is an essential feature. Modern high-speed monoplanes have very high wing loading and thus a small area of wing which makes them nearly as good as a biplane for purposes of manoeuvrability, although for display acrobatics biplanes are still popular.

CAN YOU ANSWER THESE?

See if you can answer these questions about the various manoeuvres which an aeroplane can perform –

1. What are the six degrees of freedom of an aeroplane?
2. Why is there a definite limit to the smallness of the radius on which an aeroplane can turn?
3. Two aircraft turn through 360° in the same time, i.e. at the same rate of turn, but the radius of turn of one is twice that of the other. Will they have the same angle of bank? If not, which will have the greater?

Fig. 8E   Manoeuvres
(By courtesy of the Lockheed Aircraft Corporation, USA)
A rigid-rotor helicopter performing aerobatic manoeuvres.

4. Explain the difference between a gliding and climbing turn from the point of view of holding off bank.
5. Why does an aeroplane spin?

Numerical examples on manoeuvres will be found in Appendix 3.

CHAPTER 9

STABILITY AND CONTROL

MEANING OF STABILITY AND CONTROL

The **stability of an aeroplane means its ability to return to some particular condition of flight** (after having been slightly disturbed from that condition) **without any efforts on the part of the pilot**. An aeroplane may be stable under some conditions of flight and unstable under other conditions. For instance, an aeroplane which is stable during straight and level flight may be unstable when inverted, and vice versa. If an aeroplane were stable during a nose-dive, it would mean that it would resist efforts on the part of the pilot to extricate it from the nose-dive. The stability is sometimes called **inherent stability**.

Stability is often confused with the balance or 'trim' of an aircraft, and the student should be careful to distinguish between the two. An aeroplane which flies with one wing lower than the other may often, when disturbed from this attitude, return to it. Such an aeroplane is out of its proper trim, but it is not unstable.

There is a half-way condition between stability and instability, for, as already stated, an aeroplane which, when disturbed, tends to return to its original position is said to be stable; if, on the other hand, it tends to move farther away from the original position, it is unstable. But it may tend to do neither of these and prefer to remain in its new position. This is called **neutral stability**, and is sometimes a very desirable feature.

Figure 9.1 illustrates some of the ways in which an aeroplane may behave when it is left to itself. Only a pitching motion is shown; exactly the same considerations apply to roll and yaw, although a particular aeroplane may have quite different stability characteristics about its three axes. The top diagram shows complete dead-beat stability which is very rarely achieved in practice. The second is the usual type of stability, that is to say an oscillation which is gradually damped out. The

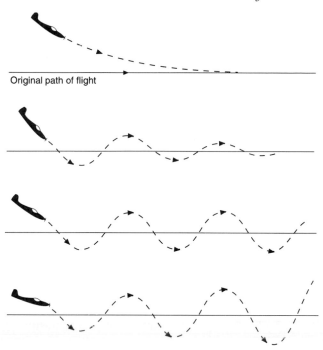

Original path of flight

Fig. 9.1 Degrees of stability

steady oscillation shown next is really a form of neutral stability, while the bottom diagram shows the kind of thing which may easily occur in certain types of aircraft, an oscillation which steadily grows worse. Even this is not so bad as the case when an aeroplane makes no attempt to return but simply departs farther and farther away from its original path. That is complete instability.

The degree of stability may differ according to what are called the **stick-fixed** and **stick-free** conditions; in pitching, for instance, stick-fixed means that the elevators are held in their neutral position relative to the tail plane, whereas stick-free means that the pilot releases the control column and allows the elevators to take up their own positions.

Another factor affecting stability is whether it is considered – and tested – under the condition of **power-off** or **power-on**. On modern aircraft the engine thrust can be comparable with or even greater than the airframe weight and therefore may significantly influence the stability.

Control means the power of the pilot to manoeuvre the aeroplane into any desired position. It is not by any means the same thing as stability; in fact, the two characteristics may directly oppose each other.

The stability or control of an aeroplane in so far as it concerns **pitching** about the lateral axis is called **longitudinal stability or control** respectively.

Stability or control which concerns **rolling** about the longitudinal axis is called **lateral stability or control**.

Stability or control which concerns **yawing** about the normal axis is called **directional stability or control**.

Before we attempt to explore this subject any further we feel it is our duty to warn the reader that the **problems involved in the consideration of stability and control of aeroplanes are considerable**. Any attempt at 'simple' explanation of such problems may, at the best, be incomplete and possibly incorrect. The reader need have no fear of the mathematics, as we shall not even attempt to tackle them, but he must be prepared, if and when he acquires greater knowledge of the subject from more advanced works, to readjust his ideas accordingly.

After this very necessary apology we will try to explain, at any rate, the practical considerations which affect stability and control.

LONGITUDINAL STABILITY

We shall start with **longitudinal stability**, since this can be considered independently of the other two. In order to obtain stability in pitching, we must ensure that **if the angle of attack is temporarily increased, forces will act in such a way as to depress the nose and thus decrease the angle of attack once again**. To a great extent we have already tackled this problem while dealing with the pitching moment, and the movement of the centre of pressure on aerofoils. We have seen that an ordinary upswept wing with a cambered aerofoil section cannot be balanced or 'trimmed' to give positive lift and at the same time be stable in the sense that a positive increase in incidence produces a nose-down pitching moment about the centre of gravity.

The position as regards the wing itself can be improved to some extent by **sweepback**, by **wash-out** (i.e. by decreasing the angle of incidence) towards the wing tips, by **change in wing section** towards the tips (very common in modern types of aircraft), and by a **reflex curvature** towards the trailing edge of the wing section.

But it is not only the wing that affects the longitudinal stability of the aircraft as a whole, and in general it can be said that this is dependent on four factors –

1. **The position of the centre of gravity,** which must not be too far back; this is probably the most important consideration.

2. **The pitching moment on the main planes**; this, as we have seen, usually tends towards instability, though it can be modified by the means mentioned.

3. **The pitching moment on the fuselage** or body of the aeroplane; this too is apt to tend towards instability.

4. **The tail plane** – its area, the angle at which it is set, its aspect ratio, and its distance from the centre of gravity. This is nearly always a stabilising influence (Fig. 9.2).

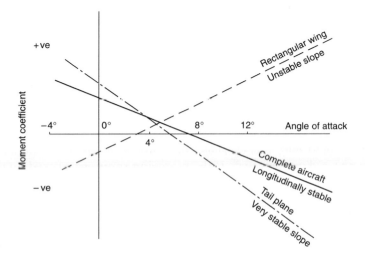

Fig. 9.2 Pitching moment coefficient about centre of gravity
Wing, tail plane and complete aircraft.

LONGITUDINAL DIHEDRAL

The tail plane is usually set at an angle less than that of the main planes, the angle between the chord of the tail plane and the chord of the main planes being known as the **longitudinal dihedral** (Fig. 9.3). This longitudinal dihedral is a practical characteristic of most types of

Fig. 9.3 Longitudinal dihedral angle

aeroplane, but so many considerations enter into the problem that it cannot be said that an aeroplane which does not possess this feature is necessarily unstable longitudinally. In any case, it is the actual angle at which the tail plane strikes the airflow, which matters; therefore we must not forget the downwash from the main planes. This downwash, if the tail plane is in the stream, will cause the actual angle of attack to be less than the angle at which the tail plane is set (Fig. 5.6). For this reason, even if the tail plane is set at the same angle as the main planes, there will **in effect** be a longitudinal dihedral angle, and this may help the aeroplane to be longitudinally stable.

Suppose an aeroplane to be flying so that the angle of attack of the main planes is 4° and the angle of attack of the tail plane is 2°; a sudden gust causes the nose to rise, inclining the longitudinal axis of the aeroplane by 1°. What will happen? The momentum of the aeroplane will cause it temporarily to continue moving practically in its original direction and at its previous speed. Therefore the angle of attack of the main planes will become nearly 5° and of the tail plane nearly 3°. The pitching moment (about the centre of gravity) of the main planes will probably have a nose-up, i.e. unstable tendency, but that of the tail plane, with its long leverage about the centre of gravity, will definitely have a nose-down tendency. **If the restoring moment caused by the tail plane is greater than the upsetting moment caused by the main planes, and possibly the fuselage, then the aircraft will be stable.**

This puts the whole thing in a nutshell, but unfortunately it is not quite so easy to analyse the practical characteristics which will bring about such a state of affairs; however the forward position of the centre of gravity and the area and leverage of the tail plane will probably have the greatest influence.

It is interesting to note that a tail plane plays much the same part, though more effectively, in providing longitudinal stability, as does reflex curvature on a wing, or sweepback with wash-out of incidence towards the tips.

When the tail plane is in front of the main planes (Fig. 5E) there will probably still be a longitudinal dihedral, which means that this front surface must have greater angle than the main planes. The latter will naturally still be at an efficient angle, such as 4°, so that the front surface may be at, say, 6° or 8°. Thus it is working at a very inefficient angle and will stall some few degrees sooner than the main planes. This fact is claimed by the enthusiasts for this type of design as its main advantage, since the stalling of the front surface will prevent the nose being raised any farther, and therefore the main planes will never reach the stalling angle.

In the tail-less type, in which there is no separate surface either in front or behind, the wings must be heavily swept back, and there is a 'wash-out' or decrease in the angle of incidence as the wing tip is approached, so that these wing tips do, in effect, act in exactly the same way as the ordinary tail plane (Figs 5C and 5D).

LATERAL STABILITY

Lateral and directional stability will first be considered separately; then we shall try to see how they affect each other.

To secure lateral stability we must so arrange things that **when a slight roll takes place the forces acting on the aeroplane tend to restore it to an even keel**.

In all aeroplanes, when flying at a small angle of attack, there is **a resistance to roll** because the angle of attack, and so the lift, will increase on the down-going wing, and decrease on the up-going wing. But this righting effect will only last while the aeroplane is actually rolling. It must also be emphasised that this only happens while the angle of attack is small; if the angle of attack is near the stalling angle, then the increased angle on the falling wing may cause a decrease in lift, and the decreased angle on the other side an increase; thus the new forces will tend to roll the aeroplane still further, this being the cause of auto-rotation previously mentioned (Fig. 8.11 on page 265).

But the real test of stability is **what happens after the roll has taken place**.

DIHEDRAL ANGLE

The most common method of obtaining lateral stability is by the use of a dihedral angle on the main planes (Figs 9.4 and 9A). Dihedral angle is taken as being the angle between each plane and the horizontal, not the total angle between the two planes, which is really the geometrical meaning of dihedral angle. If the planes are inclined upwards towards the wing tips, the dihedral is positive; if downwards, it is negative and called **anhedral** (Fig. 9B); the latter arrangement is used in practice for reasons of dynamic stability.

The effect of the dihedral angle in securing lateral stability is sometimes dismissed by saying that if one wing tip drops the horizontal equivalent on that wing is increased and therefore the lift is increased, whereas the horizontal equivalent and the lift of the wing which rises is decreased, therefore **obviously** the forces will tend to right the aeroplane.

$\beta°$ = Dihedral angle

Fig. 9.4   Lateral dihedral angle

Fig. 9A   Dihedral
Pronounced dihedral is evident on this Spitfire: a very late post-war model.

Unfortunately, it is not all quite so obvious as that.

Once the aircraft has stopped rolling, and provided it is still travelling straight ahead, the **aerodynamic** forces will be influenced only by the airstream passing over the aircraft. This will be identical for both wings and so no restoring moment will result.

What, then, is the real explanation as to why a dihedral angle is an aid to lateral stability? When the wings are both equally inclined the resultant lift on the wings is vertically upwards and will exactly balance the weight. If, however, one wing becomes lower than the other (Fig. 9.6), then the resultant lift on the wings will be slightly inclined in the

Fig. 9B   Anhedral
Like this Ilyushin IL-76, many aircraft with high-mounted swept wings are given a
significant amount of anhedral. This is primarily used to counteract the Dutch roll
tendency, a form of dynamic instability involving coupled roll and yaw motions.

Fig. 9.5   Equal lift produced by each wing – no rolling moment due to roll

direction of the lower wing, while the weight will remain vertical.
Therefore the two forces will not balance each other and there will be a
**small resultant force acting in a sideways and downwards
direction**. This force is temporarily unbalanced and therefore the
aeroplane will move in the direction of this force – i.e. it will **sideslip** –
and this will cause a flow of air in the opposite direction to the slip. This
has the effect of increasing the angle of attack of the lower plane and
increasing that of the upper plane. The lower plane will therefore
produce more lift and a restoring moment will result. Also the wing tip
of the lower plane will become, as it were, the leading edge so far as the
slip is concerned; and just as the centre of pressure across the chord is
nearer the leading edge, so the centre of the pressure distribution along

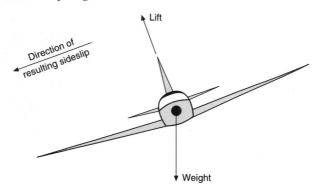

Fig. 9.6   The production of a sideslip

the span will now be on the lower plane; for both these reasons the lower plane will receive more lift, and after a slight slip sideways the aeroplane will roll back into its proper position. As a matter of fact, owing to the protection of the fuselage, it is probable that the flow of air created by the sideslip will not reach a large portion of the raised wing at all; this depends very much on the position of the wing relative to the fuselage.

Both the leading edge effect on the lower wing, and the shielding of the upper wing by the fuselage, occur on nearly all types of aircraft, and may well mean that an aeroplane has a sufficient degree of lateral stability without any dihedral angle, or too much if some of the following effects also apply. Even if there is no actual dihedral angle on the wings, these other methods of achieving lateral stability may be described as having a **'dihedral effect'**.

HIGH WING AND LOW CENTRE OF GRAVITY

If the wings are placed in a high position and the centre of gravity is correspondingly low, the lateral stability can be enhanced. When an aircraft sideslips, the lift on the lower wing becomes greater than that on the higher one. Furthermore, a small sideways drag force is introduced. In consequence, the resultant force on the wing will be in the general direction indicated in Fig. 9.7. You will see that this force does not now pass through the centre of gravity so there will be a small moment which will tend to roll the aircraft back to a level condition. This will occur even on a low-wing aircraft, but is more effective with a high wing because the moment arm is greater. For this reason a high-wing aircraft requires less dihedral than a low-wing type.

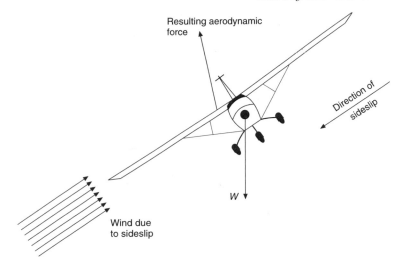

Fig. 9.7 **High wing and low centre of gravity**

SWEEPBACK AND LATERAL STABILITY

A considerable angle of sweepback (Fig. 9C) will in itself promote lateral stability, for, supposing the left wing to drop, as in the two previous cases, there will be a sideslip to the left and the left-hand wing will present, in effect, a higher aspect ratio than the right wing to the correcting airflow (Fig. 9.8). It will therefore receive more lift and, as before, recovery will take place after sideslip.

A forward sweep is sometimes used but this is not for reasons of stability (Fig. 9D).

FIN AREA AND LATERAL STABILITY

One factor which may have considerable influence on lateral stability is the position of the various side surfaces, such as the fuselage, fin and rudder, and wheels. All these will present areas at right angles to any sideslip, so there will be pressure upon them which, **if they are high above the centre of gravity,** will **tend to restore** the aeroplane to an even keel; this applies to many modern types which have a high tail plane on top of a high fin (Figs 9.9 and 9E) and such types may have anhedral on the main planes to counterbalance this effect and prevent too great a degree of lateral stability; but **if the side surfaces are low the pressure on them will tend to roll the aircraft over still more**

Fig. 9C   Sweepback
An Airbus A330 showing the sweepback of the wings.

(Figs 9.10 and 9F) and so cause lateral instability, although this must be balanced against the effect of high wing compared with the CG position.

The reader will have noticed that, whatever the method of obtaining lateral stability, correction only takes place **after a sideslip** towards the low wing.

It is the sideslip that effects the directional stability.

DIRECTIONAL STABILITY

We shall first try to consider **directional stability** by itself, if only as a means of convincing ourselves that the two are so interlinked one with the other that they cannot be disposed of separately. In order to establish directional stability we must ensure that, if the aeroplane is temporarily deflected from its course, it will, of its own accord, tend to return to that course again. This is almost entirely a question of the 'side surface' or 'fin area' which has already been mentioned when dealing with lateral stability, but here it is not a question of the relative height of this side surface, but whether it is in front of or behind the centre of gravity (Fig. 9F). When an aeroplane is flying in the normal way the airflow will approach it directly from the front, i.e. parallel to its

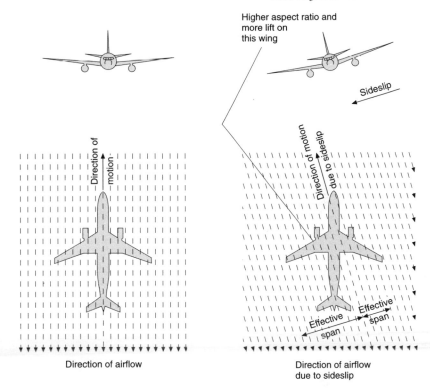

Fig. 9.8 Sweepback and lateral stability

longitudinal axis. Now imagine it to be deflected from its course as in Fig. 9.11; owing to its momentum it will for a short time tend to continue moving in its old direction, therefore the longitudinal axis will be inclined to the airflow, and a pressure will be created on all the side surfaces on one side of the aeroplane.

If the turning effect of the pressures behind the centre of gravity is greater than the turning effect in front of the centre of gravity, the aeroplane will tend to its original course.

If, on the other hand, the turning effect in front is greater than that behind, the aeroplane will turn still farther off its course. Notice that it is the turning effect or the moment that matters, and not the actual pressure; therefore it is not merely a question of how much side surface, but also of the distance from the centre of gravity of each side surface. For instance, a small fin at the end of a long fuselage may be just as effective in producing directional stability as a large fin at the end of a short fuselage. Also, there may sometimes be more side surface in the front than in the rear, but the rear surfaces will be at a greater distance.

Fig. 9.9   Effect of high fin on lateral stability

Fig. 9E   High fin
A massive high fin, and pronounced anhedral, are evident on the McDonnell
Douglas C-17.

All the side surfaces of an aeroplane, including that presented by wings
with dihedral, affect the directional stability, but to the **fin** is allotted the
particular task of finally adjusting matters and its area is settled
accordingly.

There is a very close resemblance between the directional stability of
an aeroplane and the action of a weathercock which always turns into the
wind; in fact, one often sees a model aeroplane used as a weathercock.
The simile, however, should not be carried too far, and the student must
remember that there are two essential differences between an aeroplane

---

Fig. 9D   Sweep forward (opposite)
The Grumman X-29 research aircraft with forward-swept wings. Forward-swept
wings have the same advantages in high-speed flight as swept-back wings, but give a
better spanwise distribution of lift, leading to lower induced drag.

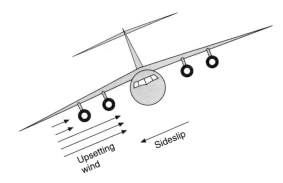

Fig. 9.10   Effect of low-slung fuselage and engine pods on lateral stability

Fig. 9F   A classic flying boat
(By courtesy of the General Dynamics Corporation, USA)
The large side area of the hull was balanced, from the lateral stability point of view,
by the dihedral on the main planes and the high fin at the rear.

and a weathercock – first, that an aeroplane is not only free to yaw, **but
also to move bodily sideways**; and secondly, that the 'wind', in the
case of an aeroplane, is not the wind we speak of when on the ground,
but the wind caused by the original motion of the aeroplane

Fig. 9G Fin and directional stability
A Sukhoi Su-27 blasting down the runway with nosewheel lifting clear of the
ground. Note the two large fins with dorsal extension below the tailplane. This
aircraft displays exceptional manoeuvrability and control at low speed.

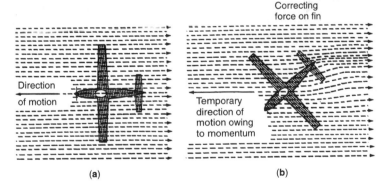

Fig. 9.11 Directional stability
(a) Before disturbance; (b) after disturbance

through the air. This point is emphasised because of the idea which
sometimes exists that an aeroplane desires to turn head to wind. If
such were the case, directional stability would be a very mixed
blessing.

LATERAL AND DIRECTIONAL STABILITY

Now we are, at last, in a position to connect these two forms of stability – the sideslip essential to lateral stability will cause an air pressure on the side surfaces which have been provided for directional stability. The effect of this pressure will be to turn the nose into the relative wind, i.e. in this case, **towards the direction of sideslip**. The aeroplane, therefore, will turn off its original course and in the direction of the lower wing. It is rather curious to note that the greater the directional stability the greater will be the tendency to turn off course in a sideslip. This turn will cause the raised wing, now on the outside of the turn, to travel faster than the inner or lower wing, and therefore to obtain more lift and so bank the aeroplane still further. By this time the nose of the aeroplane has probably dropped and the fat is properly in the fire with all three stabilities involved! The best way of seeing all this happen in real life is to watch a model aeroplane flying on a gusty day; the light loading and slow speed of the model make it possible to watch each step in the proceedings, whereas in the full-sized aeroplane it all happens more quickly, and also the pilot usually interferes by using his controls. If, for instance, the left wing drops and he applies rudder so as to turn the machine to the right, he will probably prevent it from departing appreciably from its course.

We can now explain the technique of turning an aeroplane. Suppose, when we want to turn to the left, **instead of applying any rudder we simply bank the aeroplane to the left**, as we have already seen it will slip inwards and **turn to the left**. That is all there is in it. So effective is this method that it is unnecessary to use the rudder at all for turning purposes. So far as the yaw is concerned – and a turn must involve a yaw – the rudder (with the help of the fin) is still responsible, just as (with the help of the fin) it always was. The difference is simply that the rudder and fin are brought into effect by the inward sideslip, instead of by **application** of rudder which tends to cause an outward skid. The pilot may do nothing about it, but the stability of the aeroplane puts a force on the rudder for him. It should also be emphasised that although it may be most practical, and most sensible, to commence a turn in certain aircraft without application of rudder, **such a turn cannot be absolutely perfect; there must be an inward sideslip**. The pilot may not notice it, the sideslip indicator may not detect it; but it is there just the same.

Just as a slight roll results in a sideslip and then a yawing motion so if an aircraft moves in a yawed position, as in Fig. 9.11, that is if it moves crabwise (which is really the same thing as slipping or skidding) lateral stability will come into play and cause the aircraft to roll away from the leading wing. Thus a **roll causes a yaw**, and a **yaw causes a roll**, and the study of the two cannot be separated.

If the stability characteristics of an aeroplane are such that it is **very stable directionally** and **not very stable laterally**, e.g. if it has large fin and rudder and little or no dihedral angle, or other 'dihedral effect', it will have a marked tendency to turn into a sideslip, and to bank at steeper and steeper angles, that it may get into an uncontrollable spiral – this is sometimes called **spiral instability**, but note that it is caused by **too much** stability (directional).

If, on the other hand, the aeroplane is **very stable laterally** and **not very stable directionally**, it will sideslip without any marked tendency to turn into the sideslip. Such an aircraft is easily controllable by the rudder, and if the rudder only is used for a turn the aircraft will bank and make quite a nice turn.

The reader will find it interesting to think out the other characteristics which these two extremes would cause in an aeroplane, but the main point to be emphasised is that **too much stability** (of any type) is almost as bad as too little stability.

CONTROL OF AN AEROPLANE

Where an aeroplane is stable or unstable, it is necessary for the pilot to be able to control it, so that he can manoeuvre it into any desired position.

Longitudinal control is provided by the **elevators**, i.e. flaps hinged behind the tail plane.

Roll control is provided by the **ailerons**, i.e. flaps hinged at the rear of the aerofoils near each wing tip.

Directional control is provided by the **rudder**, i.e. a vertical flap hinged to the stern post.

The system of control is the same in each case, i.e. if the control surface is moved it will, in effect, alter the angle of attack and the camber of the complete surface, and therefore change the force upon it (see Fig. 9.12). On many aircraft the roll can also be controlled by the use of spoilers. These are described in more detail later.

The elevators and ailerons are controlled by movements of a **control column** on which is mounted a handwheel which is usually abbreviated to something rather like a car steering wheel with most of the rim sawn off to leave a pair of small handgrips or 'spectacles'.

Pushing the control column forward lowers the elevators, thus increasing the lift on the tailplane and making the nose of the aircraft

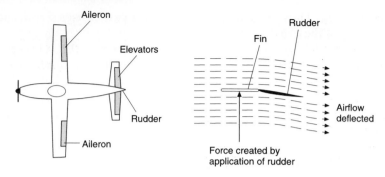

Fig. 9.12   Control surfaces

drop. Turning the handwheel anti-clockwise lowers the right hand aileron and raises the left, thus rolling the aircraft left-wing down.

In old aircraft and aircraft such as fighter aircraft, where cockpit room is restricted, there is no handwheel and instead the control column moves to left and right as well as backwards and forwards – the left-hand movement is equivalent to turning the handwheel anti-clockwise (left hand down).

The rudder is controlled by foot pedals. Pushing the left pedal forward deflects the rudder to the left and therefore turns the nose of the aircraft to the left. This can cause some problems amongst learner pilots as the movement is opposite to that of bicycle handlebars.

In each instance it will be noticed that the control surfaces are placed as far as possible away from the centre of gravity so as to provide sufficient leverage to alter the position of the aeroplane.

On modern aircraft there may also be a secondary set of inboard ailerons which are used in high speed flight where the outboard surfaces could produce excessively large rolling moments, or unacceptable structural loading or wing twist.

BALANCED CONTROLS

Although, in general, the forces which the pilot has to exert in order to move the controls are small, the continuous movement required in bumpy weather becomes tiring during long flights, especially when the control surfaces are large and the speeds fairly high. For this reason controls are often **balanced**, or, more correctly, partially balanced (Fig. 9H).

Several methods have been employed for balancing control surfaces. Figure 9.13 shows what is perhaps the most simple kind of aerodynamic balance. The hinge is set back so that **the air striking the surface in**

Fig. 9H   Balanced controls and tabs
(By courtesy of SAAB, Sweden)
Twin-jet training aircraft showing the statically and aerodynamically balanced
ailerons with geared servo-tabs; starboard tab adjustable for trimming. Elevators and
rudder are also balanced, and there is a trim tab on the rudder, and a servo-tab
(adjustable for trimming) on each elevator.

Fig. 9.13   Aerodynamic balance

front of the hinge causes a force which tends to make the control
move over still farther; this **partially balances** the effect of the air
which strikes the rear portion. This is effective but it must not be
overdone; over-balancing is dangerous since it may remove all **feel** of
the control from the pilot. It must be remembered that when the control
surface is set a small angle, the centre of pressure on the surface is well
forward of the centre of the area, and if at any angle the centre of
pressure is in front of the hinge it will tend to take the control out of the

pilot's hands (or feet). Usually not more than one–fifth of the surface may be in front of the hinge.

Figures 9.14 and 9.15 show two practical applications of this type of balance; in each some part of the surface is in front of the hinge, and each has its advantages.

Fig. 9.14   Horn balance

Fig. 9.15   Inset hinge balance

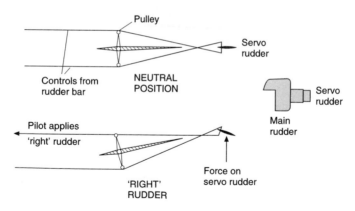

Fig. 9.16   Servo system of balance

Figure 9.16 shows the servo type of balance which differs in principle since the pilot in this case only moves the small extra surface (**in the opposite direction to normal**), and, owing to the leverage, the force on the small surface helps to move the main control in the required direction. It is, in effect, a system of gearing.

Perhaps the chief interest in the servo system of balance is that it was the forerunner of the **balancing tabs** and **trimming tabs**. The development of these control tabs was very rapid and formed an interesting little bit of aviation history.

The servo system suffered from many defects, but it did show how powerful is the effect of a small surface used to deflect the air in the opposite direction to that in which it is desired to move the control surface.

The next step was to apply this idea to an aileron when a machine was inclined to fly with one wing lower than the other. A strip of flexible metal was attached to the trailing edge of the control surface and produced the necessary corrective bias.

So far, the deflection of the air was only in one direction and so we obtained a **bias** on the controls rather than a balancing system. The next step gave us both balance and bias; the strip of metal became a **tab**, i.e. an actual flap hinged to the control surface. This tab was connected by a link to a fixed surface (the tail plane, fin or main plane), the length of this link being adjustable on the ground. When the main control surface moved in one direction, the tab moved in the other and thus experienced a force which tended to help the main surface to move – **hence the balance**. By adjusting the link, the tab could be set to give an initial force in one direction or the other – **hence the bias**.

Sometimes a spring is inserted between the tab and the main control system. The spring may be used to modify the system in two possible ways –

1. So that the amount of tab movement decreases with speed, thus preventing the action being too violent at high speeds.
2. So that the tab does not operate at all until the main control surface has been moved through a certain angle, or until a certain control force is exerted.

Tabs of this kind are called **spring tabs**.

The final step (Figs 9.17 and 9.18) required a little mechanical ingenuity, but otherwise it was a natural development. The pilot was given the means of adjusting the bias while **in the air**, and thus he was enabled to correct any flying faults, or out-of-balance effects, as and when they occurred.

On small aircraft with manual controls these tabs may be fitted to all of the primary control surfaces. The pilot can adjust their settings from within the cockpit and can thereby arrange the trim so that the aircraft will fly 'hands-off' in almost any flight conditions. On aircraft with power-operated controls, such tabs are unnecessary and the trim wheels are simply used to reset the neutral or hands-off position of either the control column or the actuator system. From the pilot's point of view this feels almost exactly like setting a trim tab.

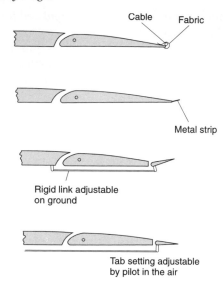

Fig. 9.17   Evolution of control tabs

MASS BALANCE

Control surfaces are often balanced in quite a different sense. A **mass** is fitted in front of the hinge. This is partly to provide a mechanical balancing of the mass of the control surface behind the hinge but may also be partly to help prevent an effect known as 'flutter' which is liable to occur at high speeds (Fig. 9.19). This flutter is a vibration which is caused by the combined effects of the changes in pressure distribution over the surface as the angle of attack is altered, and the elastic forces set up by the distortion of the structure itself. All structures are distorted when loads are applied. If the structure is elastic, as all good structures must be, it will tend to spring back as soon as the load is removed, or changes its point of application. In short, a distorted structure is like a spring that has been wound up and is ready to spring back. An aeroplane wing or fuselage can be distorted in two ways, by bending and by twisting, and each distortion can result in an independent vibration. Like all vibrations, this flutter is liable to become dangerous if the two effects add up. The flutter may affect the control surfaces such as an aileron, or the main planes, or both. The whole problem is very complicated, but we do know of two features which help to prevent it – **a rigid structure** and **mass balance of the control surfaces**. When the old types of aerodynamic balance were used, e.g. the inset hinge or horn balance, the mass could be concealed inside the forward portion of

Fig. 9.18 Control tabs

Fig. 9.19 Mass balance

the control surface and thus two birds were killed with one stone; but when the tap type of balance is used alone the mass must be placed on a special arm sticking out in front of the control surface. In general, however, the problems of flutter are best tackled by increasing the rigidity of the structure and control–system components.

Large aircraft and military types now invariably have powered controls and these are much less sensitive to problems of flutter as the actuating system is very rigid.

Fig. 9I   Tab control mechanism
(By courtesy of Piaggio, Genoa, Italy)

Perhaps it should be emphasised that the **mass** is **not** simply a weight for the purpose of balancing the control surface statically, e.g. to keep the aileron floating when the control mechanism is not connected; it may have this effect, but it also serves to alter the moments of inertia of the surface, and thus alter the period of vibration and the liability to flutter. It may help to make this clear if we realise that mass balance is just as effective on a rudder, where the weight is not involved, as on an elevator or aileron.

On old military biplane aircraft, the exact distribution of mass on the control surfaces was so important that strict orders had to be introduced concerning the application of paint and dope to these surfaces. It is for this reason that the red, white and blue stripes which used to be painted on the rudders of Royal Air Force machines were removed (they were later restored, but only on the **fixed** fin), and why the circles on the wings were not allowed to overlap the ailerons. Rumour has it that when this order was first promulgated, some units in their eagerness to comply with the order, but ignorant as to its purpose, painted over the circles and stripes with further coats of dope!

CONTROL AT LOW SPEEDS

We now turn our attention to an important and interesting problem – namely, that of control at low speeds or, what amounts to the same

Fig. 9J   Control at low speed
The massive bulk of the Antonov AN-124 in a tight turn at low speed and low
altitude. This photograph was taken from the ground!

thing, at large angles of attack. It is obviously of little use to enable a
machine to fly slowly unless we can ensure that the pilot will still have
adequate control over it (Fig. 9J).

Let us first state the problem by giving an example. Suppose, owing
to engine failure, a pilot has to make a forced landing. If he is
inexperienced – and indeed it has been known to happen to pilots of
considerable experience – he will often tend to stall his aeroplane in an
attempt to reach a distant field or to climb over some obstacle. Now the
use of slots or flaps may postpone the stall, may help him to obtain lift at
slow speeds, but they will not give him what he most needs – namely,
efficient control.

In the first place, owing to the decreased speed of the airflow over all
the control surfaces, the forces acting on them will be less and they will
feel 'sloppy'. **But this is not all.** Suppose while he is thus flying near
the stalling angle he decides that he must turn to the left, he will move
the control column over to the left (which will cause the right aileron to
go down and the left one to go up), at the same time applying left
rudder. The rudder will make a feeble effort to turn the aeroplane to the
left; but what will be the effect of the movement of the ailerons?

The effect of the right aileron going down should be to increase the
lift on the right wing, but in practice it may decrease it, since it may
increase the angle of attack beyond that angle which gives the greatest

lift. **But what is quite certain is that the drag will be considerably increased on the right wing, so tending to pull the aeroplane round to the right.** This yawing effect, caused by the ailerons, is present at nearly all angles of attack, but it becomes particularly marked near the stalling angle; it is called **aileron drag**.

Meanwhile, what of the left wing? The lift may either have decreased or increased according to the exact angle of attack, but in any case the change in lift will be small. The drag, on the other hand, will almost certainly have decreased as the aileron moved upwards. To sum it all up, the result of the pilot's attempt to turn to the left is that there may or may not be a slight tendency to roll into the left bank required for an ordinary left-hand turn, while at the same time the drag on the wings will produce a strong tendency to turn to the **right** which may completely overcome the rudder's efforts in the opposite direction (9.20). The conditions are, in fact, very favourable for a spin (both literally and metaphorically); the pilot could hardly have done better had he deliberately attempted to get the aeroplane into a spin.

Fig. 9.20   Result of an attempt to turn at large angles of attack

So much for the problem. What solution can be found? We must endeavour to ensure that when the stalling angle is reached, or even exceeded, the movement of the controls by the pilot will cause the same effect on the aeroplane as in normal flight. The following improvements would all help to attain this end –

1. **Increased turning effect from the rudder.**
2. **Down-going aileron should not increase the drag.**
3. **Up-going aileron should increase the drag.**

4. **Down-going aileron should increase the lift at all angles.**
5. **Up-going aileron should cause a loss of lift at all angles.**

A large number of practical devices have been tried out in the attempt to satisfy these conditions; most of them have been partially successful, but none of them has solved the problem completely.

Let us consider a few of these and see to what extent they meet our requirements.

(*a*) The use of **very large rudders** with sufficient power to overcome the yawing effect of the ailerons in the wrong direction.

The disadvantage is that the size of the rudder required to obtain the desired result is excessive for normal flight. Also this seems to be a method of tackling the problem from the wrong standpoint – instead of curing the disease, it allows the disease to remain while endeavouring to make the patient strong enough to withstand it.

(*b*) A **wash-out**, or decrease of the angle of incidence, towards the wing tips.

This will mean that when the centre portions of the wings are at their stalling angle, the outer portions are well below the angle, and therefore the aileron will function in the normal way. The defect of this arrangement is that the wash-out must be considerable to have any appreciable effect on the control, and the result will be a corresponding loss of lift from the outer portions of the wing in normal flight. The same effect can be obtained by rigging up the ailerons so that the trailing edge of the ailerons is above the trailing edge of the wing.

(*c*) **'Frise', or other specially shaped ailerons** (Fig. 9.21). This is a patented device, the idea being so to shape the aileron that when it is moved downwards the complete top surface of the main plane and the aileron will have a smooth, uninterrupted contour causing very little drag, but when it is moved upwards the aileron, which is of the balanced variety, will project below the bottom surface of the main plane and cause excessive drag. This method has the great advantage of being simple, and it undoubtedly serves to decrease the bad yawing effect of the ailerons, and therefore it is often used. Unfortunately, its effects are not drastic enough.

(*d*) **Differential ailerons** (Fig. 9.22). Here, again, is a delightfully simple device suffering only from the same defect that, although it provides a step in the right direction, it does not go far enough to satisfy our needs. Instead of the two ailerons moving equally up and down, a simple mechanical arrangement of the controls causes the aileron which moves upwards to move through a larger angle than the aileron which moves downwards, the idea being to increase the drag and decrease the lift on the wing with the up-going aileron, while at the same time the

Fig. 9.21   Frise ailerons   **Fig. 9.22   Differential ailerons**

down-going aileron, owing to its smaller movement, will not cause excessive drag.

(*e*) **Slot-cum-aileron control** (Fig. 9.23). The slots, which need only be at the outer portions of the wings in front of the ailerons, may be of the automatic type, or the slot may be interconnected to the aileron in such a way that **when the aileron is lowered the slot is opened**, while when the aileron is raised, or in its neutral position, the slot is closed. By this means the down-going aileron will certainly serve to increase the lift for several degrees beyond the stalling angle, nor will the drag on this wing become very large since the open slot will lessen the formation of eddies. We shall therefore obtain a greater tendency to roll in the right direction and less tendency to yaw in the wrong direction. This is exactly what is required, and the system proved to be very effective in practice.

(*f*) **Spoiler control** (Fig. 9.24). Spoilers are long narrow plates normally fitted to the upper surface of the wing though they may occasionally be fitted below as well. In the ordinary way they lie flush with the surface, or even inside it, and have no effect on the

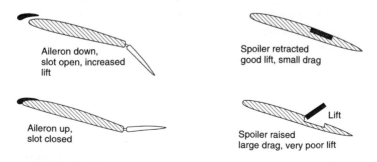

Fig. 9.23 Slot-cum-aileron control   **Fig. 9.24   Spoiler control**

performance of the aerofoil, but they can be connected to the aileron controls in such a way that when an aileron is moved up beyond a certain angle the spoiler is raised at a large angle to the airflow, or comes up through a slit, causing turbulence, decrease in lift and increase in drag. This means that the wing on which the aileron goes down gets more lift, and very little extra drag, while on the other wing the lift is 'spoilt' and the drag greatly increased. **Thus we have a large rolling effect in the right direction combined with a yawing effect, also in the right direction** – just what the doctor ordered.

This is what we aimed at, and there is the further advantage that the mechanical operation of the spoiler is easy, since the forces acting upon it are small. This method of control feels strange to the pilot who is unaccustomed to it because the loss of lift caused by the spoiler will result in a decided drop of that wing, which may be alarming when near the ground. But any such strangeness can soon be overcome and the pilot begins to realise the advantages of maintaining good lateral control, up to and beyond the normal stalling angle. The improvement in manoeuvrability is particularly noticeable when the aeroplane approaches its ceiling. But, whatever its merits, the spoiler took a long time to become popular as a means of control, though it was, and is, used extensively as an air brake.

It is rather curious that we have been describing the use of spoilers as an aid to lateral control **at low speeds**; and this indeed was their original purpose, but in many types of modern aircraft it is **at high speed** that the aileron control may result in undesirable characteristics caused by compressibility as discussed in Chapter 11. Modern airliners use a complicated arrangement of spoilers combined with more than one set of ailerons. The control system normally automatically selects the correct combination according to flight speed.

It may be noticed that the elevator control has not been mentioned in dealing with this problem; the elevators usually remain fairly efficient, even at low speeds, since the angle of attack of the tail plane is less than that of the main planes, and therefore there is not the same tendency to stall as with the ailerons. However, on high-speed aircraft, the tailplane has to be able to compensate for the rearward movement of the centre of lift of the wing and it is quite common nowadays for the whole tailplane to be movable in addition to having a hinged elevator. On propeller-driven aircraft, the extra speed of the slipstream normally adds to the effectiveness of both rudder and elevators.

Before leaving the subject of control it should be mentioned that large amounts of sweepback, and even more delta-shaped wings, cause control problems of their own, but since wings of these shapes are nearly always on high-speed aircraft, their consideration will be left to a later chapter.

POWERED CONTROLS

Nowadays all but small aircraft are usually fitted with power-actuated control surfaces which are very easy to operate even on large airliners. Because such controls offer virtually no natural resistance, they are given some form of artificial 'feel', a resistance which is designed to increase with flight speed (or more precisely with dynamic pressure) so that the control system feels like a direct mechanical linkage. Without such feel it would be quite literally possible to pull the control surfaces off at high speed.

The control column often incorporates a 'stick shaker' which operates when the aircraft approaches a stalled condition. This reproduces the shaking that normally occurs on simple mechanical systems due to the buffeting of the control surfaces caused by turbulence. It is intended to trigger the pilot's conditioned response. On some aircraft, if the pilot fails to respond correctly by pushing the column forward, the controls take over and a 'stick pusher' does the job for him! With the advent of advanced and reliable electronic devices, it has become possible to make control systems of immense complexity that respond smoothly over a very wide range of flight conditions and contain many built-in safety features. In order to prevent loss of control in the event of power failure the systems are usually duplicated, triplicated or even quadrupled.

DYNAMIC STABILITY

In this chapter we have dealt mainly with what is known as static stability. There are other forms of instability that the designer (and the pilot) has to cope with. These take the form of oscillations or deviations from the desired flight path that vary with time, and are known as dynamic instabilities. They can be a nuisance, nauseating or downright dangerous if left unchecked. An example is the spiral instability referred to earlier in this chapter.

A full description of dynamic stability is beyond the scope of this book, but the reader should be aware that the problems of stability and control are far more complex than the simple outline given here.

CAN YOU ANSWER THESE?

Questions on stability and control –

1. What would be the characteristics of an aircraft with extreme directional stability and little lateral stability?

2. What would be the characteristics of an aircraft with extreme lateral stability and little directional stability?
3. What is the object of balancing controls?
4. Distinguish between 'mass' balance and 'aerodynamic' balance.
5. Explain the difficulty of obtaining satisfactory lateral control at large angles of attack.
6. Describe some of the methods which have been tried with a view to overcoming this difficulty.
7. Explain why spoilers are sometimes used at low speeds, and sometimes at high speeds, as an aid to lateral control.

CHAPTER 10
A TRIAL FLIGHT

In the preceding chapters I have tried to explain the principles upon which the flight of an aeroplane depends; it is true that so far we have only dealt with subsonic speeds, but these are after all reasonable speeds for a start. I hope that you have been able to learn something, but however carefully you have read you may still feel that something is lacking.

There is no substitute for the real thing, and we can at least provide an approach to this by going on a flight – in our imagination – and that is what we are going to do in this chapter. During this flight we shall try to link up theory and practice, angle of attack and air speed, angle of bank and tan θ, or whatever it may be. The approach will not be exactly that of a flying instructor teaching a pupil to fly, but rather that of a demonstrator in a laboratory pointing out to students what is going on, as the results of an experiment are revealed.

The aircraft that we shall use for this purpose is a light private type similar to that shown in Fig. 10A but we will assume a retractable undercarriage. In some ways it is a little old-fashioned, having a propeller driven by a piston engine, but it is never-the-less typical of the hundreds of light aircraft that are manufactured every year. The advantages of using such an aircraft for our demonstration flight is that things happen a little more slowly, and there is less automatic control so that we are more aware of what is happening.

Our aircraft has all of the modern conveniences such as cockpit heating, radio navigation aids and a comprehensive set of instruments. These include a rate of climb indicator, sideslip indicator and artificial horizon. We have also fitted this aircraft with an angle of attack indicator and an accelerometer to show the acceleration normal to the plane of the fuselage floor. Also, since we wish to demonstrate some aerodynamic effects, we have attached strings to the wing tips, and wool tufts on the upper surface of the wings.

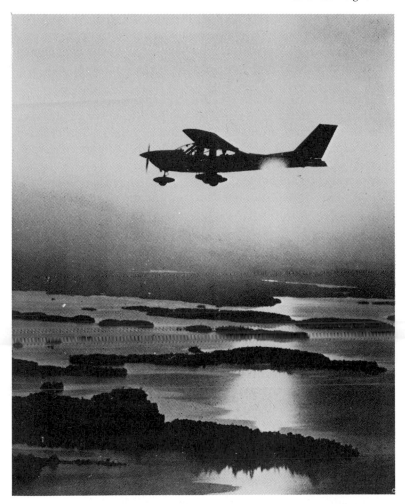

Fig. 10A   Below the clouds
(By courtesy of Cessna Aircraft Company, USA)

In this chapter we shall be dealing with practical flying so we shall use feet and knots since these are the internationally recognised units for aircraft operations.

During the flight we shall meet again all the old familiar things to which you have been introduced, angles of attack, lift, drag, wind gradients and so on – but at first you may not recognise them; the instruments will help but, even so, you will not be able to see the air coming up to meet the wings, nor the centre of pressure wandering about on the aerofoil, nor arrows to show where the forces are acting;

but all these things will be happening just the same and will have their effect on the movements of the aeroplane, so try to visualise them and to link up what you have learnt on the ground with your experiences in the air – you will find it both fascinating and instructive.

One of the first things to be checked, both in the early stages of design and before an aircraft flies, is **the position of the centre of gravity**; this, as we have learnt, affects both balance and stability. The designer has seen to this so far as the fixed parts of the aeroplane are concerned, and after construction it has been checked by weighing at two places, on the main and tail wheels of a tail-wheel type, on the main and nose wheels of a nose-wheel type, and then by calculation according to the principle of moments. But it is with the movable or changing weights that pilot and crew are concerned, and **in the certificate of airworthiness, limits have been laid down within which the position of the centre of gravity must lie**, more particularly to ensure adequate longitudinal stability. The loading of large types with passengers or cargo has to be carefully regulated and checked, but in the comparatively small aircraft used for this test there is not much choice of where to put things, and since all seats are to be occupied it can be assumed that the balance and stability will be correct without further checks.

What cannot be assumed, however, is that everything in connection with the aircraft and engine is correct; so before a flight there must be a check – first from the outside, of any damage to propeller, aircraft structure, tyres and so on – then from inside, of the movement of the controls, setting of levers, checking of fuel contents and other important items. If the aircraft has been properly maintained it is unlikely that anything will be wrong, but one cannot be too careful where flying is concerned; it may sound unbelievable that aileron controls could be rigged to work the wrong way – but **it has happened**, and moreover they have been checked (?) by the pilot and found correct, and he has then taken off, and only then comes to his senses and shown his skill by landing safely.

TAXYING

Now let us start the engine and warm it up. This is very important for until it is warmed-up it will not respond correctly to throttle adjustments and could even cut out at a critical stage. **During the warm-up** which may take a few minutes **we have to keep the brakes firmly on.**

If all is well, the brakes are taken off, the engine is opened up until the aircraft begins to move and taxying commences. This is quite straightforward as our aircraft has a steerable nosewheel. In the old days with tailwheel aircraft and particularly before the introduction of differential brakes taxying could be quite difficult in a strong cross-wind.

But here we are at the end of the runway. We stop now to run up and test the engine, if this has not already been done, and so to the final checks – sometimes called the vital actions, **and vital they are**. These include setting the flaps to a small angle as specified for this type of aircraft. With the simple camber flap used on this aircraft the lower take-off speed, and the steeper climb, more than make up for the reduced acceleration resulting from the drag of the flaps; but it is essential that they should be set at the correct small angle which is, in fact, indicated on the lever in the cockpit. The pilot now has a good look both ways to see that all is clear, gets permission from the tower to take off, opens up the engine a little so that the slipstream makes the rudder more effective and, with careful use of the brakes, turns the aircraft into the take-off position on the runway.

TAKING OFF

We are nearly into wind, which is now blowing at some 20 to 30 knots, so we shall not need so much ground speed to get off the ground. Now open the throttle steadily but surely. **But we are swinging to the right** – no instruments are needed to make us aware of that, and well do we know the reason from our previous studies. The propeller in front of us is rotating anti-clockwise (as we see it), and so is the slipstream which is striking the right-hand side of fin and rudder; reacting to the propeller torque the aircraft is trying to roll to the right. As our aircraft has a tricycle type of undercarriage with a nose wheel, this does not present a major problem. With a nose wheel there is no uneven pull on the propeller blades, the tail is already up so there is no gyroscopic effect but, even more important, the centre of gravity is in front of the main wheels and so any tendency to swing is corrected rather than intensified. With the further refinement of contra-rotating propellers we would not even get the torque and slipstream effects, and much the same applies to turbojets.

As the aircraft gathers speed, the rudder becomes more effective and as long as we are careful not to overcorrect, it will be fairly easy to keep a straight line, assuming that we are not trying to take off in a strong cross-wind.

We are now gathering speed rapidly; the pitch of the constant-speed propeller is becoming more coarse as it travels farther forward with each revolution; the air speed indicator is beginning to read now, 45, 48, 50 knots – with a slight backward pressure on the control column we could be off, but it is better to wait a little yet, 55 knots – that's it, the aircraft almost took off of its own accord.

Have a glance at the instruments at this important stage – angle of attack 8°, air speed 55 knots, the artificial horizon and the angle of attack indicator showing only a very slight upward inclination, the vertical speed indicator still almost at zero.

There are obstructions ahead, and an inexperienced or unskilled pilot is sorely tempted to exert too much pressure on the control column in order to pull the nose up to clear the obstacles – **but that is not the way to do it! We must first gather more speed**, at least up to the climbing speed for flaps down, and as the pilot retracts the undercarriage we can almost feel the aircraft leaping forward as it is relieved of that drag – 60, 65, 70 knots. Now the vertical speed indicator begins to show a definite climb and the pilot reduces power slightly, both the boost and the rpm, but still keeping sufficient power in hand for any eventuality.

See how we have climbed easily over those obstructions and as we reach 80 knots and 300 feet the pilot presses the knob to bring the flaps up from their small take-off angle; there is a nasty sinking feeling owing to the loss of lift, none too comfortable even though we have cleared the obstacles, but this is soon made up for by an even more rapid increase of speed up to **the best climbing speed (with flaps up) of about 110 knots**.

A little more angle of attack for a steady climb – **but wait, what has happened? The engine falters. What should we do – turn back? No** – that is **a very, very old habit of pilots**; but it would mean a turn through 180° near the ground, and if we made this we would be travelling down wind, and against the direction of other aircraft taking off. But our pilot knows better, he puts the nose down a little and **prepares**, if necessary, **to land in a field ahead**. But the engine is picking up, it was only a momentary falter after the tough time it had during take-off – now it is running well again and, as the pilot eases the throttle back a little, notice how the rpm of the propeller remain the same with less power from the engine.

Bumps now – look at the factories below, all those roofs in the blazing sunshine, chimneys belching forth hot air – these cause up-currents; then cool winds and down-currents, and now the shadow from a cloud for it's clouding over from the west; the atmosphere is getting very turbulent down here, so let us make a circuit and landing, and then go up above the clouds.

TURNING

Watch what happens as we do a gentle climbing turn to the left from a height of 500 feet and air speed of 110 knots; first increase the speed slightly, then turn the control wheel or grip anti-clockwise or 'left hand down'. Notice how we old-timers still call it a 'stick' even though in this case it is actually a pair of handle bars. Look at the ailerons; the right one goes down, the left one up even more (for they are differential ailerons); watch the wool tufts on top of the wing, the slip indicator and the artificial horizon. **We are slipping ever so slightly to the left,** but the pilot applies a little **left rudder and that checks it**; no slip now. Stick back a little to keep the nose up – have a look over your shoulder at the elevators.

But now we are tending to bank too much, and again to slip inwards, though this time it is because the right wing is travelling faster than the left; so the control wheel is moved slightly clockwise **to hold off bank** – and all is central again as the turn needle shows a steady Rate 1 turn to the left, the artificial horizon a slight but steady angle of bank, about 15° to 20°, and the accelerometer rather less than $+1\frac{1}{2}$; notice that it started at $+1$ when there wasn't any acceleration or extra load.

After a 90° turn the pilot takes off the bank and levels the nose – not easy to do without a slight slip or skid one way or the other – and so back to a steady climb up to **the circuit height of 1000 feet**.

Now, as we do a gentle circuit of the airfield, let us try some **straight and level flight** – one of the fundamental exercises in learning to fly, one of the fundamental conditions in designing an aeroplane.

We level out after the climb, turn another 90° onto the down-wind leg, set the power to cruising conditions and attain a speed of 140 knots, just the speed for this local flying; moreover it is **the best range speed for this aircraft**, and so excellent practice in economical flying. It is true that the best endurance speed, which requires slightly less power, is lower, but it is easier to keep the speed steady at 140. Notice that **the nose of the aeroplane is on the horizon, as is the model on the artificial horizon**. Now the pilot sets the throttle at the correct rpm for cruising, and allows the aircraft to take up its own air speed which in fact is still about 140 knots; but the aircraft is tail-heavy, so he adjusts the elevator trimming tabs by pushing the pitch trim lever a little forward. Watch the small tabs on the elevator go up; they put the elevators themselves slightly down, and the extra lift holds the tail up. Now the pilot can take his hands and feet off the controls and the aircraft flies by itself – straight and level – so there is no need to adjust rudder or aileron tabs.

Altimeter constant at 1000 feet; vertical speed indicator shows no climb or descent. **If the pilot increases power** by opening the throttle slightly notice that **the aeroplane climbs** (as shown by altimeter and vertical speed indicator) even if the same attitude is maintained – but note that **the air speed is higher**; if he keeps the same air speed but raises the nose of the aircraft, **again it will climb**, but in a steeper attitude. If he decreases power slightly, until he can just maintain height, the air speed will be less, down to about 105 knots – **that of minimum power, or maximum endurance.** Any further reduction of power means loss of height.

But we return to the throttle setting for cruising, and to a speed of about 140 knots, to complete the circuit. Notice what happens if the pilot again flies 'hands off' and a bump causes the right wing to drop; the instruments show a slight slip to the right, and a resulting slight turn to the right – owing to directional stability; then the dihedral takes effect and restores the aircraft to a level keel – but slightly off course to the right. In this way **we have tested both the lateral and directional stability** for, strange as it may seem, the going off course is all part of that. Now watch what the pilot does if a wing drops while he has control; if again it is the right wing that drops he moves the control wheel anti-clockwise and the rudder to the left and, with a slight skid, the aircraft recovers **without any appreciable deflection off course.**

Now we can make a simple test of **longitudinal stability**; put the nose down and increase speed to 150 knots, then leave the aircraft to itself and see what happens. The nose comes up, the speed drops to below 140, then the nose goes down again to above 140 – but not this time as far as 150 – then, after slight oscillation up and down, both attitude and speed return to those of straight and level flight, showing that **the longitudinal stability is excellent.**

While flying straight and level have a look at the strings attached to the wing tips, **and note the direction of rotation** – on the right wing anti-clockwise (as seen from the rear), and on the left wing clockwise.

APPROACH AND LANDING

We are still at 1000 feet, the airfield is on the left, and we are flying down wind. Before turning across wind the pilot performs **the vital actions necessary for the final approach and landing.** **Undercarriage down,** and he makes sure that it is locked down – going at this speed, now 130 knots, the extra drag tends to put the nose down a little, but that is all to the good. **Now propeller to 'high rpm'** (fine pitch), just in case of a bad approach and the need to go round again. Then, after a gentle turn through 90°, we are across wind and get a good view of the runway and, after throttling back, we reduce speed by about 10 knots and **lower the flaps a little – but not yet fully**. Notice how the aircraft takes up a natural angle of glide, but the flaps cause a slight nose-down pitching moment which the pilot corrects by re-trimming with the elevator tabs. Now, as we descend across wind, we tend to drift to the right, but a slight adjustment of direction keeps us at right angles to the runway. From 500 feet a gentle gliding turn to the left puts us in line with the runway, and we now have a straight run in; but we are tending to overshoot so **the flaps are lowered fully to the landing position** – they serve a double function by giving us a steeper glide (because they spoil the L/D ratio) and so a better view with a reduced speed, and also a lower landing speed (because they increase the C_L and the C_L max).

As we approach the boundary of the airfield, with the speed at 80 knots and 50 feet up, the pilot re-trims so that the tail will go down easily when required; at this stage he controls the speed with the elevators and the rate of descent by slight adjustments of the throttle (watch the air speed and vertical speed indicators). Now he checks the descent by moving the stick gently back, gently to keep a little speed in hand, and because we don't want g's at this stage. 75, 70, 65 knots as we flatten out and sink slowly, a few centimetres at a time. At 50 knots a slight twitch as the two main wheels hit the runway at almost the same instant, and then a slight jolt as the nosewheel touches. Now we throttle back and very gently apply the brakes. On this aircraft we do not have the luxury of reversible pitch propellers to slow us down quickly, but with such a low landing speed this is not necessary; we are after all now travelling at a speed that would not be considered fast for a family car.

OFF AGAIN

But now we have stopped; and although there seems to be plenty of room ahead to take off again, **we must obey the rules and taxy round to the previous take-off position**. Having got there everything must be done systematically as before.

This time after take-off we make for the low flying area – over those marshes, so as not to disturb people on the ground. Height 200 feet – that is quite low enough. We are still going against the wind as the pilot reduces the air speed and puts on a little flap; 90 knots, that's about right. Now look at the ground below – even with this head wind it **seems to be moving quite fast – in fact it is moving quite fast** since we are doing about 70 knots ground speed, and we know what that would feel like in a car – 130 kilometres per hour. When high up we didn't notice ground speed at all, but down here it is very different.

Now the pilot does a right turn; stick to right, a wee bit of right rudder. All well so far – but no! we seem to be slipping inwards and the pilot is taking off bank – something is very wrong. Look at the sideslip indicator – well over to the left, **so we are not slipping inwards at all, we are skidding outwards!** The pilot, **and he was not the only one, thought that we were slipping inwards** because he was looking at the ground – but all of us were deceived, deceived by our eyes. As we turned to the right the wind was first ahead, then on the left, then behind us, and we were drifting to the right across the ground, and that made us **think** that we were slipping inwards – but in fact we were making a perfect turn until the pilot spoilt it (the first time that he has let us down). Now the air speed has dropped to 60 knots, and look at the wool tufts – the air is beginning to separate over the wings – but just in time the pilot opens the throttle and we are out of that dilemma, stalling at 200 feet. **It is air speed that matters when flying, not ground speed** – after that rather badly executed turn we were going down wind, faster over the ground than before (and that was what deceived us) even though the air speed was only 60 knots.

The moral of all this is that **we only notice relative speed when we are close to a thing**. Have you ever watched a large aircraft, such as a Boeing 747, land? When it is on the approach it seems very slow; when it is on the runway it seems much faster. This is because we are looking at a background of sky on the approach and have no real reference with which to judge the speed. Because the aircraft is very large we imagine it to be closer than it really is and so it appears to be slow. As soon as we can compare it to the ground it seems to speed up.

Now, having regained an air speed of 90 knots, we turn up wind again. This time a perfect turn, but the pilot was struggling with himself, and as we looked at the ground we all thought that we were skidding outwards as we turned, but when we looked at the sideslip indicator we saw that we were not slipping at all, but **we were moving sideways – relative to the ground – that was not an optical illusion,** and if there had been a high obstacle on the outside of the turn we might have had to bank steeply to avoid it. When flying near the ground there are always **two** things to be considered – the direction and speed of the aircraft relative to the air, to keep it flying correctly – and the direction and speed relative to the ground, to avoid contact with it.

CLIMBING

But the clouds are getting thick above us, so let us go through them in a long steady climb. That's it – not too steep, but a little more power –

remember that we can climb in a more or less level attitude. Still rather steep. Look at the angle of attack, 12°; rate of climb, 1000 feet per minute; air speed 100 knots; inclination to the ground 9°. Now put the stick forward a little, until the angle of attack is 6°; inclination to the ground is now only 3°, **but rate of climb has increased to over 2000 feet per minute.** The air speed is now 110 knots, and that is the speed at which there is most surplus power, and since in many aircraft there is no angle of attack indicator it is the **air speed** for best climb that the pilot must go by. It may seem strange that by putting the nose down, and increasing speed from 100 to 110 knots, we should climb faster – but notice how all the instruments confirm this. The pilot now adjusts the trimming tabs, and the aircraft climbs away – hands off.

So into the clouds, and through the thick of them – dark and clammy, damp and cold; we can see very little outside the cockpit, yet this is a suitable time to observe **how the wisps of cloud always come from straight ahead** although as we entered the cloud we were flying cross wind; a very elementary lesson, but one that pilots take some time to learn. The author has known an experienced pilot claim that an aircraft, if left to itself, will turn into the wind owing to its directional stability; until that same day, he went up and tried it, and found that it didn't happen!

Now back into the old climb, but this time **concentrate on the instruments**; artificial horizon showing a little climb, but wings level; air speed 110 knots, but at this height an indicated speed of 110 knots is a higher true speed than that at which we started climbing; directional gyro, which has been set to correspond to the compass, reads 270° **so we are pointing west**; the top needle of the turn indicator is central, so **we are not turning**; the slip indicator is central, so there is **no slip or skid**; altimeter 5000 feet – as this time we go into the clouds relying on instruments, not instinct.

This instrument flying, with a little practice, is easier than it might seem. The aircraft will almost fly by itself – if left to itself – and **there is no need to concentrate on all the instruments all the time.** If the artificial horizon shows the aircraft to be level and climbing at the correct angle, and the turn indicator shows that we are not turning, and the air speed is right, there cannot be much wrong. But all is well this time as we climb steadily – 5500, 6000 feet – air speed 105, nose down a little to 110 again. Hands off, but there's daylight above, blue sky at last – and we are through.

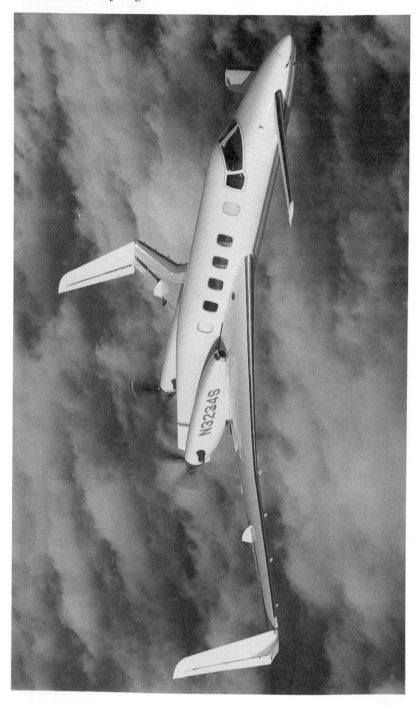

ABOVE THE CLOUDS

What a sight! There is nothing on earth to compare with this – the sun and the blue sky above – and below us a sea of rolling white clouds extending in all directions as far as the eye can see.

But let us climb to ten thousand feet – a good safe height to do some experiments. As we climb **the air gets thinner**; the engine needs more air, more boost, provided by **automatic boost control**; the mixture of air to petrol goes wrong, so **automatic mixture control**; the propeller loses its grip, **automatic pitch control**; in all these the changes in pressure and density of the air are cleverly made to do the work of providing the automatic control.

But these are mainly concerned with the engine; it is the aircraft that mostly concerns us. So let us time the rate of climb as we pass the 8000 feet mark – ten, twenty, thirty, thirty-three seconds – not very different from what it was down below. **But this rate of climb cannot go on for ever**; the engine is rated to give its best at 10 000 feet – after that the inevitable happens, **indicated speed must be lowered** until at the service ceiling it is down to about 105 knots (the best endurance speed), but the true speed of course would be much higher than this, at 40 000 feet twice as much. The rate of climb would decrease with altitude, **maximum speed would decrease**, and at the **absolute ceiling** there would be **only one speed** – both the fastest and the slowest – and only **one angle of attack** – we couldn't do anything except stagger along. But we are not out for any height records on this flight, so let us just do a gentle climbing turn – but first a little more power and a little more speed – and notice how it differs from a level turn in that the aircraft tends to over-bank – do you remember why? – and it is even more necessary to apply opposite aileron and hold off the bank. But here we are at ten thousand feet – flight level 100.

STEEP TURN

First let us try a steep turn; so far we have only tried very mild ones. The procedure is just the same, **only more so**. To turn to the right – first **more power** and **more speed**, then wheel right hand down (notice again the movement of the ailerons), right rudder to correct any slip or skid, stick back to keep the nose up. But we are losing speed

Fig. 10B   Through the clouds (opposite)
(By courtesy of the Beech Aircraft Corporation, USA)
The elegant and unconventional Beech Starship.

because the angle of attack is being increased to get more lift, so **we open the throttle a bit more.** Now the angle of bank is about 60° – it feels steeper than it sounds – but there is nothing to tell us the angle of bank except, very roughly, the artificial horizon and our own observation relative to the real horizon which is none the less useful in the form of a sheet of cloud. But look at the accelerometer, **+2, exactly what it should be for 60°,** and we all seem to weigh twice our normal weight; **rate of turn nearly 3; air speed 120; slip indicator central, no sliding about. A perfect turn.**

Now the pilot tightens the turn, as he applies even more power if available; notice that the angle of attack is now 12°, the accelerometer reads +3 (but we can almost feel that in the weight of our heads!), watch the strings at the wing tips rotating more and more violently (induced drag). Accelerometer now +4, angle of attack 14° and we can only tighten the turn by increasing this angle and reducing the air speed – but look at the wool tufts on the wing, **separation is starting – and all is going grey, we cannot see clearly** – but only for a moment as the pilot takes off the bank and levels the aircraft once more. That was what we mean by greying out, and some of us began to feel it at about +5, perhaps a salutary warning that we had gone far enough, but in any case that was the steepest turn that can be done on this aircraft – limited by engine power.

MAXIMUM SPEED OF LEVEL FLIGHT

Now for a test of maximum speed. The throttle is opened steadily, fully open – 160, 180 knots (over 90 metres per second), but we are still gaining speed – it takes time. The angle of attack is now only 1°, all the wool tufts lie straight back, and the strings on the wing tips are not rotating so violently. The angle of attack indicator shows the nose slightly down, as does the artificial horizon, but the vertical speed indicator still shows a slight rate of climb, so the pilot puts the nose down a wee bit more, and re-trims because of the downward force on the tail; 190, 200 knots (100 metres per second), **that's about the limit,** now we are losing height – **so maximum speed in level flight is 200 indicated,** which our computer shows to be **a true speed of just over 222 knots.**

Fig. 10C Above the clouds (opposite)
(By courtesy of the Grumman Corporation, USA)
Grumman F-14s flying in formation, with their variable-sweep wings in the forward low-speed position.

MINIMUM SPEED OF LEVEL FLIGHT

Now to the other end of the scale; how slowly can we fly? The pilot
throttles down a little, not much because the lowest speed can only be
achieved with full throttle – but that is not particularly good for the
engine. The stick is moved back slowly, very slowly, as the speed drops
rapidly, 190, 170 knots, and the angle of attack goes up to 6°, 8°, 12°. At
this point the vertical speed indicator still shows a climb, but with a
little more angle the climb ceases and the air speed drops to 150, 120,
100 knots. **As the angle of attack reaches 14°, then 15°, things
begin to happen rapidly and there is so much to watch** – the nose
is well up in the air, the air speed is still dropping, to 80, then 60, the
wing-tip trailers are rotating violently, the wool tufts show separation
near the fuselage, and the separation spreads outwards towards the wing
tips, and forwards towards the leading edge. The pilot tries the ailerons,
they are not very effective; he applies a little flap – that steadies the
aircraft for a while, but the speed still drops, now 55, now 50; be tries to
keep it at that, no loss of height, a little more throttle – nose up just a
little farther, 49 knots – no, that's the limit – look at the wings, the tufts
are all over the place – and we feel it too in our stomachs – the pilot tries
the **ailerons, now they have no effect at all** – he tries the **elevators,
they have no effect either** – the aircraft takes charge, the right wing
drops, almost simultaneously the nose follows it and we start to spin.
But the pilot is prepared for all this and we hardly even complete a turn
before he checks the spin, and the stall, and we are flying level again. **So
the stalling speed is 50 indicated, 58 true; and the speed range of
level flight is from 222 knots true, to 58 knots true, at 10 000 feet.**

GLIDING

With flaps fully up again we climb to 15 000 feet for some experiments
in gliding. As the pilot throttles down, right down this time, **the nose
drops automatically** – a nice steady drop, not like the stall – and we
begin to glide, very quiet and peaceful after the steep turns and stalls
with full throttle. With about 4° or 5° of flap to get the best results,
speed and angle of attacks are adjusted to give the flattest glide. The
pilot knows the best air speed, but **the angle of attack indicator
showing 4° confirms that we now have the best ratio of lift to
drag.** From the air speed of 90 knots, and the rate of descent, we can
calculate the gliding angle – **about 1 in 12** – and that is a good deal
better than the usual allowance of a kilometre for every 200 metres, only
1 in 5.

If we do a gentle turn while gliding, we find that **there is no need to hold off bank,** in fact we may have to hold it on – another reminder of what we have already learnt.

But to return to the straight glide, speed is reduced to 75, 65 knots; the aircraft takes up a **flatter attitude** (notice the fore and aft level) and **we think that we are gliding farther;** the angle of attack is 10°, but look at the rate of descent and calculate the angle of glide – **now it really is 1 in 5,** considerably steeper than the best angle of 1 in 12.

To complete this experiment the nose is lowered again and the air speed goes up to 100, the angle of attack down to 2°; this time there is no illusion that we are going to glide farther – in fact it is rather surprising that the calculation shows that the path of glide is **again 1 in 5,** but the attitude is very different as is evident on the fore and aft level, on the artificial horizon and even more by our inclination to the cloud horizon.

Well, that is enough for one day, and if we are not used to flying, our stomachs will probably agree. With our modern navigation aids we can quickly pick up the direction of the airfield, but if the visibility is good, our pilot will probably just use his knowledge of local landmarks to find his way back.

CHAPTER 11

FLIGHT AT TRANSONIC SPEEDS

INTRODUCTION

The earlier chapters of this book have dealt with flight at speeds well below that at which sound travels in air – in short, at subsonic speeds. It is true that frequent references have been made to the problems of high-speed flight, but detailed consideration of these problems has been deferred until now.

From what has already been said, it is clear that the speed of sound has an important influence on the flow, and this is what we examine first.

THE SPEED OF SOUND

When a body moves through the air at speeds well below that at which sound travels in air, there is, as it were, a message sent ahead of the body to say that it is coming. When this message is received, the air streams begin to divide to make way for the body, and **there is very little, if any, change in the density of the air as it flows past the body.** The way in which air is thus 'warned' of the presence of the body which is approaching, or of changes in the shape or attitude of that body, can be clearly illustrated in a smoke tunnel in which the air flows over an aerofoil fitted with a flap (Fig. 11.1). If the smoke streams are allowed to settle when the flap is in the closed position, and the flap is then lowered, all the streams of smoke many chords length in front of the aerofoil immediately change course, and streams which previously flowed below the aerofoil, now flow above it. Once again, if we could 'see the air' in front of an approaching aeroplane, this fact would be immediately obvious, and the air would begin to be disturbed perhaps 100 metres or more in front of the aeroplane of the approach of which it must have had some warning.

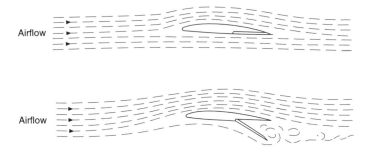

Fig. 11.1   Effect of lowering flap on airflow in front of aerofoil

What we have called a 'message' or 'warning' is really due to a wave-motion in the air set up by the areas of increased and decreased pressures around the body. **These pressures are communicated in all directions to the surrounding air by means of 'waves'.** These waves are similar to sound waves, and they travel **at the speed of sound, which is about** 340 m/s (or 1224 km/h) **in air at sea-level conditions.** There is no mystery in this relationship between pressure waves and sound waves because sound is a pressure wave set up by some local compression of the air, and the speed of sound is simply the speed of propagation of rarefactions and compressions of small amplitude in the air.

So if a body travels through the air at the speed of sound, there will be no time for the message as given by the wave to get ahead and **the air will come up against the body with a 'shock'.**

COMPRESSIBILITY AND INCOMPRESSIBILITY

It was emphasised in the early chapters that **air is compressible**; it was also emphasised, though perhaps not emphatically enough, that though it is compressible it does, in fact, **behave at ordinary speeds almost as though it were incompressible**. Of course such an assumption is not true, air is really compressible or, what is sometimes more important, expandable at all speeds and the density does change, i.e. increases or decreases, as the wings and bodies of aeroplanes move through air at quite ordinary speeds, but the point is that the error in making the assumption is so small as to be negligible, while the simplification that the assumption gives to the whole subject is by no means negligible.

As we approach the speed of sound the error in making this assumption of incompressibility can no longer be justified, the

air is definitely compressed, or expanded. We are now dealing with a compressible and expandable fluid.

It should be clear from this that **the change is gradual, not sudden**; it is all a question of deciding when the error becomes appreciable, and a rough idea of the error involved may be obtained from the following figures which represent the error in assuming the ordinary laws of aerodynamics when estimating the drag of a body moving through air at the speeds mentioned –

Speed			Error in assuming incompressibility
m/s	knots	km/h	
45	87	161	About $\frac{1}{2}$%
90	175	322	Less than 2%
134	260	483	4%
179	347	644	7%
224	436	805	11%
268	522	966	16%

It will be sufficiently clear from this that **we must begin to change our ideas at speeds considerably lower than 340 m/s**.

APPROACHING THE SPEED OF SOUND

Consider a point A sending out pressure waves. If the point is not moving, successive waves 1, 2, and 3 will form circles round the point just like the waves which radiate from a point on the surface of water when a stone is dropped into the water at that point (Fig. 11.2a). Now suppose that the point A is moving in the direction shown, but at a speed less than the speed of propagation of the waves, and that it sends out the wave 1 when it is A1, 2 at A2, 3 at A3; the picture will now take the form of Fig. 11.2b. Now increase the speed of A until it is the same as the speed of the propagation of the waves (Fig. 11.2c), and we soon see why the air ahead gets no warning that A is approaching – we get a pretty good idea too how the waves pile up and what causes the 'shock'.

Some time ago this 'shock', which occurred before the speed of sound was actually reached, and which in some instances had rather alarming effects on the flight of the aeroplane, was considered as **something to be avoided at all costs**; and the rapid increase in drag seemed a formidable enough 'barrier' at that time, if only because there seemed little hope of providing engine power to match it.

Thus it was that the transonic range of speeds, and especially the approach to the speed of sound, came to acquire rather a bad reputation

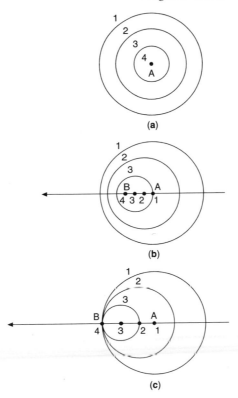

Fig. 11.2 Propagation of pressure waves
(*a*) Point stationary – sending out pressure waves 1, 2, 3, 4.
(*b*) Point moving below speed of sound.
(*c*) Point moving at the speed of sound.

in the minds of both designers and pilots, and the difficulty in solving the problems was all the more frustrating because we knew, from experience of the flight of bullets and shells, that flight beyond the speed of sound, flight at supersonic speeds, was not only possible but apparently free from some of the troubles of transonic flight.

To-day, of course, we know a great deal more about flight at transonic speeds, the 'shock' and the 'barrier' are no longer obstacles to be avoided but rather to be got over, or through, and the supersonic region is within the reach of aeroplanes as well as bullets and shells; but with all our increase of knowledge, it is interesting to note that **flight at transonic speeds still presents special problems of its own**. That is why it deserves a whole chapter to itself.

SHOCK WAVES

Let us see if we can find out a little more of what actually happens during the change from incompressible flow to compressible flow, and so discover the cause of the mounting error in making the assumption of incompressibility. Let us also investigate the 'shock', together with its cause and effects.

As the speed of airflow over say a streamline body increases, the first indication that a change in the nature of the flow is taking place would seem to be a breakaway of the airflow from the surface of the body, usually some way back, setting up a turbulent wake (Fig. 11.3). This may occur at speeds less than half that of sound and has already been dealt with when considering the boundary layer. It will, of course, cause an increase of drag over and above that which is expected at the particular speed as reckoned on the speed–squared law.

Subsonic
flow $M = 0.6$

Fig. 11.3   Breakaway of airflow

As the speed increases still further, the point of breakaway, or separation point tends to creep forward, resulting in thicker turbulent wake starting forward of the trailing edge.

When we reach about three-quarters of the speed of sound a new phenomenon appears in the form of **an incipient shock wave (Figs 11.4 and 11A).** This can be represented by a **line approximately at right angles to the surface of the body and signifying a sudden rise in pressure and density of the air, thus holding up the airflow and causing a decrease of speed of flow. There is a**

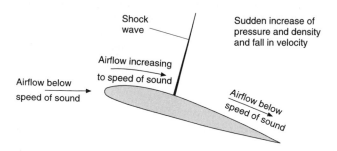

Shock
wave

Sudden increase of
pressure and density
and fall in velocity

Airflow increasing
to speed of sound

Airflow below
speed of sound

Airflow below
speed of sound

Fig. 11.4   Incipient shock wave

Fig. 11A   Incipient shock wave
(By courtesy of the Shell Petroleum Co Ltd)
An incipient shock wave (taken by schlieren photography) has formed on the upper surface; the light areas near the leading edge are expansion regions, separated by the stagnation area which appears as a dark blob at the nose.

tendency for the breakaway and turbulent wake to start from the point where the shock wave meets the surface which is usually at or near the point of maximum camber, i.e. where the speed of airflow is greatest.

OBSERVATION OF SHOCK WAVES

The understanding of shock waves is so important to the understanding of the problems of high-speed flight that it is worth going to a lot of trouble to learn as much as we can about them.

As we remarked so often in dealing with flight at subsonic speeds, it would all be so much easier to understand if we could see the air. Well, fortunately we almost can see shock waves; they are not merely imaginary lines, lines drawn on diagrams just to illustrate something which isn't actually there, they are physical phenomena which can be photographed in the laboratory by suitable optical means, and in certain conditions they may even be visible to the naked eye.

The photographic methods depend on the fact that rays of light are bent if there are changes of density in their path. In the 'direct shadow' method the shock wave appears on a ground-glass screen, or on a photographic plate, as a dark band with a lighter band on the high-pressure side. But the system that has proved most effective is known as the **'schlieren'** method, and the results obtained by it are so illuminating, and have contributed so much to our understanding of the subject, that the method itself deserves a brief description. The word 'schlieren', by the way, is not the name of some German or Austrian scientist, but simply the German word for streaking or striation, which is descriptive of the method; nor is the method itself modern, it was used a hundred years ago for finding streaks and other flaws in mirrors and lenses, just as it now finds 'flaws' or changes of density in the air.

The fundamental principle behind the schlieren method is that **light travels more slowly through denser air; so if the density of air is changing across, or at right angles to, the direction in which light rays are travelling,** the rays will be bent or deflected towards the higher density (Fig. 11.5). Notice that the bending only takes place

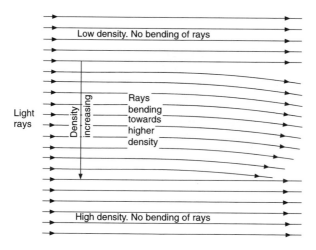

Fig. 11.5   **Bending of light rays through changing density**

when the density of the air is changing (across the path of the rays) – there is no bending when the rays pass through air density which is constant across their path, nor if the density changes along their path; in those cases the rays are merely slowed up by high density.

Figure 11.6 shows a typical arrangement of mirrors and lenses as used in the schlieren method. From a light source A the rays pass through a lens B, to a concave mirror C, which reflects parallel rays through the glass walls of a wind tunnel to another concave mirror D, which in turn reflects the rays on to a knife edge at E, where an image of the light source is formed. Rays that pass through changing density near the model in the tunnel are bent, those passing through falling density being

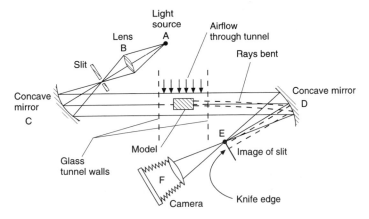

Fig. 11.6 Arrangement for schlieren photography
View from above.

deflected one way, and those passing through rising density the other way. At E either more or less light will be let through depending on whether the ray has been deflected onto or away from the knife edge by the density changes. Thus the image of the working section at F will show light or dark areas (Fig. 11A). In a film on High-speed Flight produced by the Shell Petroleum Company, to whom I am indebted for these schlieren pictures, a colour filter was used at E and in this ingenious way increasing density was shown on the screen in one colour, decreasing density in another, and unchanging density in yet another; thus giving a real live picture of changes of density as the air flowed over different shapes at different speeds.

It is interesting to note that whereas the schlieren method reveals the change of density in a certain distance, the direct shadow method shows changing rates of density change, and other methods, such as the Mach Zender and laser interferometers, the actual density in different places. So each method has its advantages, but each needs rather different interpretation if we are to realise exactly what it is showing us.

Figure 11A is a photograph of a shock wave obtained by the schlieren method, and the fact that it appears as a narrow strip at right angles to the top surface of the aerofoil shows that across the strip there is a sudden increase in density. Immediately behind and immediately in front of the strip the density is reasonably constant, though greater of course behind than in front. Colour photography reveals another small area of increasing density immediately in front of the nose of the aerofoil as we might expect, and a large area above the nose and a smaller area below in which the density is decreasing.

Schlieren pictures of airflow at speeds well below that of sound show no appreciable changes of density at all, so now we see the fundamental change that takes place as we approach the speed of sound; the air begins to reveal its true nature as a very compressible – and expandable – fluid.

EFFECTS OF SHOCK WAVES – THE SHOCK STALL

It is clear from schlieren photographs that there is **a sudden and considerable increase in density of air at the shock wave,** but there is also, as has been stated, **a rise in pressure (and incidentally of temperature), and a decrease in speed.** Most important of all perhaps is the **breakaway of the flow from the surface,** though it is sometimes argued whether this causes the shock wave or the shock wave causes it. Whichever way it is the result is the same.

As is rather to be expected all this adds up to a **sudden and considerable increase in drag** – it may be as much as a ten times increase. This is accompanied, if it is an aerofoil, by **a loss of lift** and often, due to a completely changed pressure distribution, to a **change in position of the centre of pressure and pitching moment,** which in turn may upset the balance of the aeroplane. At the same time the turbulent airflow behind the shock wave is apt to cause **severe buffeting,** especially if this flow strikes some other part of the aeroplane such as the tail plane. One can hardly avoid saying – **very like a stall.** Yes, so like the stall that it is called just that – **a shock stall.**

But the similarity must not lead us to forget the **essential difference** – no it isn't the speed, we have already made it clear, or tried very hard to make it clear, that the ordinary stall can occur at any speed; the essential difference is that the 'ordinary stall' occurs at **a large angle of attack** and, to avoid confusion, we shall in future call it the **high-incidence stall** to distinguish it from the **shock stall** which is more likely to occur at small angles of attack.

From what has already been said the reader will probably have realised that the formation of shock waves is not a phenomenon that occurs on the wings alone; it may apply to any part of the aeroplane. Even the shock stall, which may first become noticeable owing to the sudden increase of drag and onset of buffeting – it is sometimes called the **buffet boundary** – may be caused by the formation of shock waves on such parts as the body or engine intakes, rather than on the wings (Fig. 11B).

Fig. 11B   Shock waves
(By courtesy of the former British Aircraft Corporation, Preston)
Top: Lightning at M 0.98; low pressure regions above canopy and wing cause
condensation, the evaporation to the rear marks the shock wave.
Bottom: Schlieren photograph of model of Lightning at M 0.98; note the
extraordinarily close resemblance to actual flight.

SHOCK DRAG

The sudden extra drag which is such a marked feature of the shock stall has two main components. First the energy dissipated in the shock wave itself is reflected in additional drag (wave drag) on the aerofoil. Secondly, as we have seen, the shock wave may be accompanied by separation, or at any rate a thickening of and increase in turbulence level in the boundary layer. Either of these will modify both the pressure on the surface and the skin friction behind the shock wave.

So this shock drag may be considered as being made up of two parts, i.e. the wave-making resistance, or **wave drag**, and the drag caused by the thick turbulent boundary layer or region of separation which we will call **boundary layer drag**.

As has already been explained the shock wave and the thickened turbulent boundary layer or separation are like the chicken and the egg – we don't know which comes first; what we do know is that when one comes so does the other. That is not to say that they are by any means the same thing, or that they have the same effects, or that a device which reduces one will necessarily reduce the other.

MACH NUMBER

The time has come to introduce a term that is now on everybody's lips in connection with high-speed flight – **Mach Number**. This term is a compliment to the Austrian Professor Ernst Mach (1838-1916), who was professor of the history and theory of science in the University of Vienna, and who was observing and studying shock waves as long ago as 1876. Incidentally, his *Science of Mechanics*, which was published in English in 1893, throws much light on the work of Newton and others, and is well worth reading.

Fortunately the definition of Mach Number is simple. **The Mach Number (M) refers to the speed at which an aircraft is travelling in relation to the speed of sound.** Thus a Mach Number of 0.5 means that the aircraft is travelling at half the speed of sound. **Both the speed of the aircraft and the speed of sound are true speeds.**

VARIATION OF SPEED OF SOUND

There is one small complication that must be introduced into the definition of Mach Number even at this stage. **The speed of sound varies according to the temperature of the air,** and therefore we must add to the definition the fact that **the speed of sound must be** that corresponding to the temperature of the air in which the aircraft is actually travelling. People are often surprised to hear that the speed of sound in air depends on temperature alone.

The actual relationship is that the **speed of sound is proportional to the square root of the absolute temperature**. It is a solemn thought, therefore, that the speed of sound would be zero in air at the absolute zero of temperature; such a thought opens up possibilities of testing 'high-speed phenomena' at very low speeds in wind tunnels at very low temperatures. The point is that **what we are concerned with is not high-speed flight as such, but flight at speeds in the neighbourhood of the speed of sound, in other words at Mach Numbers approaching 1**. It may not be practical to experiment at the absolute zero of temperature, but **it is very practical to consider flight in the stratosphere where, at a temperature of about $-60°C$ (213 K) the speed of sound will have fallen from about 340 m/s at sea-level to about 295 m/s**.

Perhaps it should be emphasised again that this drop in the speed of sound is not really a function of the height at all; at a temperature of $-60°C$ such as may occur at sea-level in, say, the North of Canada in winter, the speed of sound would also be about 290 m/s, while in tropical climates it might be well over 340 m/s even at considerable heights.

This variation of the speed of sound with temperature accounted for the rather surprising feature of speed record attempts of some years back in that the pilots waited for hot weather, or went to places where they expected hot weather, in order to make the attempts. Surprising because it had always been considered, and was in fact true, that high temperatures act **against** the performance of both aircraft and engine. The point, of course, is that the record breakers **wanted to go as fast as possible while keeping as far away as possible from the speed of sound** – so they wanted the speed of sound to be as high as possible. Nowadays in breaking speed records the aim of the pilot is just the opposite, i.e. to get **through the speed of sound as quickly as possible** – but we are anticipating.

CRITICAL MACH NUMBER

It has already been made clear that the onset of compressibility is a gradual effect, and that things begin to happen at speeds considerably lower than the speed of sound, that is at Mach Numbers of less than 1. One reason for this is that, as explained in earlier chapters, **there is an increase in the speed of airflow over certain parts of the**

aeroplane as, for instance, over the point of greatest camber of an aerofoil. This means that although the aeroplane itself may be travelling at well below the speed of sound, the airflow relative to some parts of the aeroplane may attain that value. In short, there may be a **local increase in velocity up to beyond that of sound and a shock wave may form at this point**. This in turn, may result in an increase of drag, decrease of lift, movement of centre of pressure, and buffeting. In an aeroplane in flight the results may be such as to cause the aircraft to become uncontrollable, in much the same way as it becomes uncontrollable at the high incidence stall at the other end of the speed range.

All this will occur at a certain Mach Number (less than 1), which will be different for different types of aircraft, and which is called the **critical Mach Number** (M_{cr}) of the type.

If the reader has followed the argument so far he will not be surprised to learn that the general characteristic of a type of aircraft that has a high critical Mach Number is **slimness**, because over such an aircraft the local increases of velocity will not be very great. This was well illustrated by the Spitfire, a 'slim' aircraft that was originally designed without much thought as to its performance near the speed of sound, yet which has proved to have a critical Mach Number of nearly 0.9, one of the highest ever achieved.

We had some difficulty in deciding whether the ordinary stalling speed should be defined as the speed at which the lift coefficient is a maximum, or at which the airflow burbles over the wing, or at which the pilot loses control over the aircraft. They are all related, but they do not necessarily all occur at the same speed. So now with the critical Mach Number – is it the Mach Number at which the local airflow at some point reaches the velocity of sound? or at which a shock wave is formed? or at which the air burbles? or when severe buffeting begins (this is sometimes called the 'buffet boundary' of the aircraft)? or at which the drag coefficient begins to rise? – or, again, when the pilot loses control? I do not know – nor, apparently, does anyone else! Authorities differ on the matter, each looking at it according to his own point of view, or sometimes according to whether he wants to claim a high critical Mach Number for his pet type of aircraft. However, it doesn't matter very much; they are really all part of the same phenomenon.

Is it possible for an aircraft to fly at a Mach Number higher than its critical Mach Number? Is it possible for an aircraft to have a critical Mach Number higher than 1? These two questions may at first sound silly, but they are not. The answers to both depend entirely on which of the many definitions of critical Mach Number we adopt. If the critical Mach Number is when the pilot loses control, then he can hardly fly beyond it; but if it is when a shock wave is formed, or when the drag coefficient begins to rise, why shouldn't he? He may not even know that it has happened, any more than he knows whether he is at the maximum lift coefficient in an ordinary stall. Graphs of lift and drag coefficients are all very well, but one cannot see them on the instrument panel when flying. Supposing the pilot can maintain control through all the shock waves, increases of drag coefficient and so on, then he will find that his critical Mach Number is higher than 1 or, to be more correct, that the aircraft **has not got a critical Mach Number** in any of the senses that we have so far defined it, except for the one relating to the first appearance of supersonic flow locally.

DRAG RISE IN THE TRANSONIC REGION

The behaviour of the drag coefficient, for a thin aerofoil shape at constant angle of attack, can best be illustrated by a diagram (Fig. 11.7). This **shows that up to a Mach Number of about 0.7 the drag coefficient remains constant** – which means that our elementary principles are true – **then it begins to rise**. According to one definition the Mach Number at which it begins to rise, in this example 0.7, is the critical Mach Number. **At M of 0.8 and 0.85 C_D is rising rapidly.** Note that the curve then becomes dotted and the full line is resumed

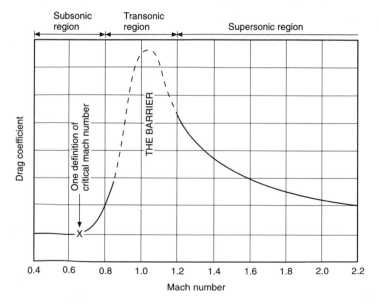

Fig. 11.7 Transonic drag rise

again at an M of about 1.2. The reason for this is interesting. For a long time, although it was possible to operate high-speed wind tunnels up to an M of about 0.85, and again at M of 1.2 or more, in the region of the speed of sound a shock wave developed right across the wind tunnel itself, the tunnel became 'choked' and the speed could not be maintained. Thus there were no reliable wind tunnel results in this region, and **the dotted part of the curve was really an intelligent guess**. This difficulty has now been overcome, and experiments have also been made by other means, by dropping bodies, or propelling them with rockets and also, of course, by full-scale flight tests, and the guess can now be confirmed. Previous experiments on shells had of course suggested that there was nothing much wrong with it. The curve is still left dotted, partly to remind us of this bit of aeronautical history, but more now to emphasise the strange behaviour of the drag coefficient in the transonic region.

After a Mach Number of about 1.2, C_D **drops and eventually, at** M **of 2 or more becomes nearly constant again though at a higher value than the original,** variously quoted as 2 or 3 times.

The diagram shows that there is a definite hurdle to be got over. But it also shows that conditions on the other side are again reasonable and that supersonic flight, as we now of course know only too well, is a practical proposition.

The reader is advised to work through Example No. 255 (page 492) in which he will plot the actual drag, as distinct from the coefficient, and it will then be clear that the drag also actually falls after $M = 1$, but that the reduction is not quite so evident in terms of drag as in terms of drag coefficient.

SUBSONIC – TRANSONIC – SUPERSONIC

We have already talked about flight at subsonic, transonic, and supersonic speeds, and it should now be clear that the problems of flight are quite different in these three regions, but the dividing lines between the regions are of necessity somewhat vague. Figure 11.7 shows the **subsonic region** as being **below a Mach Number of 0.8, the transonic region from M 0.8 to M 1.2,** and the **supersonic region above M 1.2.** There are arguments in favour of considering the transonic region as starting earlier, say at a Mach Number of about 0.7 or near the point marked in the figure as the critical Mach Number, and extending up to say a Mach Number of 1.6 or even 2.0. In terms of sea-level speeds this would mean defining subsonic speeds as being below 450 knots, transonic speeds as 450 up to 1000 or even 1200 knots, and supersonic speeds above that.

Perhaps the best definition of the three regions is to say that the **subsonic region is that in which all the airflow over all parts of the aeroplane is subsonic, the transonic region is that in which some of the airflow is subsonic and some supersonic, and the supersonic region is that in which all the airflow is supersonic**. Once again we are in trouble if we take our definition too literally. Even at very high speeds we may have local pockets of subsonic flow – just in front of a blunt nose for example. So the space shuttle would only be transonic even at the fastest point of re-entry! Also with this definition we are none the wiser as to the speeds or Mach Numbers at which each regions begins or ends; the beginnings and endings will of course be quite different for different aeroplanes.

In this chapter our main concern is with speeds in the transonic range, and particularly in the narrow range between Mach Numbers of 0.8 and 1.2. This range, as is probably already evident, presents us with some of the most baffling but fascinating problems of flight; it is the range in which most of the change takes place, the change from apparent incompressibility to actual compressibility, the gradual substitution of supersonic flow for subsonic flow; it is the range about which we are even now most ignorant.

FLIGHT AT TRANSONIC SPEEDS – DRAG AND POWER REQUIRED

There was a time when the prospects of supersonic flight seemed poor owing to the lack of engines with the necessary power to overcome the rapid rise in drag which begins at the critical Mach Number. Even without compressibility effects the drag would rise with the square of the velocity, and the power – which is Drag × Velocity – with the cube of the velocity. The effect of compressibility is to increase these values still further in accordance with the following table, which shows the approximate figures for a Spitfire –

Velocity (knots)	250	350	450	550	650
Velocity (m/s)	129	180	232	283	335
Velocity (km/h)	464	649	834	1019	1205
Power (kW) *without compressibility*	750	1900	2430	6000	9700
Actual power required (kW)	750	1900	2510	7500	22 000

Actually the problem was much more serious than this, **if we assumed that the aircraft would be driven by propellers.** As was explained in the chapter on thrust, the efficiency of a good propeller is about 80 per cent **at its best, but its 'best' is at speeds of 129 to 180 m/s, after which the efficiency falls off very rapidly** – this happens for various reasons, but chiefly because the propeller tips are the first part of the aircraft to suffer from compressibility. With the high forward speed of the aeroplane combined with the rotary speed of the circumference of the propeller disc, Mach Number troubles **begin to occur at aircraft speeds of about 180 m/s, and the result is so disastrous that the power which would have to be supplied to the propeller by the engine in order to attain the speeds given in the table would look something more like this –**

Velocity (m/s)	130	180	230	280	335
Power (kW) to be given to propeller	820	2390	6000	15 000	100 000?

When one looks at these figures one realises why it was that people who knew what they were talking about forecast not so many years ago that it would be a long, long time before we could exceed 300 m/s.

Yet they were wrong. And for one simple reason – **the advent of the gas turbine and the first flight of a jet-driven aircraft in 1941.** This made all the difference, partly because of the elimination of the propeller and its compressibility problems (it is true that there are similar problems with the turbine blades in the jet engine), but mainly because, **the efficiency of the jet increases rapidly over just those speeds, 154 to 257 m/s, when the efficiency of the propeller is falling rapidly.** The net result is that whereas the reciprocating-engine-propeller combination requires nearly twenty times as much power to fly at 500 compared with 250 knots, the jet engine only requires about five times the thrust, and it is thrust that matters in a jet engine. Further, **the weight of the jet engine is only a small fraction of that of the reciprocating-engine-propeller combination,** and at this speed **even the fuel consumption is less.**

Maybe the prophets ought to have foreseen the jet engine – but they didn't, at least not within anything like the time during which it actually appeared. Of course, there were good reasons too why they didn't foresee it, for no metal could then possibly stand up to the temperatures of the gas turbine blades.

The jet engine, then, was the first step in solving the problem of high-speed flight. And while on the subject of engines, the rocket system of propulsion takes us even a step further and no man now, who knows what he is talking about, dare predict the limits of speed that may be reached with rockets – outside the atmosphere there really isn't any limit.

It is interesting to note that lately the wheel looks like turning a full circle. Improvement in propeller design means that the tip problems can be largely overcome and a gas turbine/propeller combination promises better efficiency in the future than a turbojet at transonic speeds.

FLIGHT AT TRANSONIC SPEEDS – THE PILOT'S POINT OF VIEW

We have so far discussed the problems of approaching the speed of sound very much from the point of view of the designer – but what about the pilot? Well, if he wants to know what is going on, or what is likely to happen, **the first thing that he needs is an instrument to tell him at what Mach Number he is flying**. Various types of **machmeter** are already in existence, and no doubt they will be improved in accuracy and reliability. For a machmeter to give a reliable indication of the Mach Number it must measure, in effect, the true speed of the aircraft and the true speed of sound for the actual temperature of the air. The first is usually done via the indicated speed, which can be corrected for air density by a compensating device within the instrument, but which still includes position error. A temperature compensating device can be used to give the true speed of sound, but in modern instruments this has been eliminated, and a combination of aneroid barometer and air speed indicator gives all the correction required – except that of position error. The term 'Indicated Mach Number' is sometimes used for the reading of the machmeter, but it is an unsatisfactory term since it differs from indicated air speed in that the main correction, that of density, has already been made in the instrument itself.

In the absence of a machmeter the pilot will find that the air speed indicator is apt to give very misleading ideas – even more so than usual. This assumes, of course, that he is not one of those pilots who have already discovered that what the air speed indicator reads is not an air speed at all! Unless he is such a pilot, he may reason that if the speed of sound is around 340 m/s he cannot possibly run into trouble with say 103 m/s on the clock; a shock stall at this speed might come as a real shock.

Yet such a shock is possible because –

In the first place, **the speed of sound is** a real speed, **a true speed**.

Secondly, **it decreases with fall of temperature**, down to 295 m/s, or less, in the stratosphere.

Thirdly, **the speed as indicated is not the true speed**, and the error is more than 100 per cent in the stratosphere, so that at a real speed of 340 m/s at say 40 000 ft, the indicator will read less than 154 m/s.

Fourthly, **trouble begins** not at the speed of sound but **at the critical Mach Number**, and if this is 0.7 (a low value but by no means unknown), a shock stall may occur at 0.7 × 154, i.e. 108 m/s on the clock.

Fifthly – and a point not so far mentioned – a shock stall occurs **at an**

even lower critical Mach Number during manoeuvres, so that in a turn the 108 might be reduced to less than 103.

And there we are!

The figures are all possible, they might even be worse.

There is one compensating point; the pitot head is one of the first parts to experience the effects of compressibility, and this may cause the air speed indicator to over-read at very high speeds – but this is usually allowed for in calibration.

Height		True speed of sound		True speed at which shock stall will occur assuming M_{cr} 0.7	Reading of ASI at this speed	
feet	metres	knots	m/s		knots	m/s
0	0	661	340	463	463	238
10 000	3048	640	329	448	385	198
20 000	6096	614	316	430	315	162
30 000	9144	589	303	412	253	130
40 000	12 192	573	295	401	200	103
50 000	15 240	573	295	401	156	80

The above table may be a help to a pilot in realising what is going on; the figures are calculated on the assumption of the International Standard Atmosphere and will vary to some extent according to how much actual conditions differ from this.

The figures in the last column are the readings of the air speed indicator at which a shock stall may occur (it may even occur at lower indicated speeds because the figures given do not allow for manoeuvres), and they are apt to be rather alarming. There is, however, another way of looking at it – and one that is much more heartening.

Suppose one dives from 50 000 ft at a constant true speed of, say, 450 knots; that is at a rapidly increasing speed on the clock of –

> 176 knots at 50 000 ft,
> 225 knots at 40 000 ft,
> 275 knots at 30 000 ft,
> 328 knots at 20 000 ft,
> 387 knots at 10 000 ft,
> and 450 knots at sea-level,

strange as it may seem, **the actual Mach Numbers would be decreasing** as follows –

450/570 or 0.77 at 50 000 ft,
450/570 or 0.77 at 40 000 ft,

450/590 or 0.76 at 30 000 ft,
450/616 or 0.73 at 20 000 ft,
450/640 or 0.70 at 10 000 ft,
450/666 or 0.68 at sea-level.

This means that if the critical Mach Number were 0.7 an aircraft that **was shocked stalled at 50 000 ft would become unstalled at a height of** about 10 000 ft.

Even if the true speed were to increase during the dive, as would probably happen in practice, there might still be a drop in Mach Number.

This consoling feature of the problem is based on the assumption of rise in temperature with loss of height – if the temperature does **not** rise, that is to say, if there is an inversion, well the reader – and the pilot – can calculate what will happen!

BEHAVIOUR OF AEROPLANE AT SHOCK STALL

All this rather assumes that there is something to be feared about a shock stall, and that pilots try to avoid it. After all, there was a time when we looked upon the high incidence stall in the same way – something to be avoided at all costs. Now, however, it is practised by all pilots in the very early stages of learning. Much the same is happening to the shock stall – it is all a question of knowledge, and many aircraft currently cruise safely well into the transonic region.

By far the most important effect is a considerable change of longitudinal trim – usually, but not always, towards nose-heaviness, and sometimes first one way then the other. Unfortunately the change of trim is made even worse by the **very large forces required to move the controls**, and the **ineffectiveness of the trimmers**. There is also likely to be **buffeting, vibration of the ailerons**, and **pitching and yawing oscillations** which may become uncontrollable, and which are variously described as **snaking** (yawing from side to side), **porpoising** (pitching up and down), and the **Dutch roll** (a combination of roll and yaw).

These effects can, though, be alleviated by the use of power controls and automatic stability augmentation systems. The best way of avoiding the difficulties is to **keep an eye on the machmeter** – if there is one – and, if the worst comes to the worst, to **get into regions of higher temperature**. The best ways of getting out of trouble are to **stop going so fast** – or to go **faster**! In a climb it is easy to stop going so fast – just throttle back. That is why the safest and best research work can often be done in climbing flight. In level flight it may not be quite so easy to lose speed, especially if the controls cannot be moved; and in a dive, which is where these troubles are most likely to occur, it will be even more difficult. It is essential therefore that **all aircraft which are capable of these speeds, and which have undesirable characteristics in the transonic range, should have some kind of dive brake, or spoiler, which can safely be used at high Mach Numbers**. We have had to go very fast on aeroplanes before the need for a brake was recognised! As to going faster; well, that will take us into the region of supersonic flight which will be dealt with in the next chapter.

HEIGHT AND SPEED RANGE

It was explained in an earlier chapter that for a piston-engined aircraft owing to limitations of power the speed range of an aircraft narrows with height, until, at the absolute ceiling, there is only one possible speed of flight. This speed, however, was not the stalling speed, but rather the speed of best endurance, and the absolute ceiling was simply a question of engine power; for a given aircraft, the greater the power supplied, the higher would be the ceiling.

Now, however, **with almost unlimited thrust available** in the form of jets or rockets, or eventually perhaps atomic energy, there is **an altogether new aspect of the limitation of height at which an aircraft can fly without stalling – one way or the other. For the true speed of the high incidence stall will increase with height, while the true speed of the shock stall will fall from sea-level to the base of the stratosphere, and then remain constant**. The result, assuming a sea-level stalling speed (high incidence) of 46 m/s and a critical Mach Number of 0.8, is shown in Fig. 11.8, the shaded portions being the regions in which flight is not possible without stalling. It will be evident that if this aeroplane is to avoid both kinds of stall, it cannot fly above 23 000 m, whatever the power available, while if it flies at 23 000 m, it can only fly at the stalling speed. If it flies any slower it will stall (high incidence), and if it flies any faster it will stall (shock). Surely a modern interpretation of being between the devil and the deep sea! Figure 11.9 shows exactly the same thing from the point of view of indicated speeds. Thus, **quite apart from engine power, there is a limitation to the height of subsonic flight, and a narrowing of the speed range as the limiting height is approached**.

This curious coming together of the two stalls will occur at considerably lower heights during manoeuvres, which will cause the high incidence stalling speed to increase, and the shock stalling speed to

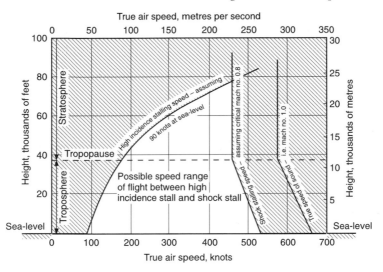

Fig. 11.8   Subsonic speed range of flight – true speeds

decrease. The following figures are quite reasonable for 40 000 ft (say 12 000 m)

High incidence stall (true air speeds)				Shock stall (true air speeds)			
Normal stall, sea-level	Normal stall at 40 000 ft	Stall at 4g, at 40 000 ft	Further increased at high M to, say,	At 4g, this may be reduced to, say	$M = 0.7$. 40 000 ft	$M = 1$. 40 000 ft	$M = 1$. Sea-level
90	180	360	380	380	400	570	661
knots $\rightarrow$	$\rightarrow$	$\rightarrow$	$\rightarrow$	$\leftarrow$	$\leftarrow$	$\leftarrow$	$\leftarrow$	knots
Increase of high incidence stalling speed $\rightarrow$				$\leftarrow$ Decrease of shock stalling speed			

One must always be careful in interpreting results of this kind. The above discussion must not be taken to imply that all aircraft are limited by shock stall or buffet at the maximum speed. This will be true for aircraft designed for economical cruise at transonic speed (as are most commercial airliners) but clearly many aircraft are designed to pass

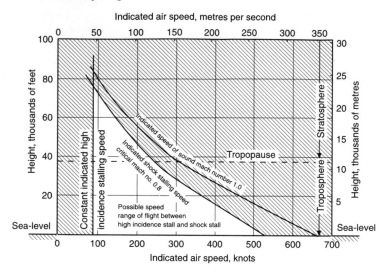

Fig. 11.9   Subsonic speed range of flight – indicated speeds

through the transonic speed range and cruise at supersonic speed – but
more of that in the next chapter.

Before we move on to supersonic flow, though, we will spend a little
time examining the ways in which knowledge of transonic flows was
acquired by experiment and how this knowledge is used in the design of
aircraft which operate at transonic speeds.

EXPERIMENTAL METHODS

In the course of this book there has been frequent comment to the effect
that theory has tended to lag behind practice as the means whereby we
have acquired aeronautical knowledge. Some critics have said that this
claim has been exaggerated, but the author still believes that it is fair
comment. It certainly isn't a new idea for this is an extract from the 14th
Annual Report of the Aeronautical Society of Great Britain (now the
Royal Aeronautical Society) – 'Mathematics up to the present day has
been quite useless to us in regard to flying' – the date of that report,
1879!

But whatever may have been the relative importance of theory and
experiment in the acquisition of knowledge about subsonic flight, even
the most theoretically-minded critic will surely agree that we have had
to rely almost entirely on experiment in solving the problems of
transonic flight, for here we have a mixture of subsonic and supersonic
flow, compressibility and incompressibility, two completely different

theories all mixed up. Curiously enough, in real supersonic flight, theory comes into its own; supersonic theory is simpler and older than that of subsonics or transonics – Newton's theories used to calculate air resistance of a body at 515 m/s give an answer nearer the truth than if they are used to calculate its resistance at 51.5 m/s.

But one of the fascinations of this subject is that the experimental methods themselves are so interesting, involving as they do their own theories quite apart from the facts that they reveal. So far as transonic and supersonic flight are concerned we have already referred to ingenious methods of photography which give us pictures not only of shock waves, but also of smaller changes of density; and by taking films by these methods we can watch the changes of density and of the shock pattern as speed is increased from subsonic to transonic, then through the transonic to the supersonic region. At subsonic speeds we have devised methods of seeing the flow of air, but the schlieren and other methods show something even more important – what happens as a result of airflow. We have learnt a very great deal by these methods, and we shall look at some more pictures later.

But though it is conceivable that by very elaborate means such photographs could be taken in flight they really require laboratory conditions, which means wind tunnels. Now wind tunnels present quite enough problems at subsonic speeds; but as we approach the speed of sound the very shock waves which we want to investigate and photograph, obstruct the flow through the tunnel (even when it hasn't got a model in it), and raise a barrier so effective that the tunnel is virtually choked and the high-speed flow cannot get through. Even if the tunnel design can be modified so as to allow the flow, the shock waves on even the smallest of models will cause very severe interference between the model and the walls of the tunnel, and so make the results of the tests valueless. This choking of wind tunnels, which was particularly difficult to overcome between Mach Numbers of 0.85 and 1.1, is the explanation of two rather curious facts in aeronautical history, that a truly supersonic wind tunnel became a practical proposition before a transonic one, and that flight at supersonic speeds took place before such speeds were reached in wind tunnels.

The problem of the transonic wind tunnel has now been largely overcome (though rather late in the day) by using slotted or perforated walls, and there are many types of supersonic tunnel in use today. Some of these are very similar to subsonic types in general outline; the extra speed has simply been obtained by more power together with suitable profiling of the duct – perhaps simply is not quite the right word, because the increase in power required, and consequent cost, is tremendous. It might be thought that fans of the propeller type would not be suitable for such tunnels; but in fact they can be used because they are situated at a portion of the tunnel where the speed is comparatively low, the cross-section of the tunnel, and therefore the propeller, being correspondingly large. The great size of the propeller, often larger than any used on aircraft, presents problems of its own, but none the less this conventional type of tunnel, which may be straight through, return flow or even open jet, is probably the most satisfactory where a large tunnel is required.

For smaller types, and higher speeds, it is more usual to employ some kind of reservoir of compressed air and, by opening a valve, to allow this to blow through the tunnel to the atmospheric pressure at the exhaust (Fig. 11.10). By arranging for the exhaust to be into a vacuum tank, even

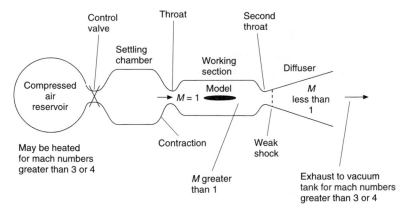

Fig. 11.10   High-speed wind tunnel: blow-through type

higher speeds can be obtained. One great disadvantage of this method is that continuous running for long periods is impossible; in fact constant speed is only achieved for a very short time. It sounds rather primitive, and in some ways it is, but Mach Numbers of 4 or more may be reached by this method, though the cross-sectional area of the working section is usually small.

More efficient, at some Mach Numbers, than this **straight blow-through** type of tunnel is the **induction or induced flow** type, in which air is blown in or injected just down stream of the working section (Fig. 11.11), thus 'inducing' a flow of air from the atmosphere through the mouth of the tunnel. Notice that in this type the compressed air does not flow over the model at all, only the induced air does. The injected flow can be the jet of a turbo-jet engine; in fact the jet engine itself can be in the tunnel. An induction type tunnel may be of

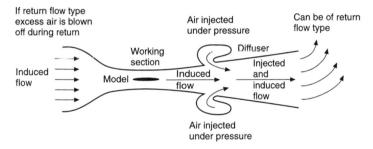

Fig. 11.11   High-speed wind tunnel: induced flow type

the return flow variety, in which case provision must be made for the excess air to be blown off from the return passage.

It is one thing to obtain a high Mach Number in a wind tunnel – it is quite another thing also to obtain a high Reynolds Number (see Appendix 2) and so eliminate scale effect. This brings in the question of high-density tunnels, low-density tunnels, cryogenic (low temperature tunnels) and even the use of gases other than air; this interesting problem will be touched on in Appendix 2, but it is really beyond the scope of this book.

A simple and fascinating way of observing patterns very similar to shock-wave patterns is by using what is sometimes called the **hydraulic analogy**. Anyone who has watched the bow wave, and other wave patterns caused by a ship making its way through water, and has also seen the shock-wave patterns in air flowing at supersonic speed, must have been struck by the similarity of the patterns. If bodies of various shapes are moved at quite moderate speeds across the surface of water (**not totally immersed**), or the water made to flow past the bodies, many shock-wave phenomena can be illustrated and, by a suitable system of lighting, thrown on to a screen. Some people say that such demonstrations are too convincing, because they make one think that it is the same thing – which of course it isn't. The patterns are similar but the angles and so on of the waves are different.

Finally, we come to full-scale or free-flight testing, and even this may be divided into two types, tests with piloted aircraft and those with pilotless aircraft or missiles. Not only are these the eventual tests, but in the case of transonic flight, in particular, they have also been the pioneer tests and most of them have been made in piloted aircraft. In the early investigations, piloted aircraft certainly had their limitations because we could only approach the speed of sound in a dive; this had rather obvious and rather serious disadvantages. It took a long time (and a lot of height lost) to reach the critical speeds and then, if the symptoms were alarming – as they sometimes were – we were unpleasantly near the ground by the time a recovery could be made; moreover the making of such a recovery was not made any easier if the symptoms were severe buffeting, or worse – a further dropping of the nose and steepening of the dive – or worse still – heaviness of the controls so severe that they could not be moved. In such circumstances the point made in an earlier paragraph that the Mach Number might be decreasing as we hurtled towards the ground was hardly sufficient consolation.

As the thrust of jet engines increased – and it was this thrust which made even the approach of the speed of sound possible – the shock stall could be reached in level flight in some types of aircraft, and later still in climbing flight. This was an altogether different proposition from the pilot's point of view and, as testing at transonic speeds lost its terrors, our knowledge increased correspondingly more rapidly.

Test on bullets and shells moving through the air at supersonic speeds have been made on ballistic ranges since before the days of practical flight, and photographs of shock waves were taken more than 60 years ago, but it was only when aircraft themselves began to approach the speed of sound that the significance of such tests in respect of aircraft design began to be realised; and it was the development of the ramjet and rocket as means of driving missiles, and the parallel development of electronic instruments which could not only guide the missiles but take readings and keep records during the flight – in some cases even transmitting them back to earth – it was these that contributed most of all to our knowledge, and to the solution of the problems of transonic and supersonic flight.

So we can sum up the experimental methods that have been used to investigate these problems as coming under the following headings –

(*a*) Photography of shock waves.
(*b*) High-speed wind tunnels.
(*c*) The hydraulic analogy.
(*d*) Free flight in piloted aircraft.
(*e*) Rockets and missiles.

Now let us see what all this has taught us.

SHOCK–WAVE PATTERNS

In an earlier paragraph we described how a shock wave is formed at a speed of about three-quarters of the speed of sound, i.e. at about $M = 0.75$. On a **symmetrical** wing at **zero angle of attack the incipient shock wave appears on both top and bottom surfaces** simultaneously, approximately at right angles to the surfaces, and, as one would expect, **at about the point of maximum camber** (Fig. 11.12*b*). On a wing at a small angle of attack, even if the aerofoil section is symmetrical, the incipient shock wave appears first on the top surface only (Fig. 11A) – again as one would expect, because it is on the top surface that the speed of the airflow first approaches the speed of sound.

Figure 11.12 shows how the shock-wave pattern changes (on a symmetrically shaped sharp-nosed aerofoil at zero angle of attack) as the speed of airflow is increased from subsonic, through the transonic range to fully supersonic flow.

Between the formation of the incipient wave (at a Mach Number of about 0.75 or 0.8) and the time when the wing as a whole is moving through the air at a speed of sound ($M = 1.0$), **the shock wave tends to move backwards, but in doing so becomes stronger and extends farther out from the surface, while there is even more violent turbulence behind it** (Fig. 11.12*c*). **At a speed just above that of sound another wave appears, in the form of a bow wave, some distance ahead of the leading edge; and the original wave, which is now at the trailing edge, tends to become curved, and shaped rather like a fish tail** (Fig. 11.12*d*). **As the speed is further increased the bow wave attaches itself to the leading edge, and the angles formed between both waves and the surfaces become more acute** (Fig. 11.12*e*). Still further increases of speed have little effect on the general shock-wave pattern – but here we are trespassing on supersonic flight, which is the subject of the next chapter.

At each wave there is a sudden increase of pressure, and density, and temperature, a decrease in velocity, and a slight change in direction of the airflow. The thickness of a shock wave, through which these changes take place, is only of the order of 2 to 3 thousandths of a millimetre – they look thicker on photographs because it is not possible to get a perfectly plane shock wave in the experiment. The changes at the shock wave are **irreversible**, which is another way of saying that the high pressure behind the wave cannot be communicated to the lower pressure in front – messages can only go down stream. It is interesting to note, however, that the incipient waves only extend a short distance from the surface, and leaks are possible round the ends of the waves; as speed increases the waves extend and there is less and less possibility of such leaks. It is interesting to note, too, that the decrease in velocity, which occurs behind the shock wave, means that when an aircraft is moving through the air, and a shock wave is formed, the air behind the shock wave begins to move in the direction in which the aircraft is travelling.

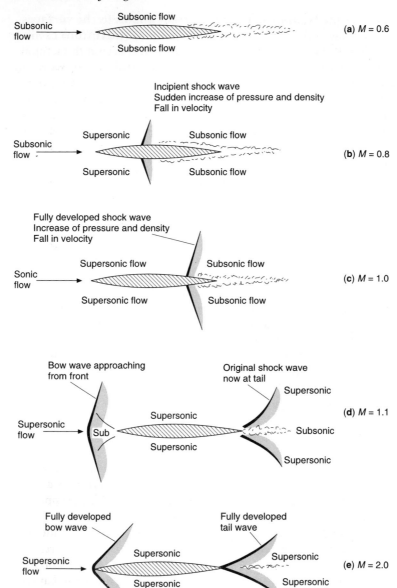

Fig. 11.12   Development of shock waves at increasing Mach numbers
(*a*) Subsonic speeds. No shock wave. Breakaway at transition point.
(*b*) At critical Mach Number. First shock wave develops.
(*c*) At speed of sound. Shock wave stronger and moving back.
(*d*) Transonic speeds. Bow wave appears from front. Original wave at tail.
(*e*) Fully supersonic flow. Fully developed waves at bow and tail.

In addition to showing the shock-wave patterns, Fig. 11.12 also indicates the areas in which the flow is subsonic or supersonic. In (*a*) at $M = 0.6$ **it is all subsonic** (clearly we are still in the subsonic region); at $M = 0.8$ **the flow immediately in front of the shock wave is supersonic**, but **all the remainder is subsonic** (we are now in the transonic region with both types of flow); at $M = 1$ **the area of supersonic flow has increased** but the **flow behind the shock wave is still subsonic** (as we shall learn later **it is always subsonic behind a shock wave that is at right angles to the flow, it can only be supersonic behind an inclined or oblique shock wave**); at $M = 1.1$ **nearly all the flow is supersonic**, but there are still small regions of subsonic flow, immediately in front of the leading edge at what is called the **stagnation point** where the flow is brought to rest, and immediately behind the trailing edge (we are still in the transonic region, but not for much longer); at $M = 2$ **the flow is all supersonic** – we are through the barrier (though to be strictly correct, unless the bow wave is actually attached to the leading edge, which will only happen if the edge is very sharp, there will still be a small area of subsonic flow at the stagnation point between the bow wave and the leading edge; and of course in the boundary layer itself the air immediately next to the surface is at rest relative to the surface, and most of the remainder of the airflow in the boundary layer is subsonic).

The figure also shows how the extent of the separated region, or thickened boundary layer tends to decrease with increasing Mach Number, and this suggests that **as wave drag becomes relatively more important, boundary layer drag becomes relatively less so.** This may also give a clue to the decrease in drag coefficient as we pass through the barrier (Fig. 11.7).

The reader should now be able to draw for himself the shock patterns, corresponding to those of Fig. 11.12 for an aerofoil inclined at a small angle of attack, and the exercise in doing so will help him to appreciate how and why shock waves are formed.

Figure 11B (on page 331) is a **remarkable example** of condensation and shock waves on an aeroplane in flight with, below, a schlieren photograph of shock waves on a model of the same aircraft. More shock waves on an aerofoil are shown in Fig. 11F at the end of this chapter.

SHOCK WAVES AND PRESSURE DISTRIBUTION

It has already been stated that at the shock stall, as at the high incidence stall, there are sudden changes both in lift and drag; and it is only to be expected that these are due in the main to changes in the pressure distribution over the wings or other surfaces. So pressure plotting has been as important a feature of research into the problems of high-speed flight as into those of subsonic flight, and the connection between the shock patterns and pressure distribution is naturally of great interest and importance.

It must be remembered that when we were originally considering the shapes of aerofoil sections in Chapter 3, the so-called **laminar flow aerofoil** (page 98) proved of great value at speeds of 140 m/s upwards. The characteristics of this section were comparative thinness and gently graduated camber, with the point of maximum camber farther back than on slow-speed types. As a result of this shape – in fact it was the purpose of it – the airflow speeded up very gradually and the distribution of decreased pressure over the upper surface was much more even than for the slower types on which there was a marked peak of suction quite near the leading edge (page 76). We naturally approached the transonic region with aerofoil sections of this type, and so in considering the pressure distribution diagrams we must expect a fairly even distribution of decreased pressure on the top surface before the shock waves appear (this will be evident in Figs 11.13 and 11.14).

Figure 11.13 shows in a very realistic way how the decrease in pressure, or suction, on the upper surface of a wing is affected by the formation of a shock wave when the wing as a whole is moving near the

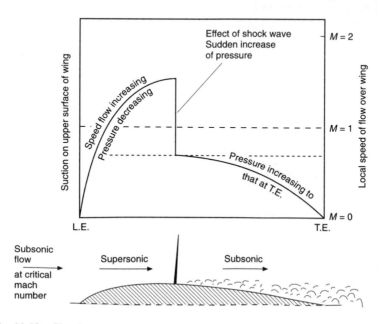

Fig. 11.13   Shock wave and pressure distribution

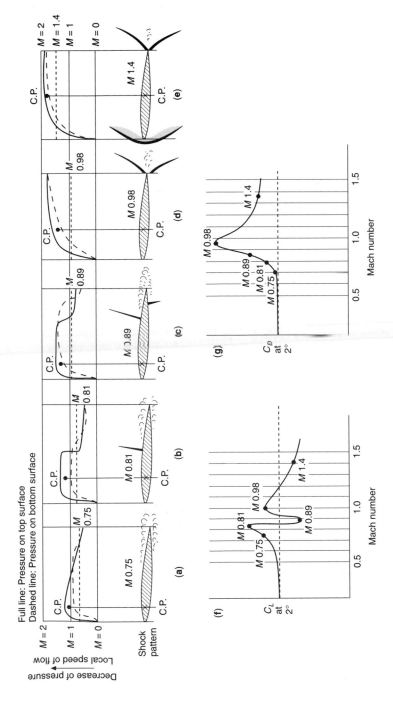

Fig. 11.14 Shock-wave patterns and pressure distribution
Symmetrical bi-convex section at 2° angle of attack.

critical Mach number. It shows the local Mach Number of flow across the surface of the wing; pressure of course depends on the speed, and hence Mach Number; the higher the local speed the less the pressure (back to Bernoulli again). The pressure scale is not given since compressibility effects complicate the issue considerably, but qualitatively it gives the right idea.

Figure 11.14 is rather more complicated, but it is worth trying to understand **because it demonstrates so clearly the practical effect of the shock waves on the pressure distribution** – this time on both top and bottom surfaces, and for a symmetrical wing section at a small angle of attack – **and the resultant effect on the lift and drag coefficient of the wing over the transonic range**. Incidentally, too, the figure gives a partial answer to the exercise suggested to the reader at the end of the previous paragraph, but perhaps he tackled that exercise before reading on!

At (a), (b), (c), (d), and (e) in the figure, which illustrate what happens at Mach Numbers of 0.75, 0.81, 0.89, 0.98, and 1.4 respectively, we see first the shock patterns at these speeds. All refer to a symmetrical bi-convex aerofoil section at an angle of attack of 2°. Above the shock pattern is shown the corresponding Mach Number distribution, across the chord; the full line representing the upper surface and the dotted line the lower one; decreased pressure is indicated upwards, as this corresponds to increasing Mach Number. It is the difference between the full line and the dotted line which shows how effective in providing lift is that part of the aerofoil section; if the dotted line is above the full line the lift is negative. The total lift is represented by the area between the lines, and the centre of pressure by the centre of area. Increasing speed, which is proportional to the decreasing pressure, is also shown upwards.

On the vertical scale on each diagram is a Mach Number of 1, i.e. the speed of sound, and the free stream Mach Number, i.e. the Mach Number at which the aeroplane as a whole is moving through the air. Thus we can see at a glance over what parts of the surfaces, upper or lower, the local airflow is subsonic or supersonic, and over what parts its speed is above or below that at which the aerofoil is travelling.

Parts (f) and (g) of the figure show the lift coefficient and drag coefficient corresponding to each diagram. Of these the lift curve is most interesting, partly because we have already seen the drag curve in Fig. 11.7, but more because the drag is not revealed by the pressure distribution to the same extent as the lift is, and in fact the pressure distribution does not show the important part of the drag that is tangential to the surface at all.

This figure should be studied at leisure; it tells a long story – too long for me to point out every detail. But let us just look at some of the more important points.

Part (*a*) of course, is the subsonic picture, except that separation has already become apparent near the trailing edge and there is practically no net lift over the rear third of the aerofoil section; the centre of pressure is well forward and, as (*f*) shows, the lift coefficient is quite good and is rising steadily; the drag coefficient, on the other hand, is only just beginning to rise.

In (*b*) the incipient shock wave has appeared on the top surface; notice the sudden increase of pressure (shown by the falling line) and decrease of speed at the shock wave. The centre of pressure has moved back a little, but the area is large, i.e. the lift is good (see (*f*)), and the drag (*g*) is rising rapidly.

The pressure distribution in (*c*) shows very clearly why there is a sudden drop in lift coefficient (see (*f*)) before the aerofoil as a whole reaches the speed of sound; on the rear portion of the wing the lift is negative because the suction on the top surface has been spoilt by the shock wave, while there is still quite good suction and high-speed flow on the lower surface. On the front portion there is nearly as much suction on the lower surface as on the upper. The centre of pressure has now moved well forward again; the drag is increasing rapidly (*g*).

Part (*d*) is particularly interesting because it shows the important results of the shock waves moving to the trailing edge, so no longer spoiling the suction or causing separation. The speed of flow over the surfaces is nearly all supersonic, the centre of pressure has gone back to about half chord, and owing to the good suction over nearly all the top surface, with rather less on the bottom, the lift coefficient has actually increased (see (*f*)). The drag coefficient is just about at its maximum.

At (*e*) we are through the transonic region. The bow wave has appeared. For the first time the speed of flow over about half of both surfaces is **less** than the Mach Number of 1.4 of the aerofoil as a whole. The lift coefficient has fallen again, because the pressures on both surfaces are nearly the same; and this time – for the first time since the critical Mach Number – the drag coefficient has fallen considerably.

SONIC BANGS

We are now all too familiar with the noises made by aircraft 'breaking the sound barrier', not to mention those unfortunate people who have suffered damage to property as a result. These so-called **sonic bangs**, or **booms**, are of course, caused by shock waves, generated by an aircraft, and striking the ears of an observer on the ground, or his glasshouses or whatever it may be; but there has been considerable argument as to the exact circumstances which result in the shock waves being heard, why there are often two or more distinct bangs, whether the second one came first, and so on.

Strangely enough many people don't seem to realise that we were familiar with sonic bangs, and their effects on us and our property, long before aircraft flew at all. A crack of the whip is probably the oldest man-made example; it may not have been responsible for breaking glasshouses, but in the hands of a circus performer it can be a pretty shattering noise. A roll or a clap of thunder is an example from nature of a series of shock waves; and one must have noticed how the bangs produced by aircraft often resemble a short roll of thunder. Explosions, too, produce shock waves, and, although a bombing raid is hardly an appropriate time to analyse such things, there must be many who were unfortunate enough to experience during the war the disastrous effects on their ears and property of the shock wave of an exploding bomb, as distinct from the damage caused by the bomb itself. Some, too, may even have noticed the rather weird way in which the different bangs arrived at different times depending on where the observer was relative to the launching and explosion of the bomb. But the nearest thing to the sonic bangs produced by aircraft are the crack of a rifle bullet, or of a shell going overhead, or of the V2 rocket of wartime memory.

If an aircraft were to fly at supersonic speed at a height of a few feet over one's head the shock waves from wings, body, tail, etc., would strike one's ears in rapid succession, so rapid that one couldn't distinguish between them, and (if one remained conscious at all) the impression, so far as noise is concerned, would probably be of a short roll of thunder. The higher the aircraft flew, the less violent would be the noise produced by the shock until it would hardly be noticed at all from the ground; the noise of the engine, and of the aircraft itself, are of course continuous noises which are quite distinct from that of the shock.

An aircraft diving towards the earth at supersonic speed, and at an angle of say 45°, then suddenly slowing up and changing direction, will 'shed' its shock waves, which will travel on towards the earth and strike any observers which may happen to be in their path. It is certainly quite clear from schlieren photographs that a bow wave **approaches from the front** as the speed of sound is approached, and, conversely, **goes ahead of the aircraft** when it decelerates below the speed of sound.

So far as effects at ground level are concerned, we know that these become **less intense with the height** of the aircraft; **more intense with Mach Number**, though not anything like in proportion; that they are affected by the **dimensions of the aircraft, increasing with its weight and volume,** and being of **longer duration according to its length;** that they are **more intense during accelerated flight** (when the shocks tend to coalesce) than in steady flight; and that they **decrease very rapidly with lateral displacement** from the line of flight of the aircraft, in fact they only extend over a certain lateral distance. All this is rather what one might expect, but the problem is complicated because shocks of different intensities may be generated by the body, wings, tail and other parts of the aircraft, hence sometimes the roll as of thunder rather than one or two sharp bangs. One factor that one might not expect is the extent to which the bangs vary **according to the conditions in the atmosphere** between the aircraft and the ground, the winds, temperature, turbulence and so on. The actual pressures created at ground level are not so great as is sometimes thought; the overpressure in the Concorde boom, for instance, is only of the order of 96 newtons per square metre.

Can anything be done about it? **No much**; if we insist on flying at these speeds. We can legislate against supersonic flight other than over the sea or thinly populated areas, but even so aircraft have to reach these areas. Moreover the tendency must be for the weight and size of such aircraft to increase rather than the reverse. Some alleviation can be obtained by control of climbing speeds; and at certain heights the speed of sound may be exceeded without creating a boom at all, the shock wave being dissipated before it reaches the ground. So really there is nothing for it but to fly as much as possible over the sea, and as high as possible, perhaps even really high – as we shall mention in the last paragraphs of this book.

Finally it should be mentioned that the publicity that has been given to sonic booms **has tended to make us forget all the other noises** created by aircraft, those from the engines, propellers or jets, and from the motion of the aircraft itself; these noises are probably **more objectionable than sonic booms** to those who live on landing or take-off paths, they are by no means confined to transonic and supersonic aircraft, and there are better prospects of reducing some of these noises as, for example, by using quieter engines.

RAISING THE CRITICAL MACH NUMBER – SLIMNESS

When increase in engine thrust – due to the rapid development of jet engines – first made transonic flight possible, research was concentrated on the problem of raising the critical Mach Number, of postponing the shock stall, of getting as near to the barrier as possible without getting into it – in short, of keeping out of trouble rather than facing it.

There are two main ways of raising the critical Mach Number. The first is **slimness**. The need for slimness will be abundantly clear from all that has been said about shock waves and their effects – and the slimness applies to all parts, the aerofoil section, the body, the engine nacelles, the fin, tail plane and control surfaces, and perhaps most of all to small excrescences (if there must be such things) on the aircraft. The aerofoil section must be of the low-drag laminar-flow type already referred to, and must have a very low ratio of thickness to chord. The Spitfire of the Second World War has already been mentioned as an example of slimness, and of a high critical Mach Number – all the more remarkable in that it was not designed for transonic speeds.

The full line in Fig. 11.15 shows very clearly the effect of thickness/chord ratio on the critical Mach Number for a straight wing (the dotted line will be referred to in the next paragraph); at a t/c ratio of 10 per cent this wing has a critical Mach Number of only just over 0.8, at a t/c ratio of 8 per cent it is raised to 0.85, and at 4 per cent it is over 0.9. Not long ago the t/c ratios of wings for fighter aircraft were from 9 to 12 per cent, but they have now been reduced to 7 or 8 per cent, and may yet be still further reduced to a figure as low as 3 per cent, though there are of course very great design and manufacturing difficulties in producing such thin wings. In thus speaking of thin wings it is important to keep in mind that what really matters is not the actual thickness, but **the ratio of thickness to chord**.

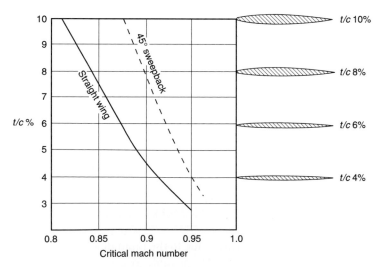

Fig. 11.15   **Critical Mach Number and** t/c **ratio**

RAISING THE CRITICAL MACH NUMBER – SWEEPBACK

The second main way of raising the critical Mach Number (and this applies only to the wings, tail, fin, and control surfaces) is **sweepback** –

not just the few degrees of sweepback that was sometimes used, rather apologetically and for various and sometimes rather doubtful reasons, on subsonic aircraft, but 40°, 50°, 70° or more.

Sweepback of this magnitude not only delays the shock stall, but reduces its severity when it does occur. The theory behind this is that it is only the component of the velocity across the chord of the wing ($V \cos \alpha$) which is responsible for the pressure distribution and so for causing the shock wave (Fig. 11.16); the component $V \sin \alpha$ along the span of

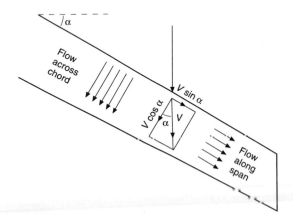

Fig. 11.16   Sweepback – components of velocity

the wing causes only frictional drag. This theory is borne out by the fact that when it does appear the shock wave lies parallel to the span of the wing, and only that part of the velocity perpendicular to the shock wave, i.e. across the chord, is reduced by the shock wave to subsonic speeds. As the figure clearly shows, the greater the sweepback the smaller will be the component of the velocity which is affected, and so the higher will be the critical Mach Number, and the less will be the drag at all transonic speeds of a wing of the same t/c ratio and at the same angle of attack.

Experiment confirms the theoretical advantages of sweepback, though the improvement is not quite so great as the theory suggests. The dotted line in Fig. 11.15 shows how a wing swept back at 45° has a higher critical Mach Number than a straight wing at all values of t/c ratio, the advantage being greater for the wings with the higher values of t/c. Figure 11.17 tells us even more; it shows that sweepback not only increases the critical Mach Number, but it reduces the rate at which the drag coefficient rises (the slope of the curve), and it lowers the peak of the drag coefficient – and 45° of sweepback does all this better

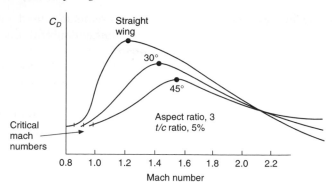

Fig. 11.17   Effects of sweepback

than 30°. Incidentally this figure also shows that, above about *M2*, sweepback has very little advantage – but that is another story and, in any case, aeroplanes cannot fly at *M2* without first going through the transonic range.

Figure 11.18 shows various plan forms of swept-back wings.

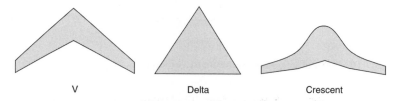

Fig. 11.18   Swept-back wings for transonic speeds

Of course, as always, there are snags, and the heavily swept-back wing is no exception. There is **tip stalling** – an old problem, but a very important one; in the crescent-shaped wing (Fig. 11.18) an attempt has been made – with some success – to alleviate this by gradually reducing the sweepback from root to tip. $C_L$max is low, and therefore **the stalling speed is high**, and $C_L$max is obtained **at too large an angle to be suitable for landing** – another old problem, and one that can generally be overcome by special slots, flaps or suction devices. There are also control problems of various kinds, and the designer doesn't like the extra bending and twisting stresses that are inherent in the heavily swept-back wing design. But whatever the problems sweepback seems to have come to stay – at least for aircraft which are designed to fly for any length of time at transonic or low supersonic speeds.

CONTROL PROBLEMS

Reference has already been made to the unpleasant things that may happen to aircraft as they go through the speed of sound – violent changes of trim, up or down, oscillations, buffeting, and so on. In such circumstances the first essential from the pilot's point of view is that he should have complete control; and so be master of the aircraft and its movements.

Unfortunately this is by no means easy to provide – and for several reasons.

Consider, for instance, an ordinary tail plane and elevator. At subsonic speeds an elevator depends for its effectiveness **on the complete change of flow** which occurs **over both tail plane and elevator** when the elevator is raised or depressed; the actual forces which control the aircraft are in fact much greater on the tail plane than they are on the elevator itself. Now as soon as a shock wave is formed on the tail plane – and the most likely place for its formation is at the hinge between tail plane and elevator – **a movement of the elevator cannot affect the flow in front of the shock wave**, so we have to rely entirely on the forces on the elevator itself to effect control. **But the elevator will be in the turbulent flow behind the shock wave, and so may itself be very ineffective in producing the control forces required**.

At higher speeds, when the shock wave moves back over the elevator, the opposite trouble may occur, and the **forces on the elevator may be so great that it becomes almost immovable**.

The answer to these troubles has been found in the **all-moving slab type of tail plane** (Fig. 5H) – sometimes given the rather curious name of a **flying tail** – and in making this **power-operated**. In this way we have what at the same time is an adjustable tail plane and an elevator.

The same principle can be applied to the ailerons – by having **all-moving wing tips** – and, more rarely, to the rudder, by having a combined fin and rudder.

But this is not the only high-speed control problem. Another is the possibility of **reversal of the controls**, owing to the distortion of the structure by the great forces on the control surfaces. This is most likely to happen on the ailerons. Suppose, for instance, that the starboard aileron is lowered in order to raise the starboard wing – if the force on the aileron is very great it will tend to **twist the wing in such a way as to reduce its angle of attack**, and so **reduce the lift** instead of increasing the lift on that wing; and the net result is that the wing will fall instead of rise. The aileron is acting on the wing just as a control tab is intended to act on a control surface; but that is little consolation to the pilot, and it doesn't take much imagination to realise his horror when a movement of the control column has exactly the opposite effect to that intended! Nor is it an easy problem for the designer to solve since the extra weight involved in providing sufficient stiffness, especially on thin and heavily swept-back wings, may be prohibitive. So as well as requiring power-generated controls, special high-speed ailerons may be needed – these are situated well inboard where the structure is stiffer instead of at the tips. Another solution is to use 'spoilers' instead of ailerons. These are small flaps which come out of the top surface of the wing and disrupt the local flow, thus reducing lift on that wing and giving rolling control.

The introduction of power-operated controls has in itself caused a new problem in that the pilot no longer 'feels' the pressure resisting the movement of the controls; this feel was always a safety factor in that it made the pilot conscious of the forces he was applying, and in fact there was some advantage in that there was a limit to what he could do to the aeroplane owing to the sheer limitation of his strength. So important is this matter of feel that when power-operated controls are used it has been necessary to incorporate artificial or synthetic 'feel'; and this is made even more real by grading it so that it varies not only with the movement of the control surface but with the density of the air and the air speed, in other words with the dynamic or stagnation pressure, $\frac{1}{2}\rho V^2$ or q – it is sometimes called 'q-feel'. Quite apart from the safety aspect, this synthetic feel gives the pilot a sense of control over the aeroplane which restores something of the art of flying.

But this is not the only problem resulting from the use of powered controls. If the power fails, and if there is no means of reverting to manual operation, the control may lock solid and the pilot be denied the use of rudder, elevators or ailerons. The answer to this is to introduce a safety factor by having **more than one control surface**, each having a separate power control; thus the Concorde has **two rudders**, one above the other, and **six elevons** (combined elevators and ailerons, which is the usual arrangement on delta or highly-swept wings). The Russian counterpart even has eight elevons.

Again on delta and highly-swept wings there is sometimes an interesting use of **spoilers**, this time as an aid to longitudinal control. Large aircraft flying at high speed, with their considerable inertia, are slow to respond to the elevators, just as they are to the ailerons. **If one set of spoilers is fitted inboard** (i.e. close to the fuselage) on each wing, and so **forward of the aerodynamic centre**; and another set is fitted **well outboard** and so, owing to the sweepback, **behind the aerodynamic centre**; and these are then linked to the elevator control in such a way that the fore and aft sets can be operated differentially, they will cause a movement of the centre of pressure which will aid the elevator control, just as when they are operated differentially on the port and starboard wings they assist the ailerons in providing lateral control, as described above.

AREA RULE

We should by now realise that if the drag is to be kept to a minimum at transonic speeds, bodies must be slim and smooth, and have 'clean lines'. What is the significance of clean lines? Well, it is often said to be in the eye of the beholder, what looks right is right – yes, but it depends on who looks at it; and a little calculation, a little rule, formula, or whatever it may be will often aid our eyes in designing the best shapes for definite purposes. The **area rule** (Fig. 11C) is simply one of these rules, and put in its simplest form it means that **the area of cross-section should increase gradually to a maximum, then decrease gradually**; in this sense a streamline shape obeys the area rule, though for transonic speeds, and indeed for high subsonic speeds, the maximum cross-sectional area should be about half-way, rather than one-third of the way back, this giving a more gradual increase of cross-sectional area with an equally gradual decrease. The body in Fig. 11.19 obeys the area

Fig. 11C   Transonic area rule
The transonic British Aerospace Buccaneer which finally saw action in the Gulf War shortly before retirement. The bulge in the rear fuselage is for purposes of area rule.

Fig. 11.19   Area rule – plan view of fuselage without wings or tail

rule – but it hasn't got any wings. If we add a projection to a body, such as the wings to a fuselage, we shall get a sudden jump in the cross-sectional area – and that means that the area rule is not being obeyed. What then can we do? – the answer is that we must **decrease the cross-sectional area of the fuselage as we add the cross-sectional area of the wings** in such a way that **the total cross-sectional area of the aeroplane increases gradually**. Similarly behind the point of maximum cross-sectional area it is **the total cross-sectional area** that must be gradually decreased.

It will be realised that the application of this rule gives a **waist** to the fuselage where wings or other parts such as the tail plane are attached (Figs 11.19 and 11.20). It will be realised, too, that sweepback – in addition to its other advantages – is to some extent an area rule in itself so far as the wings are concerned, the cross-sectional area being added gradually, and so the waisting of the fuselage will be less marked with swept-back wings than with straight wings.

Fig. 11.20   Area rule – effect of wings and tail on plan view of fuselage

VORTEX GENERATORS

Many devices are used by the designer to control the **separation** or **breakaway of the airflow from the surface of the wing** – all these devices, in one way or another, over one part of the wing or another, have this in common, that **they are intended to prevent or delay this breakaway**. How? Well, that depends to some extent on the device, and we will consider **vortex generators** first (Fig. 11D).

The fundamental reason for the breakaway is that the boundary layer becomes sluggish over the rear part of the wing section, flowing as it is against the pressure gradient. The formation of a shock wave makes

Fig. 11D Vortex generators
The Buccaneer again, showing vortex generators on the outer wing. This 'fix' was
used to improve wing flow attachment on many early swept-wing aircraft.

matters worse, the speed in the boundary layer is still subsonic which
means that pressure can be transmitted up stream, causing the boundary
layer to thicken and, if the pressure rise is too steep, to break away from
the surface. Now vortex generators are small plates or wedges,
projecting an inch or so from the top surface of the wing, i.e. three or
four times the thickness of the boundary layer. **Their purpose is to
put new life into a sluggish boundary layer**; this they do by
shedding small lively vortices which act as scavengers, making the
boundary layer turbulent and causing it to mix with and acquire extra
energy from the surrounding faster air, thus helping it to go farther
along the surface before being slowed up and separating from the
surface. In this way **the small drag which they create is far more
than compensated by the considerable boundary layer drag
which they save, and in fact they may also weaken the shock
waves and so reduce shock drag also; and the vorticity which
they generate can actually serve to prevent buffeting of the
aircraft as a whole** – a clever idea indeed, and so simple. The net
effect is very much the same as blowing or sucking the boundary layer,
but the device is so much lighter in weight and simpler. The greater the
value of the thickness/chord ratio the more necessary does some such
device become.

There are various types of vortex generator; Fig. 11.21 illustrates the

Fig. 11.21   **Vortex generators: bent-tin type**

bent-tin type, which may be co-rotating or contra-rotating. The plates are inclined at about 15° to the airflow, and on a wing are usually situated on the upper surface fairly near the leading edge.

OTHER DEVICES TO PREVENT OR DELAY SEPARATION

There are several other devices which have been used to prevent or delay separation of the boundary layer, and so allay the rapid increase in drag at the sonic barrier, or the buffeting, or violent changes in trim which are liable to occur as a result of shock waves or separation, or some or all of these troubles.

A thickened trailing edge is sometimes employed; this causes vortices which have much the same effect as those created by vortex generators, though naturally the effect is not felt so far forward on the wing surface.

On heavily swept-back wings **fences** (Figs 11.22 and 3F) are often fitted; these are vanes of similar height to vortex generators, but running fore and aft across the top surface of the wing, and designed to check any spanwise flow of air along the wing, for this in turn is likely to cause a breakaway of the flow near the wing tips and so lead to tip stalling, particularly on swept wings.

Another problem with highly swept wings is the tendency for the flow

Fig. 11.22   **A boundary layer fence**

Fig. 11.23   Leading edge saw-tooth or dog-tooth

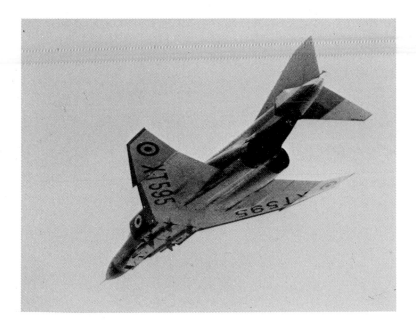

Fig. 11E   Saw-tooth or dog-tooth
(By courtesy of McDonnell Douglas Corporation, USA)
The Phantom (RAF version), showing clearly the dog-tooth on the leading edge; the outer wings have 12° dihedral; there are blown leading and trailing edge flaps; the slab tail has 15° anhedral and a fixed slot; the rudder is inter-connected with the ailerons at low speeds.

Fig. 11.24   Leading edge flap, double-slotted trailing edge flap and air brake

to separate in the tip region first. This causes all sorts of problems, for example large changes in pitching moment. This effect may be reduced by introducing a **notch** or **saw-tooth** in the leading edge (Figs 11.23 and 11E). The notch also generates a strong vortex which controls the boundary layer in the tip region.

Leading-edge droop and **leading-edge flaps** are becoming quite common features of high-speed aircraft, but these are to prevent separation of the flow at the low-speed end of the range, i.e. at large angles of attack, and so help to solve one of the main problems of aircraft designed for transonic and supersonic speeds, that of making them fly safely slowly. A **permanent droop** is called leading-edge droop or

Fig. 11F   Shock waves
Shock waves from an aerofoil at incidence to the flow. Note the stronger leading edge shock on the underside. The wave in the top left-hand corner is a reflection from the wind-tunnel wall.

droop-snoot; when it is adjustable it is called a leading-edge flap. Either may be combined with trailing-edge flaps and other devices, and Fig. 11.24 illustrates a combination of leading-edge droop, double slotted trailing-edge flaps, and air brakes – all helping to the same end.

But to conclude the problems of flight at transonic speeds on an optimistic note, it can generally be said that once one has **a good transonic shape, it remains good,** and the flow around it changes little between subsonic and transonic speeds.

CAN YOU ANSWER THESE?

1. Is the speed at which sound travels in water higher or lower than that at which it travels in air?
2. Does the speed of sound change with height – if so, why?
3. At what part of a wing does a shock wave first form?
4. What is the buffet boundary of an aircraft?
5. What is a Mach Number, a critical Mach Number, and a Machmeter?
6. How does the appearance of a shock wave on a wing affect the pressure distribution over the wing?

CHAPTER 12

FLIGHT AT SUPERSONIC SPEEDS

INTRODUCTION

As explained in the previous chapter subsonic, transonic and supersonic flight merge one into the other, and it is not easy to define where one ends and the other begins. But the change from subsonic to transonic does at least involve some outward and visible signs – in the laboratory there is the formation of shock waves, and in flight there is the shock stall with its varying effects according to the type of aircraft – whereas the change from transonic to supersonic is not accompanied by any such signs whether in the laboratory or the air, so the dividing line between the two is even more vague. In general we can only say that supersonic flight begins when the flow over all parts of the aeroplane becomes supersonic. But at what Mach Number does that happen? Does it in fact ever happen? Are there not always likely to be one or two stagnation points? And what about the boundary layer where the flow near the surface is certainly subsonic? Perhaps after all it is better to say at about $M1.2$, or 1.5, or maybe 2? Fig. 12A illustrates an aircraft designed for supersonic speeds.

SUPERSONIC SHOCK PATTERN

We have already had a look at the supersonic shock pattern in Fig. 11.12e, and except that the angles of shock waves become rather more acute as the speed increases there is very little change in this pattern over the supersonic range of speeds. To see why the angles of the shock waves change, we must understand the meaning and significance of the **Mach Angle**.

MACH ANGLE

Figure 11.2 illustrated the piling up of the air in front of a body moving at the speed of sound, and explained how the incipient shock wave is formed. This incipient shock wave is at right angles to the direction of the airflow, and this means as near as matters at right angles to the surface of a body such as a wing.

Now suppose a point is moving at a velocity V (which is greater than the speed of sound) in the direction A to D (Fig. 12.1). A pressure wave sent out when the point is at A will travel outwards in all directions at the speed of sound; but the point will move faster than this, and by the time it has reached D, the wave from A and other pressure waves sent out when it was at B and C will have formed circles as shown in the figure, and it will be possible to draw a common tangent DE to these circles – this tangent represents the limit to which all these pressure waves will have got when the point has reached D.

Now AE, the radius of the first circle, represents the distance that sound has travelled while the point has travelled from A to D, or, expressing it in velocities, AE represents the velocity of sound – usually denoted by a – and AD represents the velocity of the point V.

So the Mach Number $M = \dfrac{\text{Speed of point}}{\text{Speed of sound}} = \dfrac{V}{a}$

(as illustrated in the figure this is about $2\frac{1}{2}$).

The angle ADE, or α, is called the **Mach Angle** and by simple trigonometry it will be clear that

$$\sin \alpha = \frac{a}{V} = \frac{1}{M}$$

in other words, **the greater the Mach Number the more acute the angle α.** At a Mach Number of 1, α of course is 90°.

Fig. 12.1   Mach angle

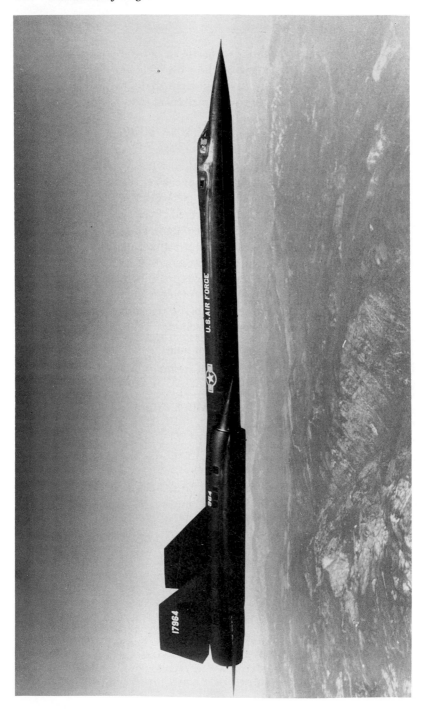

If the moving point is a solid 3-dimensional body, such as a bullet, a complete cone – called the **Mach Cone** – will be formed, the angle at the apex being 2α. If the moving point represents a straight line such as the leading edge of a wing, a **wedge** will be formed, again with an angle 2α at the leading edge.

The tangent line DE is called a **Mach Line**, and it clearly represents the angle at which small wavelets are formed; the velocity of the airflow can even be calculated by measuring the angle on photographs of the wavelets.

Again the hydraulic analogy may be useful, since similar effects are seen when a ship passes through water or a thin stick is placed in a fast-moving stream of water. Only the region within the wedge formed by the bow waves is affected by the stick; the water outside this region flows on as if nothing was there. And the faster the flow, the sharper is the angle of the wedge.

It might be thought that the Mach Line represented the inclination of the shock waves – **but this is not so**. Disturbances of small amplitude travel at the speed of sound, but **shock waves, which are waves of larger amplitude, actually travel slightly faster than sound**, and therefore **they form at a rather larger angle to the surface**. This fact is difficult to explain without going into the mathematics of fluid flow, which is quite beyond the scope of this book, but the following explanation of how shock waves are formed may help us to understand how their slope is determined.

Imagine a supersonic flow of air over a flat surface. This surface can never be perfectly smooth, and may be considered as consisting of a very large number of particles or slight bumps. At each of these bumps a Mach Line will be formed; its angle to the surface depending on the speed of the flow in accordance with the formula $\sin \alpha = 1/M$.

If the speed of flow remains constant, the Mach Lines will all be parallel as in Fig. 12.2*a*.

If the speed of flow is accelerating, the Mach Lines will diverge as the angle becomes more acute with the increasing speed (Fig. 12.2*b*).

But if the speed of flow is decelerating the Mach Lines will converge, add up as it were, and form a more intense disturbance or wave, one of greater amplitude (Fig. 12.2*c*).

Fig. 12A   Flight at supersonic speed (opposite)
(By courtesy of the Lockheed-California Company, USA)
The SR-71 Blackbird was capable of flight at Mach numbers in excess of 3. Delta wings with high leading-edge sweep, lifting chines on the forward fuselage, and two turbo-ramjet engines: bypass turbojets that effectively function as ramjets in high-speed flight.

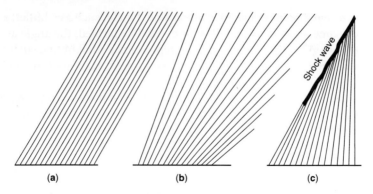

Fig. 12.2   Mach lines and shock wave
(*a*) Speed of flow constant.
(*b*) Accelerating flow.
(*c*) Decelerating flow.

This is one way of explaining how a shock wave is formed at all, but it also gives some indication of how its slope is determined. Unfortunately it is not very convincing from this point of view, and it could even be argued that the slope of the shock wave is less steep than **some** of the Mach Lines. Also, is a shock wave formed because the air is slowing up, or is it the shock wave that slows up the flow?

So perhaps we must fall back on the argument that shock waves travel faster than sound, and even more on the **fact** that the shock wave **is** at a steeper angle than the Mach Lines, for very fortunately that it is a fact can easily be seen on photographs; e.g. in Fig. 11B it is clear that the **bow and tail main shock waves are at a coarser angle than the small wavelets which result from small pressure disturbances due to the surface roughness.**

Other examples of the practical effect of large disturbances travelling faster than the speed of sound are the hearing of sonic bangs before the noise of the aircraft, and the way in which the shock wave of an explosion is followed by the other noises.

But has all this got any practical significance in aircraft design? Yes, as a matter of fact, it has; but to understand what the practical significance is we must study the nature of supersonic flow.

SUPERSONIC FLOW

There are fundamental differences between supersonic flow and subsonic flow, and perhaps these differences are best illustrated by the different ways in which the two kinds of flow turn corners, or − what comes to much the same thing – pass through contracting or expanding ducts.

Although we may not have put it in that way we have already studied this in the case of subsonic flow; and perhaps we may sum up the results by saying –

1. That subsonic flow **anticipates** the corner or whatever it may be, and so **the pattern of the flow changes before the corner is reached.**

2. That the **change of flow** takes place **gradually** on **curved paths.**

3. That at what we might call a **concave corner**, or in a **contracting duct**, the **flow speeds up** and the **pressure falls.**

4. That at what we might call a **convex corner**, or in an **expanding duct**, the **flow slows down** and the **pressure rises.**

There are of course complications (as, for instance, if we overdo the suddenness of the change and the flow breaks away from the surface), but in the main we have established these four principles, and have seen numerous examples of how they are applied in practice.

Now let us look at supersonic flow.

We have already made it sufficiently clear that the first principle is different – and we have explained why it is different. **Supersonic flow does not, and cannot, anticipate a corner or anything else that lies ahead, because there is no means by which it can know that it is there.**

What, then, of the other three principles?

COMPRESSIVE FLOW

Let us consider first what happens when supersonic flow meets what we have called a concave corner, or putting it more practically, a sharp, small-angled wedge. One way of describing this kind of corner is to say that if the flow were to go straight on it would intersect the body (Fig. 12.3).

Figure 12.4 shows what happens. The flow will in fact go straight on

Fig. 12.3   Compressive flow – concave corner

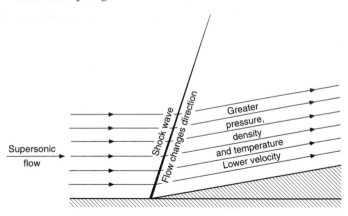

Fig. 12.4   Compressive flow

until it hits something – but what it hits will not be the wedge itself, but the **shock wave** which is formed by the slowing up of the flow as a result of the point of the wedge being inserted in the flow, and the consequent converging of the Mach Lines (perhaps that wasn't such a bad explanation after all).

In this type of flow there will be an **inclined** or **oblique shock wave**. Now it has already been explained that a shock wave at right angles to the flow causes a sudden reduction in the speed of flow, but a shock wave oblique to the flow causes **both a reduction in the magnitude of the velocity**, and a **change in its direction**. The change of direction is a result of the fact that it is only the component of the velocity at right angles to the shock wave which is reduced; the other component (along the shock wave) remains unchanged in passing through the shock wave. This is illustrated in Fig. 12.5, and it is clear that the new direction of flow will be parallel to the new surface.

So the flow has turned the corner; **the change of direction was sudden** and **occurred entirely at the shock wave**. The flow after the corner is at a reduced velocity (though it may still be supersonic), the lines of flow are closer together, the pressure is higher, the density is higher (the air is compressed, possibly quite appreciably), and the temperature is higher. The Mach Lines, at the lower speed, will be more steeply inclined to the new surface.

Supersonic flow most commonly compresses through a shock wave; and at the leading edge of a wing, or the nose of a body, or at the mouth of a contracting duct, there is – as at this wedge – no gradual change of pressure as with subsonic flow, but a **sudden rise in pressure, density, and temperature, and a sudden fall in velocity**. This type of flow is called **compressive flow**. By very careful

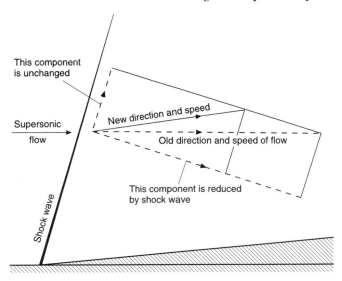

Fig. 12.5 **Change of direction and speed of flow**

design it is possible to obtain a gradual compression by avoiding the conditions where the Mach lines coalesce (Fig. 12.2(c)). The shock compression is, though, much more usual.

EXPANSIVE FLOW

There is, however, another way in which a supersonic flow can turn corners.

To understand this let us consider what happens at a convex corner, i.e. one at which, if the flow were to go straight on, it would get farther away from the surface (Fig. 12.6).

Figure 12.7 shows the result, though it must be admitted that it

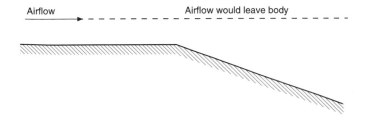

Fig. 12.6 **Expansive flow – convex corner**

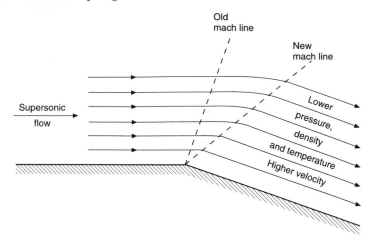

Fig. 12.7   Expansive flow

doesn't really indicate the reason for what happens – this unfortunately can only be given in a mathematical treatment of the subject.

As we can see from the figures the supersonic airflow, on meeting a corner of this type, is free to expand; this it does, becoming **more rarefied, i.e. decreasing in density** in the **decreased pressure**, and the **lines of flow are therefore farther apart**, and the **temperature also falls** as is usual in an expanding flow. The speed on the other hand **increases**.

So far it seems very similar to compressive flow – except that all the opposites happen! There is, however, another fundamental difference which is illustrated on the figure though it may not be immediately obvious. The dotted lines indicate the slopes of the two Mach Lines, the first one for the velocity of flow before the corner, the second one after the corner; notice that the second one is at a more acute angle to the new surface than the first one is to the original surface, i.e. that the angle between the Mach lines is greater than the change of angle of the surface – this of course is because the velocity after the corner is greater than before the corner. But more important than this is to notice that between the Mach Lines **the flow changes gradually on a curved path**, not suddenly as at a shock wave, because the Mach lines no longer converge as in Fig. 12.2c, but, on the contrary, they now diverge.

This gradual change of flow helps to emphasise the fact that the Mach Lines are nothing like shock waves; for there is some danger that even the dotted lines in the figures may suggest that they are. If we go back to the original explanation of Mach Lines, it will be realised that, in comparison with shock waves, they are small weak waves that may appear anywhere along the surface, not just at corners, and that the flow passes through them without sudden changes in its direction or physical properties.

This type of flow is called **expansive flow**, and the phenomenon at the corner which causes the flow to change is sometimes called an **expansion wave**.

It should be noted that although the change at an expansion wave is gradual when compared with that at a shock wave, it still takes place over a very short time and distance compared with subsonic flow in which things happen long before – and long after – the corner is reached.

It should be noted, too, that although there is a limit to the angle through which supersonic flow can be turned at one expansion wave, it is possible to turn it through a large angle by a succession of expansion waves. In fact, of course **a curved surface is an infinite series of corners**, and over a convex curved surface there will be a succession of expansion waves, and the changes in direction of the flow, and in pressure, density, and temperature will be **even more gradual**, though still, unlike subsonic flow, being confined to the passage over the surface itself, and not before or after. So, although supersonic flow **can** turn sharp corners, it should not be thought that it cannot also be persuaded to pass over curved surfaces, and this is a very good thing because even supersonic aeroplanes sometimes have to fly slowly and curved surfaces are very much better for low speeds.

SUPERSONIC FLOW OVER AN AEROFOIL

We are now in a position to look at the supersonic flow over an aerofoil – but what is to be the shape of our supersonic aerofoil?

Since straight lines and sharp corners seem to be at least as good as curves, the simplest aerofoil section for supersonic speeds would seem to be a flat plate inclined at a small angle of attack, and there is no doubt that if it were possible to give adequate strength to such a plate it would be the obvious answer.

If the plate were thin enough the flow would be undisturbed at zero angle of attack – and of course there would be no lift. At a small angle, on the top surface there would be an expansion wave at the leading edge and a shock wave at the trailing edge, and on the bottom surface a shock wave at the leading edge and an expansion wave at the trailing edge, the flow being as in Fig. 12.8.

On the top surface, owing to the expansion wave at the leading edge, the flow would be speeded up and there would be a decreased pressure;

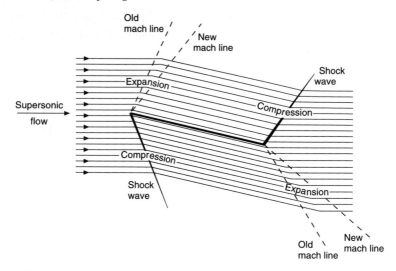

Fig. 12.8   Supersonic flow over flat plate at small angle of attack

on the bottom surface, owing to the shock wave at the leading edge, the flow would be slowed down, and there would be an increased pressure. So there would be lift – and drag.

But a flat plate is clearly not a practical proposition, so let us have a look at a shape that is – the double-wedge.

First let us see what happens at zero angle of attack (Fig. 12.9).

The pattern of shock waves and expansions will, of course, be the

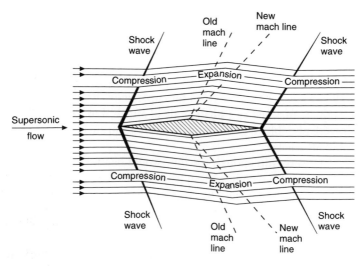

Fig. 12.9   Supersonic flow over double-wedge at zero angle of attack

same on both top and bottom surfaces. At the nose (which corresponds exactly to the small-angled wedge of Fig. 12.4) there will be shock waves, and the consequent increases in pressure, density, and temperature, and a decrease in velocity; at the point of maximum camber there will be the expansions, and the corresponding changes over the rear portion of the aerofoil; at the trailing edge, though not perhaps quite so obvious, there will again be the wedge effect, shock waves on both surfaces, and again the changes of pressure, density, temperature, and velocity.

At very small angles of attack, for reasons that should now be quite clear, the bow shock wave on the upper surface becomes less intense and that on the lower surface more intense; the tail shock wave, on the other hand, becomes more intense on the upper surface and less intense on the lower surface. But one of the most interesting, and perhaps surprising, features of the flow is that there is no upwash in front of the aerofoil (how can there be when the airflow doesn't know that the aerofoil is coming?) – and no appreciable downwash behind the aerofoil; the deflection of the air (the eventual cause of the lift) is only between the shock waves. The pressure distribution over the aerofoil accounts for both lift and drag – as it did with the flat plate, which also caused neither upwash nor downwash. After all, a speedboat travelling through the water causes a considerable depression (and rising) of the water as it passes, but it does not leave a permanent dent!

When the angle of attack reaches that at which the front portion of the top surface is parallel to the approaching airflow (an important condition because it gives the maximum lift/drag ratio for this type of aerofoil) the bow shock wave on the upper surface, and the tail shock wave on the lower surface, both disappear – as one would expect (Fig. 12.10).

At a still larger angle – but the reader may like to draw this for himself. Eventually, as the angle of attack is increased, the bow wave will become detached, as it always is in front of a blunt nose.

The reader should have no difficulty in sketching the flow patterns for other shapes (such as those in Figs 12.20, 12.21, and 12.22), and at various angles.

CONVERGENT–DIVERGENT NOZZLE

Let us now consider the flow through a contracting-expanding duct, a convergent-divergent nozzle, a de Laval nozzle – at any rate a constriction in a duct; or, as we called it in the low speed case, a venturi meter.

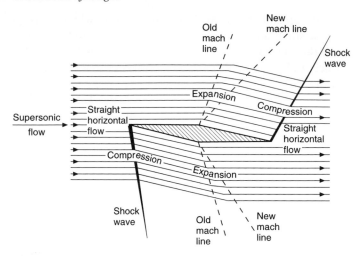

Fig. 12.10   Supersonic flow over double wedge at angle giving maximum
L/D

It is true that there are variations in the shapes of these devices according to what they are required to do, and according to the speed of flow with which they have to deal, but all have this in common – that they converge from the inlet to a throat, then diverge to the outlet.

We are familiar by now with the picture which illustrates subsonic flow through a venturi tube (Fig. 2.19); in the contracting portion the streamlines converge, the airflow accelerates, the pressure falls and, although the change in density is small (so small that we could then afford to neglect it), it is a decrease, i.e. the air is rarefied. In the expanding duct, beyond the throat, the streamlines diverge, the airflow decelerates, the pressure rises, and such small change in density as may occur is an increase, i.e. the air is compressed.

What happens if we gradually speed up the flow through the venturi tube? The reader can probably answer that from what we have discovered first about transonic flow, then about supersonic flow.

The first thing that will happen is that when the airflow, at the throat, reaches the speed of sound, **a shock wave will be formed just beyond the throat** just as it was on the camber of an aerofoil; so instead of the gradual rise in pressure and fall in speed beyond the throat, **there will be a sudden rise in pressure, a sudden fall in speed, and the flow behind the shock wave will become turbulent** (Fig. 12.11). The front portion of the venturi will still function reasonably like a venturi, i.e. the velocity will increase and the pressure fall, but the rear portion will no longer serve its proper purpose.

Fig. 12.11 Incipient shock wave in a venturi tube

It is interesting to note that if the upstream flow is subsonic then at the throat the speed can never be greater than the speed of sound. If the flow does reach this speed it may then accelerate to supersonic speed downstream of the throat (Fig. 12.11). In this condition the duct is said to be choked. If the downstream pressure is reduced the supersonic region extends and the shock moves out of the divergent section.

If the flow in the duct ahead of the contraction is supersonic, we find that the flow behaves in the opposite way! This time the speed reduces in the convergent section and increases in the divergent section, reaching a minimum value at the throat (Fig. 12.12). The increase in speed is still, however, accompanied by a decrease in pressure and vice versa.

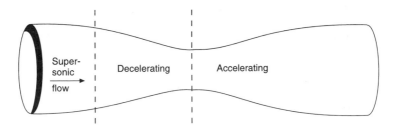

Fig. 12.12 Contracting-expanding duct – supersonic upstream flow

EXPANDING–CONTRACTING DUCT

The fact that a venturi tube has an effect at supersonic speeds opposite to that at subsonic speeds leads one to wonder whether **we could not get the venturi effect at supersonic speeds by having a duct shaped the opposite to that of a venturi tube, i.e. by first expanding and then contracting – and the answer, of course, is yes.**

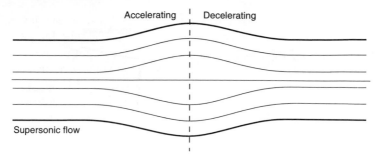

Fig. 12.13   Supersonic flow through an expanding-contracting duct

Figure 12.13 illustrates supersonic flow through an expanding-contracting duct.

The purpose of ramjets and jet engines is to provide the thrust to propel the aeroplane or missile, and it can only do this if the velocity of outflow from the engine is greater than the velocity of the aeroplane or missile through the air. The air enters the ramjet or turbojet at the inlet where it arrives with the velocity of the aeroplane; if this is above the speed of sound we can by a clever arrangement of a centre body in the inlet (Fig. 12.14) cause shock waves to be formed here and so put up the

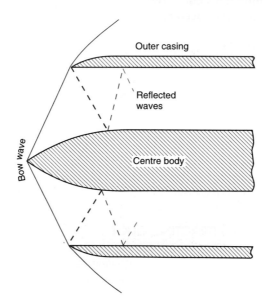

Fig. 12.14   Centre body at the inlet of a ramjet or turbine
Since the angle of the bow wave will depend on the Mach Number, the centre body must be movable to be fully effective.

pressure which, in the case of the turbojet, is further increased by the compressor itself. The air then speeds up in the expanding duct, and the burning of the fuel adds still further to its energy. When the gases leave the jet pipe a system of shocks and expansion wave will form in the emerging jet if the pressure is not matched to that of the atmosphere at exit, resulting in losses and consequent inefficiency.

But we are not yet beaten. If we now add a divergent nozzle to the contracting duct (Fig. 12.15) we get at the throat an expansion wave

Fig. 12.15   A de Laval nozzle

which is reasonably gradual and, after it, a decrease of pressure more gradually to atmospheric, together with an increase of velocity – which is just what we wanted. It is in this form that the convergent-divergent nozzle is sometimes referred to as a de Laval nozzle after the famous turbine engineer of that name.

SUBSONIC AND SUPERSONIC FLOW – A SUMMARY

Now, I think, we are in a position to sum up the essential differences between subsonic and supersonic flow in contracting and expanding ducts, and in similar circumstances such as over aerofoils or other bodies.

	In a Contracting Duct	*In an Expanding Duct*
Subsonic Flow	**Flow accelerates** **Air rarefies slightly** **Pressure falls**	**Flow decelerates** **Air is compressed slightly** **Pressure rises**
Supersonic Flow	**Flow decelerates** **Air is compressed** **Pressure rises**	**Flow accelerates** **Air is rarefied** **Pressure falls**

In short, everything in supersonic flow is exactly the opposite to subsonic flow **with one important exception – in both cases**

increasing speed goes with decreasing pressure. So Bernoulli's principle, which at low speeds is really the conservation of energy, has still some significance (though modifications are needed before it can be applied quantitatively in compressible flow).

Notice, too, that what happens in supersonic flow is what we said in an earlier chapter was what common sense might lead us to expect – a decrease of speed and compression at the throat of a venturi. It is some measure of our learning and understanding of the subject if by now this is no longer common sense!

BOUNDARY LAYER AND SUPERSONIC FLOW

It may be noticed that in all this we have said little or nothing about the boundary layer, and it was the boundary layer that caused all the trouble in subsonic flow when it came to corners, and it was the boundary layer that was so important in transonic flow when the incipient shock wave was formed, and for which we had to think of such devices as vortex generators.

The truth is that the **boundary layer is relatively unimportant in supersonic flow**; it is **thin** and the **viscous forces within it are relatively small**. This largely accounts for the ability of supersonic flow to turn sharp corners.

Curiously enough, at the even higher speeds which will be mentioned later in this chapter, the boundary layer thickens again, and once more becomes significant. So the supersonic region is especially privileged in this respect, and in many respects the theory of the flow is simpler than over any other range of speeds.

SUPERSONIC WING SHAPES – PLAN FORM

In flight at subsonic speeds the shape of the aerofoil section is more important than the plan form of the wing, but at supersonic speeds it is the plan form which is the more important (Fig. 12B). On the other hand, the more one studies the seemingly endless variety of both aerofoil section and plan form that are not only possible but seem to have proved successful in supersonic flight, the more one is forced to the conclusion that neither shape matters very much; supersonic flow is more accommodating than subsonic flow, less fussy in what it encounters, and although, compared with subsonic flow, the lift coefficient is less, the drag coefficient greater, and the L/D ratio in consequence lower, the actual values of C_L, C_D, and L/D, and the position of the centre of pressure seem to be little affected by the shapes of either the cross-section or the plan form of the wing.

Let us consider first the plan form. It will be remembered that in the transonic region there was advantage in a considerable degree of sweepback of the leading edge because it delayed the shock stall, the increase of drag, buffeting, and so on – in other words, it raised the critical Mach Number. It is often stated that there is no advantage in sweepback after the critical Mach Number has been passed, and that straight wings are better for supersonic flight. This might be true if the **only** effect of sweepback was to delay the critical Mach Number – but actually it does more than this.

Consider, for instance, the plan shapes A, B, C, and D (Fig. 12.16);

Fig. 12.16　Supersonic wings – plan shapes

with the possible exception of B, all these have been used on high-speed aircraft. At the apex of each are shown the Mach Lines for a Mach Number of about 1.8, and it will be noticed that the leading edges for these shapes all lie within the Mach Cone, and this in turn means that the airflow which strikes the wing has been affected by the wing before it reaches it; if, as is probable, there are also shock waves at the nose of the aeroplane, or at the apex of the wing, the whole of the leading edge of the wing will be behind these shock waves and so will encounter an airflow of speed lower than that of the aeroplane. This airflow may not be actually subsonic, but at least the resolved part of it at right angles to the leading edge, or across the chord, is likely to be. So although a swept-back wing is better than an unswept wing in the transonic region, it may retain some of its advantages even into the supersonic region – and this applies particularly to thick wings which are naturally more prone to the formation of shock waves.

Of course, if we are to keep within the Mach Cone the sweepback must increase with the Mach Number, until eventually the delta shape may be more appropriately described as an arrow-head shape (Fig. 12.17).

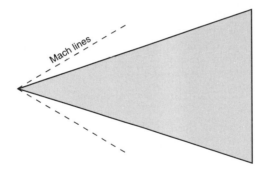

Fig. 12.17   Arrow-head delta wing for high Mach Numbers

But whenever we discuss the advantages of sweepback we must never forget its disadvantages which are largely structural; the twisting and bending stresses on a heavily swept-back wing give many headaches to the designer and mean extra weight to provide the strength. But there is also the old bogey of tip stalling and lateral control near the stalling – and landing – speed. Shapes A and B are better structurally than C and D, they are better, too, from the point of view of tip stalling; they also have an interesting, though perhaps rather concealed, advantage in that owing to the long chord the wing can be **thick** (which means a good ratio of strength to weight), yet still **slim** as regards **thickness/chord ratio** (which is what matters as regards shock drag). Have C and D then no advantages? – it would be strange if they hadn't, because they are in fact more common in practical aeroplanes than B, until recently than A also, **but A is becoming increasingly popular in modern designs for supersonic transport**. The advantage of C and D lies chiefly in lower drag (in spite of the point mentioned above), and so in better lift/drag ratio; they are also more suitable for the conventional fuselage and control system (if that is an advantage), and for engine installation.

One rather unexpected bonus resulting from the use of delta wings, or others with extreme taper and sweepback of more than 55° or so, comes from the stall itself; this is a leading edge stall which starts at the wing tip and progresses gradually inboard, the separation bubble is then swept back with the leading edge and shed as a trailing vortex, tightly rolled up and **with a very low pressure at its core**. The low pressure acts on the forward facing parts of the upper surface of the wing giving a **'form thrust'** (in effect a negative drag) and a **lift boost**; moreover the

Fig. 12B   Supersonic configuration (opposite)
(By courtesy of British Aerospace Defence Ltd, Military Aircraft Division.)
The Eurofighter 2000 with delta wing and canard foreplane for control.

flow in the core is **stable and causes little buffeting,** unlike the separation vortex on wings with sweepback of less than 50°. This is, in fact, an effective way of producing lift. Concorde uses it at both sub and supersonic speeds. The use of fences, saw teeth and vortex generators can, at best, only give partial mitigation of the resulting stalling phenomena such as the buffeting, wing drop and pitch up.

But whatever the pros and cons of sweepback there is no doubt that there is a lot to be said for the straight rectangular wing for really high supersonic speeds (Fig. 12C). With the small aspect ratio, and tremendously high wing loading associated with such speeds, the wings are very small anyway, and from the strength point of view a rectangular wing, or a wing that is tapered for structural reasons rather than for aerodynamic reasons, will probably win the day.

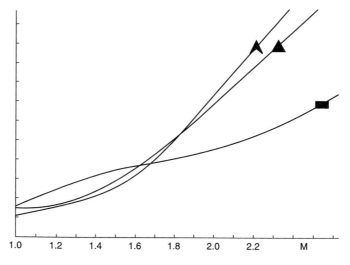

Fig. 12C   Wing shapes and drag
How the drag increases with Mach Number for straight, highly-swept and delta wing shapes.

Figure 12.18 shows how only small portions of a rectangular wing at supersonic speeds (the shaded areas) can know of the existence of the tips, and these portions will tend to exhibit the normal characteristics of 3-dimensional subsonic flow, wing-tip vortices, etc., while the flow over the remainder of the wing will be straight 2-dimensional flow as if the wing was of infinite span and there were no wing tips. This leads to a rather obvious suggestion – cut off the shaded portions.

This, in fact, is sometimes evident in design, but the arguments one way and another, for sweep at leading or trailing edge, (or both, or

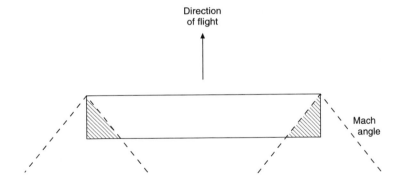

Fig. 12.18 Rectangular wing at high Mach Numbers

neither), for delta and arrow plan forms, for tail first or wing first, for tail to be larger or smaller than the wing, even to decide which is the wing and which the tail – that these arguments are endless is clearly evident from the numerous shapes and configurations which have been tried or suggested for missiles or supersonic aeroplanes.

The fundamental difficulty, for aircraft rather than missiles, is to provide wings that are suitable not only for supersonic flight, but also for subsonic and transonic flight. After all, supersonic aeroplanes have to take off and land; and they also have to pass through the transonic region. The real answer – so far as plan form is concerned – is surely in **variable sweep** (Fig. 12.19); advocated many years ago by an eminent British inventor but, as so often happens, left to others to put into practice; the Americans had serious teething troubles with their first real effort in this direction, the General Dynamics swing-wing F1-11, of

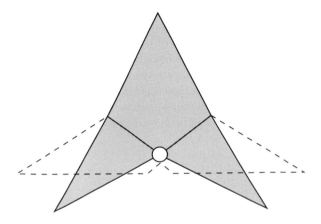

Fig. 12.19 Variable sweep

which so much was expected. Perhaps as a result of this disappointment, the Boeing idea of a swing-wing supersonic transport was also abandoned. Nor, in the meantime, were the French initially successful with their Mirage G (Fig. 12D), but accumulated experience pays off and the variable sweep concept has been adopted for a number of aircraft including the Tornado.

The design process is finely balanced though, and a great number of solutions to the problem of supersonic flight abound, such as the simple long-nosed delta configuration of the Anglo–French Concorde (Fig. 12E).

SUPERSONIC WING SHAPES – AEROFOIL SECTIONS

What then of the shape of the aerofoil section?

The proviso here is that it must be thin – or to be more correct, that it must have **a low thickness/chord ratio** – but apart from this it doesn't seem to matter very much. **Straight lines are as good as, or better than curved surfaces**; and there is **no objection to corners**, even sharp corners – within reason.

A flat plate would make an excellent supersonic wing section, but would not have the necessary stiffness or strength; and the easiest way to make it a practical proposition is to thicken it somewhere in the middle, and thickness in the middle leads naturally to the double-wedge, or rhombus shape, which we have already discussed (Fig. 12.9), and which is as good as any other supersonic aerofoil section.

It makes little difference whether the thickest point is half-way back, or more or less; there is little change of drag for x/c ratios between 40 per cent and 60 per cent (Fig. 12.20), and the lift and centre of pressure positions are not affected at all. But we have always got to consider flight at subsonic speeds and, from this point of view, maximum thickness should be at 40 per cent of the chord rather than farther back; from this point of view, too, it may pay to round the corners slightly.

A variation of the double-wedge is the hexagonal shape (Fig. 12.21). This gives greater depth along the chord and so greater strength, and also makes the leading edge rather less sharp, which has advantages both as regards strength and, as will be considered later, aerodynamic heating.

A bi-convex wing is also quite good (Fig. 12.22), and this is better than the others at subsonic speeds. A bi-convex wing has about the same drag as a double-wedge with maximum thickness rather outside the best range, i.e. at about 25 or 75 per cent of the chord.

Enough has been said about supersonic aerofoil sections to make it clear why the sections in Appendix 1 are all of subsonic type; there

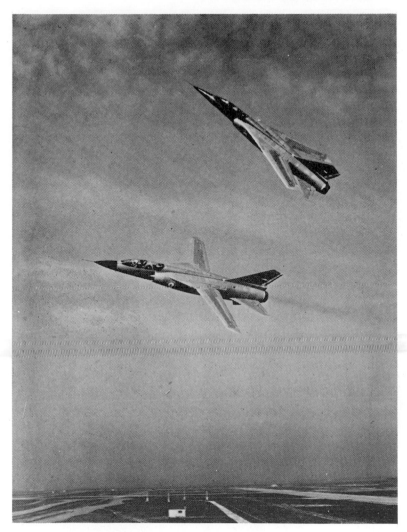

Fig. 12D   Variable sweep
(By courtesy of Avions Marcel Dassault, France)
The Mirage G, with wings folded and 70° sweepback, wings extended and 20°
sweepback; maximum speed in level flight Mach 2.5, landing speed 110 knots; no
ailerons, lateral control (wings back) by differential action of slab tail plane
supplemented when wings are spread by spoilers on wings.

would not be much point in giving the contour of a flat plate, or even of
a double-wedge – and moreover there is little difference between the lift
and drag coefficients of all reasonable shapes, and still less difference in
the positions of centre of pressure.

Fig. 12.20   *x/c* ratio

Fig. 12.21   **Hexagonal wing section**

Fig. 12.22   **Bi-convex wing section**

Theory predicts a maximum value of L/D of 12.5 for a wing with a thickness/chord ratio of 4 per cent at a Mach Number above about 1.3. (Note that in this statement there is **no reference to the shape of the wing, or where is the greatest thickness**.) This value is inferior to L/D ratios for subsonic wing shapes (only about half), but it is reasonably economical when everything is taken into consideration. The lift coefficient is the same for all the shapes, and although it is smaller than those of subsonic aerofoils this does not matter at high speeds; where it does matter is that it means high stalling and landing speeds, which in turn mean long runways, and devices such as tail parachutes to help reduce the speed after landing. A leading-edge flap, or a permanent droop at the leading edge (sometimes called a droop-snoot), will appreciably lower the landing and stalling speed of a supersonic aerofoil

Fig. 12E   **Concorde (opposite)**
(By courtesy of the British Aircraft Corporation)

section. As with plan shape the only way of making an aerofoil suitable for subsonic, transonic, and supersonic flight is to make it **variable in shape**; but in this case we know that it can be done because, in fact, it has long been done – so the only question is the best way of doing it.

Figure 12.23 shows the pressure distribution, and position of the centre of pressure, for three shapes at a small angle of attack. Comment would be superfluous.

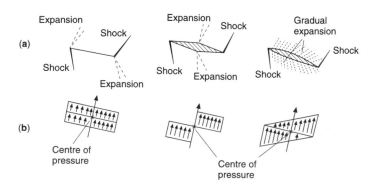

Fig. 12.23   Supersonic pressure distributions
(*a*) Pattern of shocks and expansions.
(*b*) Pressure distribution and centre of pressure.

SUPERSONIC BODY SHAPES

The considerations which decide body shapes for supersonic speeds are similar to those which apply to wing sections. Bodies should be slender, but there are limits in practice owing to the need for room inside the body, for stowage, etc. All in all, the optimum fineness ratio is about 6 to 8 per cent.

Also for reasons of stowage and body capacity there are advantages in curves rather than straight lines, and in a rounded nose and tail (Fig. 12.24*a*). From the drag and speed point of view the nose should be sharp pointed, and often is; but there are disadvantages – the pilot's view is bad (hence the droop-snoot as on the Concorde), the sharp point is useless for stowage, and the transmission of radar pulses is unsatisfactory. The tail portion can be cut off, like the rear of a bullet, without much loss of efficiency; and this is necessary in any case when the jet or rocket efflux is at the rear of the body (Fig. 12.24*b*).

A body or fuselage of some kind is clearly necessary if the aircraft is to carry pilot, passengers, mail, or goods, and if the wings are to be thin – but **are wings really necessary** at supersonic speeds? Bodies can easily

Fig. 12.24   **Supersonic body shapes**

be designed to give lift (whatever their shape they will give lift at a small angle of attack), but cannot the thrust be used to provide the lift? In earlier chapters we talked of flying wing; why not a flying body?

Well, of course, a rocket can be nothing more nor less than a flying body – and more will be said about rockets in the next chapter – but even rockets need guidance and, within the atmosphere at least, guidance and control are best achieved by fixed and movable surfaces. There is also, so far as aeroplanes are concerned, the not unimportant point of getting back to earth.

But the reason for mentioning this problem is something quite different. **One advantage in having a wing at supersonic speeds is that the presence of the wing improves the lift on the body** – there is interference between wing and body, but it is **useful interference**; and it is mutually useful, because the body produces an upwash which improves the lift of the wing.

There are great possibilities in the exploitation of beneficial interference at supersonic speeds, and it is something which we may hear a lot more about in the future. Another example of it has already been mentioned in connection with putting a centre body at the inlet of a ramjet or jet engine. A suggestion has even been made of beneficial **biplane** effects, by eliminating external shock bow waves, and using the shock between the wings to good effect as in an engine intake – perhaps making the biplane the engine. Who knows?

A form of area rule is still effective in reducing shock drag at supersonic speeds, but its application is rather more complicated. Since the shock waves and Mach Lines are now oblique, instead of being at right angles to the flow, the **'area' which must change smoothly** is not that at right angles to the line of flight, but **in planes parallel to the Mach Lines**. And unfortunately the inclination of the Mach Lines depends on the Mach Number at which the aircraft is flying, so the shape of the aircraft can only be correct for a particular Mach Number.

KINETIC HEATING

We all know that **friction increases temperature**, an example of deterioration of energy from the highest to the lowest form (from mechanical energy to thermal energy), a natural process – and skin friction in the flow of fluids is no exception. We all know, too, that an increase of pressure, as in a pump, raises temperature; another example of the same process – and the stagnation pressure on the nose of a body or wing is no exception. So when an aeroplane moves through the air it gets hot; some parts more than others, some owing to the temperature increase created by skin friction, some owing to that created by pressure.

When, then, do we first come up against this? The answer is – when we first fly! But it isn't serious? No – like many other things, **it isn't serious at low speeds**. It has been said that aeroplanes made of wax melt at 300 to 400 knots, those made of aluminium at 1600 to 1800, those of stainless steel at 2300 to 2400 knots. Aeroplanes are not made of wax (wind tunnel models sometimes have been), but some are made of aluminium alloy, and some of stainless steel, and of other metals (such as titanium) and their alloys, and just because of this very problem. Nor can we afford to go anywhere near the melting point; metals are weakened long before that – and what about the passengers, and crew, and freight?

Bullets and shells certainly travel at speeds where heating is significant, but within limits it doesn't matter whether bullets and shells get hot or not. Also their flight is not usually of very long duration, and it takes time for the surface to heat up. But meteors and satellites re-enter the atmosphere without any means of braking – and we know what happens to them, they get frizzled up. It is true that most manned space-craft have survived re-entry and more will be said about that in the next chapter. Let us at least consider what we can do to reduce heating effects.

A very simple formula $(V/100)^2$, where V is the speed in knots, gives a very fair approximation to the temperature rise in degrees Celsius. So what is merely a rise of 1°C at 100 knots, or 4°C at 200 knots, becomes 36°C at 600, 100°C at 1000, and 400°C at 2000 knots. That is how we discovered that the aeroplane made of wax would melt! Figure 12.25 shows rather more accurately local surface temperatures that may be reached under certain conditions at Mach numbers up to 4; these have been calculated from the formula $t/T = (1 + M^2/5)$ where t is the stagnation temperature, i.e. the temperature of air moving at a Mach Number of M being brought to rest, and T is the local temperature of the air; the figures relate to 8500 m where the local temperature is $-40°C$. The temperatures shown in the graph apply to a laminar

Fig. 12.25   How the surface temperature rises with the Mach Number
The graph relates to a height of 28 000 ft (8500 m) where the local temperature of the surrounding air is −40°C.

boundary layer; the temperatures are rather higher for a turbulent boundary layer. Moreover, at Mach numbers above 2 these surface temperatures may be reached in a matter of seconds, and certainly within a minute or two, unless there is some method of insulation.

Many devices have been tried, and no doubt many more will be tried, in an effort to counter this heating problem. These devices may be classified under the following headings –

(*a*) To insulate the structure from the heat.
(*b*) To use materials which can stand the high temperatures without serious loss of strength.
(*c*) To encourage radiation from the surfaces and so reduce the temperatures.
(*d*) To circulate a cooling fluid below the surface.
(*e*) Refrigeration by any of the normal methods.

As regards materials for the aircraft structure light alloys are suitable for Mach numbers up to 2, or even higher for short periods. Between $M2$ and $M4$ titanium alloy may be the answer, but above 3 or $3\frac{1}{2}$ stainless steel is probably better as being more readily available.

It must be remembered that the crew, the equipment and the fuel must be protected as well as the structure itself, so there is no point in using materials which will stand the high temperatures, unless there is also refrigeration to keep the interior of the aircraft cool.

Perhaps the most ingenious idea is to apply the heat to a suitable working fluid (hydrogen has been suggested), and to eject the expanding fluid through a suitable nozzle, and so propel the aircraft! Ingenious and fascinating – drag produces heat, heat produces thrust to help overcome the drag. In principle it is not impossible.

An interesting aspect of surface heating is the effect of shape. It is the speed of flow adjacent to the boundary layer which is the deciding factor in the temperature rise – and to some extent, of course, the nature and thickness of the boundary layer itself – and the speed of flow depends on the shape of the body. But there is more in it than that. A rise in temperature is created owing to skin friction, and owing to the stagnation pressure, **but it is also created by shock waves**, and whereas the main effect of skin friction in the boundary layer is to raise the temperature of the surface, the main effect of the shock waves is to raise the temperature of the air – and that doesn't matter very much. So from the point of view of keeping down surface temperatures it is better to have wave drag than boundary layer drag. This conclusion isn't very helpful with regard to aircraft in which we try to reduce every kind of drag to a minimum, but it is a most important consideration in designing bodies for re-entry to the atmosphere from space, bodies in which we want drag, but we don't want heating of the surfaces.

Another influence of the heating problem on shape is in the avoidance of sharp edges, which might seem desirable from the flow point of view, but which would be particularly susceptible to local temperature rise and consequent weakening of the material.

Kinetic heating is already a limiting factor in the speed of certain types of aircraft, and it provides a very formidable problem in regard to the re-entry into the atmosphere of space-ships and even of long-range missiles such as will be considered in the next chapter.

STABILITY AND CONTROL PROBLEMS

In the previous chapter we considered some of the problems of stability and control at transonic speeds. Some of these still apply in the supersonic region, and control surfaces should be of the all-moving slab type, and fully power-operated. But whereas in the transonic range this applied mostly to the tail, in the supersonic range it can be applied also in all-moving wing tips to replace conventional ailerons, and even to an all-moving fin and rudder. Since the main plane may be nearly as small as the tail plane, it, too, may be movable to give pitch; in fact in some missiles it is not easy to decide which is the main plane and which the tail plane.

But there are also new problems at supersonic speeds because the inertia forces are so great that it is practically impossible to provide the inherent natural stability of the kind that is associated with such devices as dihedral and fin area. In order to be effective against the inertia forces the surfaces would have to be so large that the cost in weight and drag would be prohibitive – and this applies particularly at great heights where there is so little air density.

Another difficulty, which is especially applicable to military aircraft, is that pilot and crew have so much to do in looking after the equipment that they must be relieved as far as possible of flying the aeroplane.

Of course we have long been familiar with the automatic pilot, but the modern conception is very different from this – nothing more nor less, in fact, than **synthetic stability** and **automatic control**. Is the pilot then necessary at all? Strictly speaking, probably not, the aircraft can be controlled from the ground like a guided missile. But pilots can still do some things that instruments cannot, they can monitor the automatic systems, tell them what to do, investigate any failure and, if necessary, take over control.

CONCORDE

It seems fitting to conclude a discussion on the problems of flight, from subsonic to supersonic, with some comments on the design of the Concorde, because controversy on its cost, on its sonic boom and on its commercial viability have tended to obscure the cleverness of its design. Readers who have followed the arguments put forward in this book will surely be fascinated by some of its outstanding features. It would possibly be going too far to describe one of the main features of such a costly and sophisticated piece of equipment as **simplicity** yet, as we shall see, there is some truth in such a description.

The now familiar ogee (or double curve) plan shape of the wing gives both **a large chord** (27.7 m), **with its advantages**, and **a large span** (25.6 m), **with its advantages**.   •

The large chord means that although the wing is **thin, very thin**, from the aerodynamic (t/c) point of view (only 3 per cent at the root and 2.15 per cent outside the engine nacelles), and so has **low wave drag**; yet at the same time it is deep enough (83 cm) to give the required **strength and structural stiffness**, usually a difficulty with slender swept-back wings. The main advantage of the large span is the **reduction of vortex, or induced, drag** at all speeds.

The **slimness of wings and body** (the Boeing 747 Jumbo Jet is longer, higher and much fatter than the Concorde), and the limitation of speed to just over Mach 2, **have kept down the temperature rise** and made it possible to use **aluminium alloys**, with which we have long been familiar, for most parts of the structure, instead of having to experiment with the more costly, and heavier, stainless steel or titanium alloys. The temperature rise in the structure is also reduced by using the fuel as a heat sink. The maximum landing weight is 1068 kN, and the maximum landing wing loading 4.786 kN/m², **less than that of many comparable aircraft**.

The ogee plan shape has another advantage in that the **stalling angle is so large that it is unlikely to be reached in any ordinary condition of flight**; this is because the shape leads to the formation of leading edge vortices (without any vortex generators!), and so improves the flow in the boundary layer and gives smooth changes of lift and pitching moment with angle of attack.

On the wing there are **no flaps, no slots, no tabs, no spoilers, no saw teeth, no fences**, nor any other devices usually required for such a large speed range as from 65.4 m/s (at 18° angle of attack) to a true speed of 649 m/s. **The only moving surfaces on the wing are the six elevons** (combined ailerons and elevators) which control both rolling and pitching – and very effectively too. The rudder has two sections, but is otherwise simple and conventional.

The large 'leading edge' vortices are useful when landing as lift continues to increase to large values with increasing angle of attack, the **'lift boost'** and the **'form thrust'** already mentioned in connection with highly-swept wings, and the large area of the delta wing gives a considerable **cushioning effect** when near the ground (two reasons for dispensing with flaps).

Perhaps one of the most interesting features, taking us back to one of the earliest ideas for adjusting trim is the **movement of fuel between tanks**, automatically between the main tanks for adjusting the centre of gravity during cruising flight and, under the pilot's control, from forward tanks to a rear tank under the fin during acceleration to supersonic flight when the aerodynamic centre moves back, and vice versa when returning to subsonic flight.

Pilots report that it is **only by looking at the machmeter** that they know when Concorde **is going supersonic**.

FLIGHT AT HYPERSONIC SPEEDS

In concluding this subject we must not leave an impression that once we have conquered the problems of supersonic flight, we have finished. Far from it – no sooner do we learn to get through one region of speeds than another, with new problems to solve, opens up before us.

Rather strangely, too, there does not seem to be much argument about where the change takes place in this case – **above a Mach Number of 5** we talk of **hypersonic speeds** instead of supersonic speeds. It doesn't pay to try to see the sense of this rather extraordinary use of the English language, or perhaps we should say of Latin and Greek, which makes hypersonic superior as regards speed to supersonic; nor shall we get very far if we try to discover just why the Mach Number of 5 is so significant; actually it is a change in the theory of the flow that decides the issue, and that is far beyond the scope of this book.

To us hypersonic flight is simply supersonic flight – only more so. Here we are deep into kinetic heating effects with all the associated problems; at a Mach Number of 5 at about 61 km the temperature given by the formula is over 1000°C and by Mach 15 it has risen to over 10 000°C. The actual temperatures are found to be rather lower, but not so much lower as to give any real comfort!

Mach Lines are inclined at very acute angles; Fig. 12.26 shows the shock waves and Mach Lines over a double-wedge section at a Mach Number of 10, and Fig. 12.27 is sketched from a photograph of a bullet moving at the same Mach Number. These suggest an arrow type of aircraft as being most suitable (Fig. 12.28).

Fig. 12.26   **Hypersonic shock and expansion waves**

Fig. 12.27   **From a photograph of a bullet moving at Mach 10**

Fig. 12.28   Arrow-head: shape of the future?

One feature of hypersonic flow is a thickening of the boundary layer and an increased importance of the nature of the flow within the boundary layer.

Aerofoil shape seems to matter even less than in supersonic flight; lift and drag coefficients tend towards a constant value as the Mach Number increases.

An interesting aspect of this part of the subject is the wide variety of experimental methods used to investigate it – arc-heated jets, gun tunnels, shock tubes, shock tunnels, hot shot tunnels, models moved by rockets or guns, and ballistic ranges; names that are all to some extent descriptive of the methods employed, each of which would need a chapter to itself. Mach Numbers of 15 can be achieved in still air on ballistic ranges, and even more if the projectile is fired against the airflow in a wind tunnel.

But even hypersonic flow is not the end; at Mach Numbers of 8 or 9 something entirely new begins to happen – molecules, first of oxygen, then at even higher speeds of nitrogen, dissociate, or split into atoms and ions, thus changing the very nature of the air and its physical properties (another phenomenon that is experienced by space-ships on re-entry into the atmosphere, and which affects radio communication with them). At this stage there are possibilities of the control of the flow by electro-magnetic devices.

And so it goes on. Nothing, so far, suggests an end. Figure 12.29 shows how speed records have gone up – and up – and up.

LIFTING BODIES

So far as aeroplanes are concerned, the North American X15 was quoted in the last edition of this book as the nearest approach to a piloted hypersonic aircraft – indeed, if we accept $M5$ as the threshold of hypersonic flight, **it was hypersonic** for it achieved a speed of 1860 m/s and a height of 95 940 m as long ago as 1962 – this speed corresponds to a **Mach Number of over 6**. It was launched from a mother craft at 13 720 m and 224 m/s; it had a rocket engine giving a thrust of 260 kilonewtons, and rocket nozzles for space control, but fuel

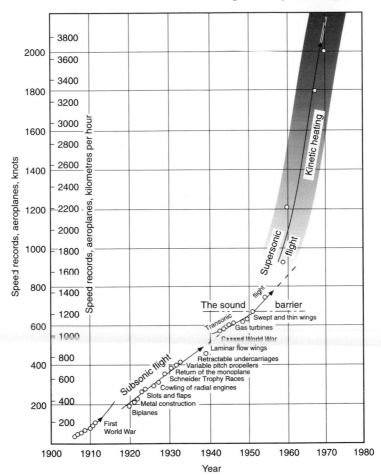

Fig. 12.29 Where do we go from here?

for only 68 seconds at full power. It landed, as an aeroplane, at 134 m/s some 320 kilometres from launch.

The experimental programme with the X15 has now been superseded by a similar programme with what are called **lifting bodies**, with little if anything in the way of wings, with powerful control surfaces and, as their name implies, with bodies so shaped as to give a certain amount of lift at high speeds. As with the X15 these are launched from a mother craft, have rocket power available for a very limited period, and this can be used first to attain considerable height and then, after a fast and steep descent – usually without power – towards the earth, a circling approach at some 15 m/s for the last 6000 m or so, followed by a steep glide, using speed brakes as required, and finally a touch of rocket power to help the round out, and a fast landing at about 100 m/s.

These lifting bodies offer some advantages for future shuttle and sub orbital vehicles.

CAN YOU ANSWER THESE?

1. What is a Mach Angle? a Mach Cone? a Mach Line?
2. Do shock waves travel at the speed of sound? If there is a difference, what effect does it have?
3. What is an expansion wave?
4. Why are sharp leading edges used on supersonic wings?
5. What influence is the heating problem having on the materials used for aircraft construction?
6. What are the main features of the wing used on the Concorde?

For numerical questions on flight at transonic and supersonic speeds see Appendix 3.

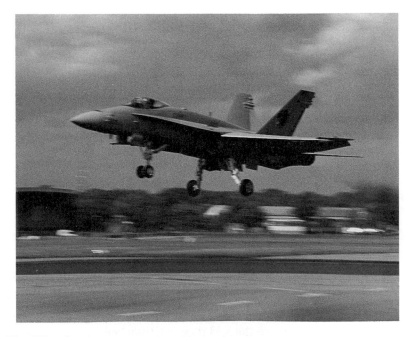

Fig. 12F   Another supersonic configuration
The McDonnell Douglas F-18 Hornet which features a wing with only moderate rearward sweep on the leading edge, and a small forward sweep on the trailing edge. The sharply swept strakes also generate lift, and control the flow over the main wing.

CHAPTER 13

SPACE FLIGHT

BALLISTICS AND ASTRONAUTICS

At first glance it may seem that space flight (Fig. 13A) is purely concerned with ballistics and is completely divorced from aerodynamics, but as we hinted in Chapter 1 at the speeds and heights of modern flight **aeroplanes behave to some extent like missiles, and missiles, given the required thrust and speed, can actually become satellites.** Moreover, both missiles and artificial satellites must pass through the earth's atmosphere when they are launched, and there experience aerodynamic forces, and – what is far more important – the most hopeful means of returning them safely to earth again is to use aerodynamic forces to slow them up and control them.

In short, **the subjects of aerodynamics, ballistics, and astronautics have merged into one subject, the mechanics of flight**, and no apology is needed for the inclusion of this chapter in a book on that subject.

THE UPPER ATMOSPHERE

The atmosphere with which we have been concerned in the flight of aeroplanes – i.e. the troposphere and the stratosphere – is sometimes called **the lower atmosphere**; the remainder is called **the upper atmosphere** (Fig. 13.1).

In the lower atmosphere the temperature had dropped from an average of $+15°C$ (288 K) at sea-level to $-57°C$ (217 K) at the base of the stratosphere, and had then remained more or less constant. The pressure and density of the air had both dropped to a mere fraction of their values at sea-level, about 1 per cent in fact. One might almost be

Fig. 13A   Shuttle lift off
(By courtesy of NASA)

tempted to think that not much more could happen, **but such an
assumption would be very far from the truth.**

There is a lot of atmosphere above 20 km – several hundred
kilometres of it, we don't exactly know, it merges so gradually into space
that there is really no exact limit to it – but a great deal happens in these
hundreds of kilometres. The temperature, for instance, behaves in a
very strange way; it may have been fairly easy to explain its drop in the
troposphere, not quite so easy to explain why it should then remain
constant in the stratosphere, but what about its next move? For from
217 K **it proceeds to rise again** – in what is called the **mesosphere** –
to a new maximum which is nearly as high as at sea-level, perhaps
271 K; then, after a pause, **down it goes again to another minimum**

Fig. 13.1   The upper atmosphere
The figures given are based on the US Standard Atmosphere, 1962, which was
prepared under the sponsorship of NASA, the USAF and the US Weather Bureau.

at the top of the mesosphere. Estimates vary of just how cold it is at this height (only 80 kilometres, by the way, only the distance from London to Brighton), but all agree that it is lower than in the stratosphere, lower, that is to say, than anywhere on earth, perhaps 181 K ($-92°$C). But its strange behaviour doesn't stop at that and, once more after a pause at this level, as the name of the next region, the **thermosphere**, suggests, **it proceeds to rise again**, and this time it really excels itself **rising steadily, inexorably** to over 1200 K at 200 km, nearly 1500 K at 400 km, and still upward in the exosphere until it reaches **over 1500 K at the outer fringes of the atmosphere**.

An interesting point about these temperature changes in the upper atmosphere is their effect upon the speed of sound which, as we learned in Chapter 11, rises with the temperature, being proportional to the square root of the absolute temperature. The interest is not so much in the effects of this on shock waves, or on the flight of rockets, but rather in that one method of estimating the temperatures in the upper atmosphere is by measuring the speed of sound there.

While these strange and erratic changes of temperature have been taking place the density and the pressure of the air have fallen to values that are so low that they are almost meaningless if expressed in the ordinary units of mechanics; at a mere 100 km, for instance, **the density is less than one-millionth of that at ground level**.

It is believed that at these heights there may be **great winds**, of hundreds, perhaps even a thousand kilometres per hour. The air above about the 70 km level is 'electrified' or ionised, that is to say it contains sufficient free electrons to affect the propagation of radio waves. For this reason the portion of the atmosphere above this level is sometimes called the **ionosphere**, which really overlaps both the mesosphere and the thermosphere. Then there are the mysterious **cosmic rays** which come from outer space, and from which on the earth's surface we are protected by the atmosphere, but beyond this we know very little about them except that they may be the most dangerous hazard of all since they affect living tissues. Then there are the much more readily understandable **meteors**, 'shooting stars' as we usually call them, but actually particles of stone or iron which have travelled through outer space and may enter the earth's atmosphere at speeds of 100 kilometres per second, and which have masses of anything from a tiny fraction of a gram up to hundreds of kilograms. The larger ones are very rare, but some of these have actually survived the passage through the atmosphere without burning up, and have 'landed' on the earth causing craters of considerable size – these are called **meteorites**.

To prospective space travellers all this may sound rather alarming, but there are some redeeming features. The winds, for instance, wouldn't even 'stir the hair on one's head' for the simple reason that **the air has practically no density, no substance**. For the same reason the extreme temperatures are not 'felt' by a satellite or space-ship (what is felt is the temperature rise of the body itself, caused by the skin friction at the terrific speeds; it is this which burns up the meteors, it is this which has eventually caused the disintegration of many man-launched satellites on re-entering the atmosphere – but all this has little or nothing to do with the actual temperature of the atmosphere). Then, as regards the very low densities and pressures, no-one is going to venture outside the vehicle, or walk in space, or even put his head out of the window to see whether the wind stirs the hair on his head, unless he is wearing a space-suit, and we have long ago learned to pressurise vehicles because this is required even for the modest heights in the lower atmosphere. Moreover, the strong outer casing of the vehicle which is required for pressurising will in itself give protection at least from the small and common meteors, and to some extent even from the cosmic rays, the greatest unknown. So altogether the prospect is not as bad as it might at first seem to be.

THE LAW OF UNIVERSAL GRAVITATION

Now let us consider the motion of bodies in this upper atmosphere, and in the space beyond.

If we throw a stone or cricket ball up into the air it goes up to a certain height, stops, and then comes down again. If we throw it vertically upwards it comes down on the same spot as that from which we threw it and, if we neglect the effects of air resistance, it returns with the same velocity downwards as that with which we threw it upwards. Moreover, again neglecting air resistance, we can easily calculate how high it will go because we know that (at first, at any rate) it loses velocity as it travels upwards at the rate of 9.81 m/s or, very roughly, 10 m/s every second, and gains it again at the same rate as it comes downwards.

But we ought to know better by now than to neglect air resistance? Yes, we certainly ought to know better, but unfortunately it wasn't only air resistance that we were neglecting in the simple examples in Chapter 1 – though, in fairness, that was our worst error, and our other omissions really were negligible in the circumstances that we were then considering. This is no longer true of the circumstances of this chapter for some of which air resistance really can be neglected, but other things that we calmly assumed most certainly cannot.

Newton would probably have been less surprised than we were when artificial satellites began to circle the globe – for these are but examples of the laws he enunciated.

When he saw the apple drop – assuming that story is true – he wondered why it did so, and eventually decided that it was because there was a mutual force of attraction between the apple and the earth; so it wasn't just a case of the apple dropping, it was the apple and the earth coming together, due to this mutual force of attraction, when there was no longer anything to hold them apart. And if the apple and the earth, why not any mass and any other mass? And so eventually, by observing the facts, and by reasoning, he came to realise that **there is a force of attraction between any two masses, and that this force is proportional to the product of the masses, and inversely proportional to the square of the distance between them. This is the law of universal gravitation,** perhaps the most important of all the physical laws, the law that governs the movement of bodies in space (whether they be natural or artificial), the law that Newton enunciated 300 years ago.

ESCAPE FROM THE EARTH

Consider a stone thrown vertically from the earth's surface with an initial velocity. Because the weight of the stone, caused by the gravitational attraction between the stone and the earth, opposes the stone's motion its velocity will be reduced with time (i.e. the stone has an acceleration g **towards** the earth). With a normal initial velocity gravitational acceleration will stop the stone at a certain height above the earth's surface and it will then reverse direction and accelerate **towards** the earth, eventually arriving at its launch point with its launch velocity but in the reverse direction (if air resistance is neglected).

For moderate initial velocity it is sufficiently accurate to assume that the weight (and hence g) is constant. What happens, though, if the initial velocity is large enough for a great height to be reached before the stone's direction of travel is reversed?

The weight of the stone, in accordance with the law of gravitation, **is inversely proportional to the square of the distance between the masses.** So supposing that the stone has a mass of 1 kilogram it will, at the surface of the earth, have a weight of 981 N. But what if it is moved away from the earth's surface altogether? What if it is thrown upwards 1 kilometre, 100, 1000, 6000 kilometres? Let us pause here for a moment because the radius of the earth is not much more than 6000 km, 6370 km in fact, so at a distance of 6370 km from the earth's surface, the force of attraction, i.e. the weight of the stone, being inversely proportional to the square of the distance from the centre of the earth – now doubled – will only be 1/4 of its weight at the earth's surface; similarly, at 12 740 km it will be 1/9, at 19 110 km only 1/16, and so on (see Fig. 13.2). Notice that in the figure the distances are given from the centre of the earth, and not from the earth's surface; for the earth is a very small thing in space, and if we are to understand the mechanics of space we must think more and more of the mass of the earth as concentrated at its centre.

The mass of the stone of course **does not change**, but as the weight changes so also does the acceleration (g) in proportion – this is just Newton's Second Law again. So the rate at which the stone loses speed on the outward 'flight', though starting at 9.81 m/s^2, gets less and less as the distance from the centre of the earth increases. This makes it more difficult to calculate how far the stone will go with a given starting

Fig. 13.2   How weight varies with height
R = radius of earth, i.e. approx. 670 km

velocity, but it has an even more interesting and important effect than this. For think of the stone returning to earth again; at great distances the rate at which it picks up speed will be very small, but the rate will increase until at the earth's surface it reaches the definite and finite value of 9.81 m/s². This, it will be noticed, is a **maximum** rate of increase, and it can be shown mathematically that even if the stone starts from what the mathematicians call infinity (which means so far away that it couldn't come from any farther!) **the velocity reached will also have a definite and finite maximum value**, which is in fact 11.184 km/s (about 40 250 km/h). So, if a stone is 'dropped' onto the earth from infinity, it will hit the earth at 11.184 km/s; and, by the same token, if it is thrown vertically from the earth at 11.184 km/s it will travel to infinity – **and never return**. This velocity is called the **escape velocity. If it is thrown with any velocity less than this, it will return.**

What happens if it is thrown from the earth at a velocity greater than the escape velocity? Or is this not possible? Yes, it is not only possible, but in a sense it has been done though not quite in this simple way. And all that happens is that it still has a velocity away from the earth when it reaches infinity – and so will go beyond infinity – but since infinity is the limit of our imagination perhaps it will be best to leave it at that. The reader may have noticed that to simplify things, we have only considered the earth's attraction on the stone, and the rest of the universe has been left out! But still the principle is illustrated.

It is important to remember that although the force of gravity, the weight of the stone, gets less and less as it travels farther and farther from the earth it never ceases altogether (at least not until the stone reaches infinity which is only another way of saying 'never'). It is often stated, quite incorrectly, that 'escape' from the earth means getting away from the pull of the earth. This we can never do, the earth 'pulls' on all other bodies **wherever they are** – that after all, is the universal law of gravitation. Why then do astronauts talk about **'weightlessness'?** – and even demonstrate it? – we shall soon see.

In the meantime, it will be noticed that we have already introduced a new unit of velocity, the kilometre per second. Our reason for this is simply one of convenience; in this part of the subject we have to deal with very high speeds, and it is easier to remember these speeds, and even to think what they mean, as kilometres per second, than as so many thousands of knots, or metres per second. At the same time we must remember that our old friend g is still in m/s², so if we wish to use any of the standard formulae of mechanics we must be careful to convert the velocities into metres per second.

PROJECTILES AND SATELLITES

Among the many assumptions so far made one of the most impracticable has been the idea of leaving the earth's surface at speeds of about 11 km/s.

But we all know the answer to this problem now, and it lies in rocket propulsion; by this means the acceleration is so comparatively gentle that it can even be withstood by human beings, at any rate in the lying down position.

When we consider the use of rockets to propel bodies to great heights or into a space a new complication is introduced in that owing to the great rate of fuel consumption **the very mass of the projectile decreases rapidly,** so we have the double effect of mass decreasing with fuel consumption and weight decreasing with distance from the earth. Moreover, in multi-stage rockets, which are the only practical means of achieving the velocities required for launching into space, there is a further decrease in mass each time a stage is completed and a part of the rocket is detached. The final mass that becomes a satellite, or goes off into space never to return, is but a small fraction of the mass at take-off.

In the interests of fuel economy turbojets, or better still ramjets, may be used for the flight of projectiles **while they pass through the earth's atmosphere, but in space rockets are the only means of powered propulsion,** and all journeys in space are dependent on rockets **and the law of universal gravitation.**

In order to get our ideas straight we have so far considered the motion of missiles in a straight line – straight up and down from the earth's surface. We did the same thing in Chapter 1 in dealing with ordinary mechanics, but then we graduated to the much more interesting motion on curved paths; this is what we are going to do now.

What happens if instead of throwing the stone vertically we throw it horizontally? – still neglecting the effects of air resistance.

It will start with no vertical velocity, but will immediately begin to acquire a downward velocity at the rate of roughly $10 \, \text{m/s}^2$ – meanwhile it will retain its horizontal velocity. After 1 sec it will have fallen about 5 m, after 2 sec 20 m, and so on. If its initial horizontal velocity was 100 m/s, and if it was launched from a height of 30 m, its path of travel would be something like (*a*) in Fig. 13.3, or if it was launched from twice the height, like (*b*). If its initial horizontal velocity was 200 m/s, its path of travel would be more like (*c*) or (*d*) respectively.

It will be quite clear from these figures that the **distance it will travel over the ground before striking the ground depends on the height at which it is projected, and the horizontal velocity with**

Fig. 13.3   Bodies launched horizontally – paths of flight

which it is projected (the velocity being more important than the height). If it is projected at ground level it won't get any distance before hitting the ground whatever its horizontal velocity; on the other hand, if it is projected from considerable height, and at a considerable horizontal velocity, it will travel a considerable distance horizontally before reaching the ground. The path of flight, or trajectory, is a mathematical curve called a parabola.

But now we have been guilty of making yet another assumption – has the reader noticed it? In the figures, and in our reasoning, we have assumed **that the earth is flat**. Not just that it is free of hills and dales (these won't affect its flight path but they may obviously affect the point at which it hits the ground), but that the earth is itself a flat plane instead of being spherical or very nearly so. Such an assumption has no practical significance in the flight of stones or cricket balls – but is all-important in the flight of high-speed projectiles. The curvature of the earth will affect the path of the projectile, and the distance it travels before striking the ground, for two reasons – first, because if the projectile didn't 'fall' towards the earth it would go off at a tangent and so get farther and farther away; secondly, because although the force of gravity acting on the body is always 'vertical' in the sense that it is always towards the centre of the earth, the direction of the vertical will change in space and this will change the shape of the curve – in mathematical terms it means that the trajectory is an ellipse rather than a parabola. These effects are shown in the figure (Fig. 13.4) – hopelessly

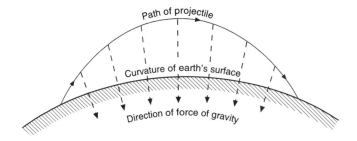

Fig. 13.4   Path of projectile – curved earth

exaggerated of course in the case of a stone or cricket ball, or even an ordinary shell fired from an ordinary gun, but not by any means exaggerated for modern rocket-propelled high-speed ballistic missiles and not, in this figure, even going as far as the man-launched satellite or space-ship.

It should now be clear that **the greater the speed with which a projectile is launched from a given height above the earth's surface in a horizontal direction, the larger will be the curve it describes and so the greater will be the distance it travels before it hits the surface; in other words, the greater will be the range.**

The actual range, of course, depends on the direction of launch as well as the speed and height but at first it is easier to consider only horizontal launches. And by 'launching' we mean that the projectile is given a velocity and then left to itself – and so becomes subject to the mechanics of ballistics. If it continues to be rocket-propelled almost anything may happen!

If we take 'launch' to have this meaning, then the launch conditions can be achieved at the end of the 'launch cycle' involving, usually, rocket propulsion between the earth's surface and the launch point where the rocket is switched off and the body left to the mercy of ballistics.

Let us assume then that we are launching a projectile horizontally from a height of 800 km above the earth's surface. This besides being a nice round figure is well outside any appreciable effects of the earth's atmosphere, and although it is even farther beyond the ceiling of 'aeroplanes' it is within the reach of multi-stage rockets. To those of us who are accustomed to think of heights in thousands of feet or metres it sounds a great height (eight hundred thousand metres), but in terms of space travel it is practically nothing, only one eighth of the radius of the earth away from the surface, about one four hundred and eightieth of the distance to the moon, and one hundred thousandth of the distance to the sun. At this height the weight of a body and the acceleration due to gravity are reduced by about 20 per cent of what they were at the earth's surface.

So much then for conditions at our launching platform, let us now see what happens as we increase the velocity of a horizontal launch. At first the projectile will simply get farther and farther round the earth, but always coming down to earth again at some point less than half way round (Fig. 13.5*a*). **Then at a certain velocity** (about 7.16 km/s or 25 800 km/h) **a most exciting thing will happen** (or at least it would happen if it were not for our old enemy air resistance) – **the projectile will just miss the surface on the far side of the earth and will then gain height again,** and, **perhaps most exciting of all, will circle the earth and come round to where it started – and will then repeat the performance** – and so on. **The projectile has become a satellite – it is travelling round the earth under its own steam as it were** (Fig. 13.5*b*).

Unfortunately it can never happen quite like this because although it was clear of air resistance at the launch, on the far side it would have come right through the atmosphere to ground level and so would burn up owing to the heat created or, even if it could be shielded in some way from this, it would lose speed and fall to the earth.

But we have only got to increase the launching velocity a little further and the projectile will then miss the far side of the earth by an appreciable margin, and when this is say 300 km, it will miss most of the atmosphere and so continue to circle the earth **on an elliptical orbit,** clearing it by 800 km on one side and 300 km on the other – in any such orbit the point at the greatest distance from the centre of the earth is called the **apogee,** and the point nearest the centre of the earth the **perigee. Our projectile is now a practical satellite** – practical, but still not very probable.

Even at a height of 300 km there is some atmosphere, and there probably is even at 800 km for that matter, so this satellite will lose speed every time it dips into the atmosphere, and so will gradually lose the energy given to it at the launch and will sooner or later come down to earth on a spiral path.

But we need not be disheartened, because a further increase of launching speed will further increase the clearance on the far side, and so gradually eliminate this problem until, at a launching speed of about

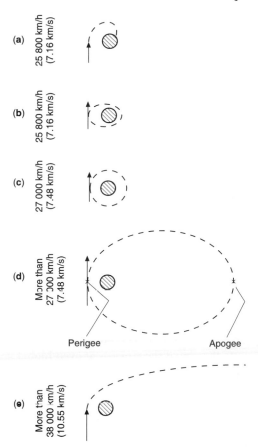

Fig. 13.5   Earth satellites
Speeds refer to horizontal launches from 800 km above the earth's surface, in the interests of clarity this distance has been exaggerated in comparison with the radius of the earth.

7.48 km/s we reach another interesting stage at which the satellite – we can no longer call it a projectile – **travels round the earth on a circular path**, 800 km from the earth's surface, and there is no longer any distinction between the apogee and the perigee (Fig. 13.5*c*). The launching speed at which this occurs is called the **circular velocity**.

After this long story the reader will probably be able to guess what happens with further increase of launching speed. Yes, the circle again becomes an ellipse but the **apogee**, or farthest point, **is now on the far side of the earth and the perigee is the point of launch** (Fig.

13.5*d*). Still further increases of speed make the ellipse more and more elongated with the apogee getting farther and farther from the earth. It will not be difficult to understand why this type of orbit is frequently used; it practically eliminates the problems of air resistance, it is not dependent on an exact launching speed and it allows scope for travel at the apogee to great distances and so, for instance, for passing beyond the moon or other planets, or hitting them.

Is this the end of the story? Not quite. Strange as it may seem the ellipse cannot be stretched indefinitely, and at a launching speed of about 10.7 km/s – different speeds at different heights – **the ellipse becomes an open curve, a parabola, and the satellite travelling on this open curve escapes from the earth for ever,** and becomes a satellite of the sun (Fig. 13.5*e*). Yes, **this is the escape velocity again,** and it only differs numerically from the previous one because we are now launching from 800 km instead of from ground level. **So the escape velocity is the highest velocity at which a body can be launched in any direction and be expected either to orbit the earth or return to it again** – above that speed, speaking vulgarly, we have had it. Above that speed, too, the path of travel changes from a parabola to a different open curve called a hyperbola, but this is a subtle change for a subtle reason, and it need not worry us.

This is not quite the end of the story as at even higher speeds there comes a point where the object can escape from the solar system and carry on out into space indefinitely.

THE MECHANICS OF CIRCULAR ORBITS

So much for the story of what happens – what is the explanation of it all? In the particular case of the circular orbit the satellite is very like a stone on the end of a string stretching from the centre of the earth to the satellite; the satellite is all the time trying to go off at a tangent but is being given an acceleration towards the centre by the centripetal force which is of course the force of gravity. So, near the earth's surface, if we neglect air resistance, the centripetal force will be the weight of the satellite, and the acceleration towards the centre will be $9.81 \, \text{m/s}^2$. Notice that a body circling the earth is accelerating towards the centre at the same rate as a body falling straight towards the earth. So we can easily calculate the circular velocity near the earth's surface because the acceleration $= v^2/r$ (see page 14).

Now r is the radius of the earth, say 6370 km (6 370 000 m), so $v^2/r = g$, i.e. $v^2/6\,370\,000 = 9.81$

$$\therefore v^2 = 62\,490\,000$$
$$v = 7905\,\text{m/s}$$
$$= 7.9\,\text{km/s approx}$$
$$= 28\,440\,\text{km/h approx}$$

How long will the satellite take to make a complete circuit of the earth at this speed?

Circumference of earth $= 2\pi \times 6370\,\text{km}$

So time of circuit at $28\,440\,\text{km/h}$

$$= (2\pi \times 6370)/28\,440$$
$$= 1.41 \text{ hours or about}$$
1 hour 25 minutes

It will be noticed that the circular velocity we have calculated, i.e. 7.9 km/s, is higher than the circular velocity at 800 km from the earth's surface, i.e. 7.48 km/s; but there is no mystery in that and we can easily work it out for ourselves by replacing the earth's radius of 6370 km by 7170 km, and reducing g by 20 per cent, i.e. to about 7.85 m/s². The value of v^2, and so of v, will then be less because whereas the value of g is reduced by 20 per cent, the value of r is only increased by $12\frac{1}{2}$ per cent – this, in turn, is because the value of g depends on the force of gravity, which is inversely proportional to the square of the distance r.

And what will be the time of a complete circuit **at 800 km** from the earth's surface? The distance is greater, the speed less, so the time of orbit will be greater. Work it out and you will find that it is about **1 hr 40 min**. Similarly **at 1600 km** the circular velocity is about 6.9 km/s and the time of orbit **nearly 2 hours**.

A distance of **35 400 km** from the earth gives a particularly interesting circular orbit because the time of a complete circuit is **24 hours**; so a satellite travelling at this speed – in the right direction, of course – **remains over one spot on the earth**; a communication satellite such as is used for transmitting TV and radio signals from one part of the earth to another.

Then at about **385 000 km** the circular velocity is a mere 3700 km/h (just over 1 km/s), and **the time of orbit 28 days** – but on that circuit we already have a satellite that surpasses in many ways any so far launched by man – **the moon**.

And now we can answer an obvious question – why doesn't the moon fall on to the earth? Because it is revolving round the earth at just such velocity and radius that the centripetal force is provided by the gravitational attraction, in other words, **in a sense** it is 'weightless' – and this applies to all those bodies in orbit, whether circular or not, and to all the people and things inside them. **Strictly speaking they are**

not weightless at all; in fact **it is their weight**, the force of attraction between them and the earth (or moon) which they are orbiting, **that keeps them in orbit** and prevents them from going off at a tangent. They merely seem to be weightless, and that is why a man can get out of a space-ship while in orbit, and continue in orbit himself, just like the space-ship, without any fear of 'falling' back to earth or to anywhere else. Although he may be travelling at several kilometres per second **he has no sense of speed, and apparently no weight – he just floats**, and has no difficulty in keeping near the space-ship which is also **just floating!** More correctly the man, and the space-ship, and all the other things in orbit, **are falling freely, are accelerating towards the earth** because of the attraction of the earth – in short **because of their weight – so much for weightlessness!** In the same sense the moon is falling towards the earth, though it never gets any nearer!

Notice that the reader can calculate all these circular velocities and times of orbit for himself, including that of the moon. For at all distances from the centre of the earth, the condition for a circular orbit is that the acceleration towards the centre shall be the 'g' or acceleration of gravity at that distance; this we might call g_d, and it must be equal to v^2/d.

But g_d is also proportional to the force of gravity, which is inversely proportional to the square of the distance.

Since at the earth's radius r, the acceleration is g,

g_d will be $g \times r^2/d^2$

Therefore for circular velocity at any distance d,

$v^2/d = g \times r^2/d^2$

i.e. $v^2 = gr^2/d$.

Figure 13.6 shows circular orbits at different distances from the centre of the earth.

It shows how a whole system of bodies can circle the earth, of their own free will as it were (once they have been put in orbit), and how **the farther out the orbit the slower is the speed**. It rather reminds one of the way in which Sir James Jeans once described the solar system as being like the traffic in Piccadilly Circus, with 'the traffic nearest the centre moving fastest, that farther out more slowly, while that at the extreme edge merely crawls – at least by comparison with the fast traffic near the centre.'* But there is **an important distinction** between the solar system – the work of nature – and bodies orbiting the earth – the

---

* The quotation from *The Stars in their Courses* by Sir James Jeans is given by courtesy of the Cambridge University Press.

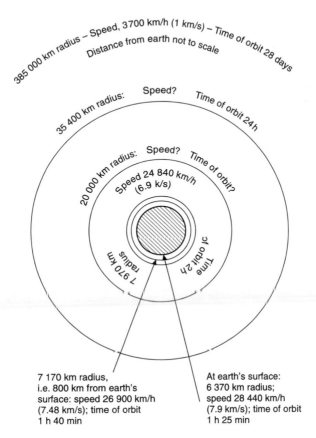

The ● moon

385 000 km radius – Speed, 3700 km/h (1 km/s) – Time of orbit 28 days
Distance from earth not to scale

35 400 km radius: Speed?   Time of orbit 24 h

20 000 km radius: Speed?   Time of orbit?

Speed 24 840 km/h (6.9 k/s)

7 970 km radius

Time of orbit 2 h

7 170 km radius, i.e. 800 km from earth's surface: speed 26 900 km/h (7.48 km/s); time of orbit 1 h 40 min

At earth's surface: 6 370 km radius; speed 28 440 km/h (7.9 km/s); time of orbit 1 h 25 min

Fig. 13.6   Circular orbits at different distances from centre of the earth
Note. The speed at 35 400 km radius, and the speed and time of orbit at 20 000 km radius, have been left for the reader to work out for himself.

work of man (except for the moon); in the solar system, again to quote Sir James Jeans, there is only **'one-way traffic'**, and the orbits of the planets round the sun are mostly circular, or very nearly so, whereas the **man-made satellites orbit the earth in various directions** and, as we shall soon discover, **some of their paths are very far from circular.**

When talking of the interesting possibilities of a 24 hour circuit we mentioned the direction of rotation. This would be all important in this case because if the satellite was rotating round the equator in the same direction as the earth's rotation it would stay over the same spot on the earth's surface but if it was travelling in the opposite direction – well, what would it do? Would it go twice round in a day? or would it merely appear to do so? or what?

But the fact that the earth is rotating will of course affect all launches, because it means that **we are launching from a moving platform**. The surface of the earth at the equator is travelling at a speed of rather over 1600 km/h owing to the spin of the earth on its axis, so a body launched in the same direction, i.e. towards the east, will already have the advantage of this speed and so will need 1600 km/h **less extra speed** to achieve circular velocity, escape velocity, or whatever it may be. Towards the west it will need 1600 km/h **more extra speed**. There can also be circuits of the earth in other planes altogether, e.g. over the poles, and in these cases the effect of the earth's rotation on launching and orbiting is more complicated.

It is not always realised, and it is interesting to note, that since the earth's surface at the equator is travelling at about 1600 km/h all bodies on the earth are in a sense trying to be satellites, and to go straight on instead of following the curvature of the earth. Thus there are two reasons why a body of the same mass weighs less at the equator than at the poles, first because it is farther from the centre of the earth so the true gravitational attraction is less, and secondly, because a proportion of the gravitational force has to provide the centripetal acceleration. How much is this centripetal force? Is it appreciable? Well, work it out for yourself. Take the actual velocity as 1690 km/h, the radius of the earth as 6370 km, g as 9.81 m/s^2, and you will find that the centrifugal force on a mass of 1 kg is about 0.018 N.

ESCAPE VELOCITY AND CIRCULAR VELOCITY

As already established the velocity for a circular orbit is given by the formula –

$$v_c^2 = gr^2/d$$

where r is the radius of the earth, and d the distance from the centre of the earth.

At the earth's surface $d = r$

∴ $v_c^2 = gr.$

Can the escape velocity be calculated equally simply? Very nearly so. If we go back to the idea of a body being projected vertically from the earth's surface with sufficient kinetic energy to enable it to do work against the force of gravitation all the way to infinity, then if the escape velocity is denoted by the symbol v_e the

kinetic energy of a mass m will be $\frac{1}{2}mv_e^2$.

This kinetic energy must be sufficient to provide the energy needed to lift m from the earth's surface to infinity. At a particular height above the earth, the energy needed to lift m one more metre is equal to its weight which equals mg at the earth's surface, and decreases all the way to infinity when it will be zero. Since the weight changes and since the change is not a simple ratio but inversely as the square of the distance, it needs the principles of calculus to estimate the total work done, but the answer is very simple; it is the same as the weight at the earth's surface $mg \times$ the radius of the earth, i.e. mgr.

So $\qquad \frac{1}{2}mv_e^2 = mgr$

or $\qquad v_e^2 = 2gr$

but $\qquad v_c^2 = gr$

$\therefore \qquad v_e = v_c \times \sqrt{2}$

i.e. Escape velocity $= \sqrt{2} \times$ Velocity of Circular Orbit at that Radius.

Thus there is a simple relationship between all escape velocities and all circular velocities at a given radius from a mass such as that of the earth, or moon, or sun – **the escape velocity is 1.41 $\times$ the circular velocity, or 41 per cent more.**

ELLIPTICAL ORBITS

So much for circular orbits – now what about the elliptical orbits which, as we have already explained, are much more common for artificial satellites? They are also more common in nature, the orbit of the moon round the earth being one of the few examples of a nearly circular orbit.

Mathematically it is a little more difficult to calculate what happens during an elliptical orbit when both the velocity and the radius are constantly changing, but the reader who has followed the arguments so far should have no difficulty in understanding the principles involved. Returning to horizontal launches from a height of 800 km it will be remembered that at a speed of launch of 7.48 km/s the orbit was circular, below and above this speed it was elliptical, though at the escape velocity of 10.7 km/s it became an open curve, a parabola, and then above this a hyperbola. The real criterion is how much energy, kinetic energy, the body has when it is launched; because it is this energy which enables it to do work against the force of gravity – it is really much more like the case of the stone that was thrown vertically upwards than might at first appear. In vertical ascent the kinetic energy given to the body at the launch enables it to do work against gravity and so gradually acquire potential energy; at the highest point reached all the kinetic energy has become potential energy; then as the body falls again the potential energy is lost and kinetic energy regained until on striking the ground the original kinetic energy has all been regained (neglecting air resistance of course). The body travelling on a curved path **also has to work against gravity**, is **also accelerating all the time**, and **downwards**, just like the body on the vertical path, and in fact, **at the same rate** – once that is understood, all is clear.

The only difference then is that on curved paths the body must retain **some** of its kinetic energy throughout the circuit if it is to continue on its orbit. At a launching height of 800 km it has at the start both kinetic and potential energy; if the launching speed is less than the circular velocity of 7.48 km/s, but sufficient to ensure that it doesn't come down to earth before reaching the far side, then by the time the body reaches the far side of the earth it will have dropped in height, i.e. **lost some of its potential energy**, but by the same token will have gained some kinetic energy (just like the falling stone), and so will be travelling faster – at over 7.9 km/s in fact if it is near the earth's surface. Then as it returns to the starting point it will gain potential energy, rising again to 800 km, and lose kinetic energy, until the proportions (and values) are the same as they were at the launch. In all this we are still neglecting air resistance, and it is air resistance which in practice makes an orbit of this kind impossible; for as soon as the body travelling round the earth, and getting lower all the time, meets appreciable air density, **it will experience drag**, and in working against this will **lose kinetic energy** and so the speed necessary to take it round the far side of the earth. Thus it will fall to the ground before getting half way round or the heat generated by kinetic heating will cause it to burn up.

As the launching speed gets nearer the circular velocity, the body will keep clear of appreciable atmosphere all the way round, and the orbit becomes more practical. But until the circular velocity is reached the launching point will be the apogee, and the perigee will be on the far side of the earth, where the body will be nearer the earth's surface, and where the velocity will increase until, at the circular velocity, it remains constant all the way round.

As the launching speed is increased above the circular velocity, the body will have more kinetic energy than is necessary to keep the 800 km of height, and so it will gain height and potential energy, at the same time losing speed and kinetic energy, until it passes round the far side of the earth – now the apogee – at lower velocity and greater height. And so it will go on as the launching speed is still further increased, the ellipse becoming more and more elongated, the apogee getting farther and farther from the earth, and the residual velocity at the apogee getting less and less.

As the launching velocity approaches the escape velocity of 10.7 km/s, the body at the apogee will only just have sufficient kinetic energy to enable it to get over the top, as it were, and return again to earth. If launched at the escape velocity, or above, it won't even be able to do this, and it will go off into space on its parabola, or hyperbola as the case may be, at the mercy of the gravitational attraction of some other body, probably the sun. This new force of attraction will restore its kinetic energy and velocity as it embarks on a completely new orbit, possibly of enormous size like that of the earth round the sun (radius of orbit about 150 million kilometres).

Since nearly all practical orbits are in the form of ellipses it is interesting to consider some of the properties of these curves. The reader may know that an ellipse can be drawn on paper with a pencil and a piece of string of fixed length attached to two pins (Fig. 13.7). In other words, an ellipse is the locus of a point moving so that the sum of the distances from these two points (the pins) is constant. These points are called the foci of the ellipse – in a circle, which is only a particular case of an ellipse, they coincide – and it will be noticed that in the figures illustrating the paths of satellites **the centre of the earth is always one of the foci of the ellipse**.

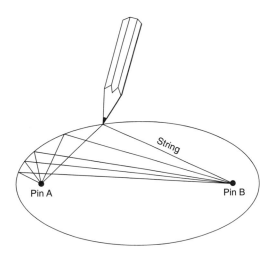

Fig. 13.7   Drawing an ellipse

At launching speeds below the circular velocity the centre of the earth is the focus farthest from the launch, at the circular velocity it is of course the centre of the circular orbit, and at higher launching speeds it is the focus nearest to the launch, the other focus getting farther and farther away as the launching speed is increased. For launches at the escape velocity the distant focus has gone to infinity, and only one focus is left at the centre of the earth – a parabolic curve has only one focus. Above the escape velocity, too, the hyperbolic curve has only one focus (see Fig. 13.8); strictly speaking, a hyperbola has two foci, but this is because there are two parts of the curve, back to back as it were, and we are here only concerned with one part of the curve.

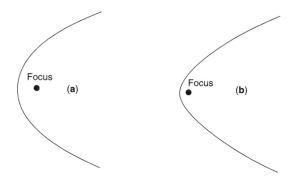

Fig. 13.8 (*a*) a parabola, (*b*) a hyperbola

It is not of practical importance in understanding the mechanics of projectiles and satellites, but it may be of interest to any reader who has not studied these curves mathematically to know that they are all derived from a cone (they are the intersections of a plane and the surface of a cone) and are sometimes called 'conic sections' (Fig. 13.9).

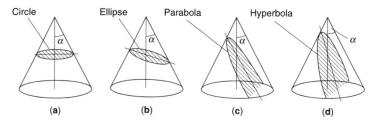

Fig. 13.9   Conic sections
In (*a*) the plane cuts the cone *at right angles* to the centre line, forming a circle. In (*b*) the plane is at an acute angle to the centre line, but the angle is *greater than* α, forming an ellipse. In (*c*) the angle is *equal to* α, forming a parabola. In (*d*) the angle is *less than* α, forming a hyperbola.

GOING TO THE MOON?

Now let us get just a little nearer to the true state of affairs by realising at least the existence of the moon. How will this affect the stone that is thrown vertically from the earth?

Well, it will affect it quite simply, in principle at any rate, in that there will now be three masses all attracting each other – the stone, the earth and the moon.

The moon is about 385 000 km away from the earth. Just for the moment let us suppose that it remains over the point on the earth's surface from which we throw the stone, and that we increase the starting velocity of the stone until it reaches distances of 6370 km, 12 740 km, and so on. As before, the weight will decrease with the distance from the earth's surface, **but now rather more so** because **a new factor has been introduced, the attraction between the moon and the stone.** The mass of the moon is only about 1/81 of the mass of the earth, so the force of attraction at the same distance will only be a fraction of that of the earth (1/81 in fact), but as we know the moon's attraction is a very real thing, even at the earth's surface, for it is largely responsible for the tides.

We have already imagined so much that we may as well go one step further and imagine that our stone travels in a straight line between earth and moon. As it gets nearer to the moon the attraction of the earth will decrease and that of the moon increase, until a point is reached where the two attractions are the same. Since the force is inversely proportional to the square of the distance, the distance from the centre of the moon at which this occurs will be $1/\sqrt{81}$ of the distance from the centre of the earth, or approximately one ninth (Fig. 13.10); roughly say 39 000 km from the moon and 346 000 km from the earth. **So if a stone is launched with just sufficient velocity to reach this point, it will stay there** – and will once more be 'weightless', this time perhaps more correctly so, though again we notice that it is just a question of the forces being balanced. In fact the balance is too delicate, and the stone **will not in fact stay at this neutral position,** because some other heavenly body will attract it and tip the balance, and it will fall either onto the earth or the moon. **If the stone is launched with a velocity slightly greater than that required to reach this neutral point, it will still be travelling towards the moon at this point, and since as it passes the point the attraction of the moon will become greater than that of the earth the stone will pick up speed and fall on to the moon.**

It has already been emphasised that all bodies attract all other bodies, and that therefore one can never really escape from the gravitational

Fig. 13.10   Neutral point between earth and moon

attraction of all the bodies in the heavens. But the motion of a body in space becomes extremely complicated if the forces of attraction on it of even three bodies (such as the sun, the moon, and the earth) are taken into account, and for this reason it is convenient – and not very far wrong – to consider a **zone of influence** for each body, this being a sphere in space in which that particular body has a greater gravitational effect than a larger body. The earth, of course, is well within the sun's zone of influence – its motion round the sun is in fact controlled by the force of attraction between it and the sun – but on the other hand, bodies near the earth's surface, although attracted by the sun, come much more under the influence of the earth's attraction than that of the sun (which is perhaps just as well, since otherwise we would all be off to the sun). The attraction of the earth remains greater than that of the sun for a distance of about a million kilometres, so the earth's zone of influence relative to the sun is a sphere of about one million kilometres radius (Fig. 13.11). Similarly the moon's zone of influence relative to the earth is a sphere of about 39 000 km radius – as we discovered in the last paragraph, though we didn't give it that name.

   Well, we have described one way of getting to the moon – to go

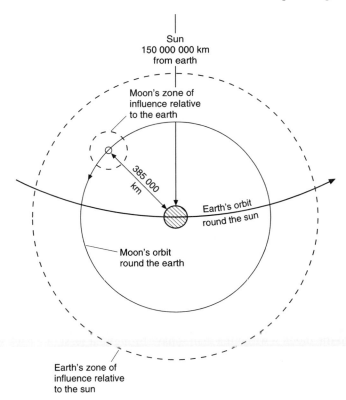

Fig. 13.11   Zones of influence
Not to scale.

straight there in fact – but it is not quite so easy as it sounds because of the great accuracy needed both in aim and launching speed. As regards aim, we have made things much too easy in our imagination; in reality the earth is travelling round the sun, and is spinning on its own axis, while the moon is travelling round the earth. But even more interesting is the sensitivity to exact velocity.

The distance of 385 000 km to the moon is much less than the distance to infinity, only a minute fraction of it, and to the neutral point even less, **but it requires nearly as much energy to reach this distance as it does to reach infinity** because the attraction of the earth at this distance has been so reduced that nearly all the serious work in overcoming the earth's gravity has already been done; or, putting it another way, whereas 11.184 km/s is needed to send the stone to infinity, about 11.168 km/s is needed to send it to the neutral point from which it will drop on the moon. So if launched **at less than**

11.168 km/s it will return to earth; at about 11.168 km/s it will go to the moon; at 11.184 km/s it will go to infinity (it could hit the moon on the way); and at over 11.184 km/s – well, we have mentioned that before.

The mass of the moon being so much less than that of the earth **its escape velocity is only about 2.4 km/s**, and that is the speed at which the stone would arrive on the moon if it 'fell' from infinity; it will for all practical purposes be the same if it falls the 39 000 km from the neutral point. This is about the muzzle velocity of a shell as it leaves a long-range gun, and so the landing on the moon will not be a very soft one, and this is the minimum speed at which the stone can arrive unless there is some means of slowing it down; there is no 'atmosphere' to do this, and the only hope is to break the fall **by rockets fired towards the moon**.

ANOTHER WAY TO THE MOON

But man has been to the moon – more than once – and has come back again; perhaps even more remarkable the Russians have sent spacecraft to the moon – without a man – have brought at least one back with samples, and have driven a moon-bug about on the surface! How has it been done? For the answer we must go back to circular and elliptical orbits. For a horizontal speed of launch of 10.46 km/s (just below the escape velocity), from a height of about 800 km, gives an elliptical orbit which will strike the moon, and at velocities of launch slightly above this, orbits will pass round both the earth and the moon (Fig. 13.12).

But the moon isn't such an easy target as all that! The shape and size of the elliptical orbit is very sensitive to the exact direction and velocity of launch, and moreover the moon is itself travelling at rather over 3700 km/h, whereas the speed of the satellite at its apogee will only be about 700 km/h. Also, if the launch from the earth is made in an easterly direction – to take advantage of the earth's rotation and consequent circumferential speed of 1600 km/h – the satellite at 700 km/h will be chasing the moon at 3700 km/h in the same direction; so it will be a case of the moon hitting the satellite rather than the satellite hitting the moon – not that it matters which hits which, but it does mean that the satellite should be launched in the other direction and so approach the moon from the front, as it were, instead of chasing it.

In practice the initial launch must be made from ground level (Fig. 13B), and not from an altitude of 800 km, and it has been calculated that if the satellite is guided only during the launching phase, and if the

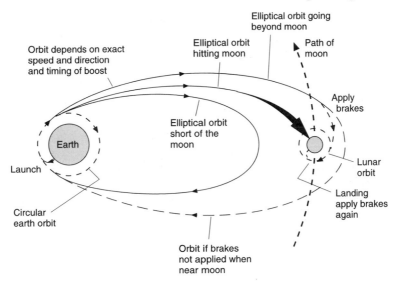

Fig. 13.12   Sending space-craft to the moon
Not to scale.

angle of launch is exactly correct, **there must not be an error of more than 23 m/s in the launching speed of 11 125 m/s;** or if the velocity is exactly correct **the angle of launch must be accurate to within 0.01°.** If the satellite is to pass round the moon and the earth the accuracy must be even greater, so much so that some guidance after launch is a virtual necessity.

In view of the accuracy needed, not to mention the expense and man-power involved, it would be a mistake to imagine that flights to the moon or other planets have become, or are ever likely to become commonplace. None the less the experience so far gained has resulted in what might be called a **standard procedure** consisting of –

1. **The launch** to orbital height and speed.
2. One or more **orbits of the earth.**
3. Rocket boost to **required speed and direction for the moon.**
4. Reverse burst of power to **slow down, and put into orbit round the moon.**
5. Separation of lunar module, and more reverse power to **give a soft landing on the moon.**
6. **Lift-off from the moon,** and **into orbit** to join up again with the command module.
7. Rocket boost to **required speed and direction for the earth.**

Fig. 13B   Lift off for the moon
(By courtesy of the General Dynamics Corporation, USA)

    8. Reverse burst to slow down and **put into orbit round the earth.**
    9. **Re-entry, splash down** and **pick-up**.

These phases have been described in detail in the Press, on radio and television, and in numerous articles and books; our purpose here is simply to indicate how the principles of mechanics apply to these various phases.

    First then, **the launch**.

LAUNCHING A SPACE-CRAFT

A projectile, whether it is launched for the purpose of escaping from the earth, or landing on the moon, or becoming a satellite, or simply travelling over the earth's surface to some other place, must first pass through the atmosphere.

The most important effect of this is **that we cannot neglect air resistance**, as we have so calmly done throughout this chapter (though with constant reminders). And the practical effect of air resistance is to reduce speeds, so **the actual speeds of launching within the atmosphere must all be higher than those we have given**. How much higher? From ground level, something of the order of 10 or 12 per cent, e.g. if the escape velocity is 40 250 km/h (11.134 km/s), the actual velocity of launch at ground level would have to be about 45 000 km/h (12.5 km/s), and for a circular velocity of 29 000 km/h (8.05 km/s), say 32 000 km/h (8.9 km/s). This naturally makes accuracy more difficult to achieve.

But there is a further difficulty. The launching speed cannot be attained at ground level. As has already been explained, the body on the first part of its flight is propelled by rockets; if it is required to reach great heights by multi-stage rockets. So in contrast with a shell fired from a gun there is time – and distance – in which to gather speed; and by so deciding and regulating the thrust of the rockets in relation to the mass of the projectile, and taking into account the drag due to air resistance, the acceleration can be moderated sufficiently to prevent damage to the missile itself and its mechanisms, and if passengers are to be carried, even to human beings. This moderation of the acceleration is of course an advantage, but it also makes it extremely difficult to calculate just what the speed, direction, and height of the vehicle will be when it is finally launched, i.e. when the fuel of the last launching rocket has been exhausted.

No one who has thought of this problem, even in the very elementary form such as we have attempted to explain in this book, can be anything but amazed at the accuracy that has actually been achieved in the launching of space-craft.

As greater heights are reached there is less density of air, and so the drag decreases in spite of ever-decreasing speeds. Eventually the rocket power is shut off, the last stage of the launching rocket is jettisoned, and the projectile, or space-craft, or whatever it may be, travels on its elliptical path under the force of gravity until it begins to descend and again approaches the earth's atmosphere. The distance it travels during this ballistic phase – under its own steam, one might almost say! – will depend on the velocity it had achieved and the direction in which it was travelling when the rocket power ended. It may be hundreds or thousands of kilometres, it could be round the earth and back again, or several times round; there is no fundamental difference between a missile, a satellite and a space-craft, they differ only in the speed, direction and height of launch.

The fact that the final launch takes place at considerable height does, at least, provide partial justification for our earlier neglect of air resistance when considering their motion. It is true that in thinking of launches at a height of 800 km we may have been guilty of going rather far though, as explained at the time, it had the advantage that we really could neglect air resistance, and so the speeds we gave for that height were reasonably correct. Typical figures for an actual launch (Fig. 13.13) are to a height of 60 km and a speed of 6000 km/h (1.67 km/s) at the end of the first stage, 200 km and 14 500 km/h (4.03 km/s) at the end of the second stage, and 500 km and 28 000 km/h (7.78 km/s) at the end of the third stage. The take-off is vertical, the path is then inclined at say 45°, then when the velocity is sufficient there is a period of coasting or free-wheeling between the second and third stages to the required height (which will become the perigee if the missile is to be a satellite) where the path will be horizontal, then the third stage rocket boosts the velocity to that required for orbit. The more this exceeds the circular velocity, the more distant will be the apogee. The perigee of nearly all

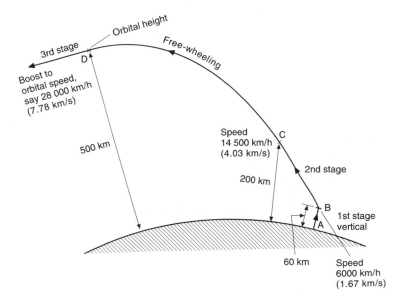

Fig. 13.13   Typical flight path for launching of space-craft
Not to scale.

the early satellites was less than 800 km, but the apogee varied from just over 800 km for Sputnik 1 up to – well, to the moon and beyond.

The second stage, orbiting the earth, has already been considered in some detail, and there is little to add. This is the aspect of space flight of which we have had most experience, and there are now literally hundreds of 'bodies' of various shapes and sizes and masses orbiting the earth, and on a variety of orbits, and hundreds more that have finished their flights and have been burnt up on re-entering the atmosphere. There have also been a number of manned orbits of the earth, and it is only a matter of time before space stations are set up which can be permanently manned in 'shifts' by shuttle services, put together and enlarged up there, and used for a variety of purposes, some peaceful – others perhaps not so peaceful.

So far as going to the moon is concerned the first orbits are more or less circular and then, at the third stage, at exactly the correct part of the orbit, **a burst of power is given to boost the speed and put the space-ship on its journey to the moon.** Although this journey is often represented in diagrams as a straight line it is in fact merely an elongated elliptical orbit designed to pass near the moon, so the astronauts still experience the sensation of 'weightlessness'. Mid-course and other corrections, if required, can be given by short bursts of rocket power; since there is no air resistance the thrust required to make such changes is not very great. As in all elliptical orbits **the speed will decrease as the apogee is approached,** but by then the space-ship will have passed the neutral point, will be attracted by the moon and will again pick up speed, but now new problems arise and we must consider how orbits of the moon differ from those round the earth.

ORBITING THE MOON

In order to understand this we must consider how the moon differs from the earth. It is, of course, **much smaller,** its diameter (3490 km) being rather more than 1/4 that of the earth, and **its mass,** which is more important from the point of view of satellite orbits, **about 1/81 that of the earth. The weight of a body on the moon's surface is about one sixth of its weight on earth** – if this puzzles the reader let him work it out, remembering that weight is the force of attraction which is proportional to the two masses multiplied together and inversely proportional to the square of the respective distances, i.e. the radii of the moon and the earth. **The acceleration of gravity** on the moon is also, of course, **about one sixth of that on earth,** i.e. just over $1.6\,\text{m/s}^2$. But the most interesting difference – and it is the result of the smaller mass of the moon and the lesser weight of bodies near the moon – is that **the velocities for moon satellites,** circular velocity, escape velocity, etc., **are much lower than for earth satellites; the escape velocity at the surface of the moon is only about 2.4 km/s,** the circular velocity being 2.4/1.41 or 1.7 km/s. Another important point is that owing to the lack of air resistance it is possible for a satellite to circle the moon **very close to its surface**.

But if the prospective satellite has been fired to meet the moon, the relative speed between satellite (700 km/h) and moon (3700 km/h) will be much too great, and so, unless the body actually hits the moon, it will merely go past it and escape. Thus, on first thoughts, it would seem that a body fired from the earth cannot become a moon satellite – this is true **so long as there is no propulsion in the reverse direction,** in other words **braking**; and the moon has no atmosphere to act as a brake, so the only practical means of persuading a satellite to orbit the moon, under the influence of the moon, is to provide for a rocket to act as a brake on its speed as it gets into the moon's sphere of influence (Fig. 13.12).

This, in fact, is how the space-craft is put into orbit round the moon – a burst of power slows it down to the correct speed for the orbit required which, as already explained, can be much nearer the moon's surface than orbits of the earth (Fig. 13C).

Thus far there have been quite a number of flights, manned and otherwise, and there have also been several soft, and some not so soft, landings of craft conveying instruments designed to send messages back to earth; but the **actual landing** of men on the moon has not been achieved so often that it can be considered as a matter of standard procedure. In the successful attempts so far made a small part of the space-craft, the lunar module, has been detached at that part of the orbit which, as calculated by computer, will result in a landing at the desired point on the moon's surface. By a small burst of reverse power the lunar module is again slowed down to bring it closer to the moon, while the command module continues on its circular orbit. As the lunar module 'falls' onto the moon, fortunately not so fast as it would onto the earth, but quite fast enough to be uncomfortable, a final reverse rocket thrust is fired to enable the module to land gently on the surface – owing to the lack of air no parachutes, or any kind of air brake, are of avail in controlling the fall (Fig. 13D).

THE RETURN FLIGHT

The moon lift-off, by another burst of rocket power, is again made just that much easier than from earth owing to the reduction in the mass of

Fig. 13C   Orbiting the moon
(By courtesy of the Boeing Company, USA)
Scale model of a lunar orbiter; the lunar landscape (taken by Lunar Orbiter 1 on the far side of the moon) is authentic, and shows the sharp break between sunshine and shadow.

the space-ship which has thrown off the multi-stage rockets, owing too to the lesser force of gravity, the lack of air resistance, and the lower speed required for orbit; all this is just as well because the rockets that developed the tremendous thrust for lift-off from earth are no longer available, and by now much of the reserve fuel and power has been

Fig. 13D   Moon landing craft
(By courtesy of the Bell Aerospace Division of Textron Inc, USA)
Astronaut Charles Conrad Jr in a lunar landing training vehicle during a simulation
flight for the Apollo 12 mission.

expended. Even so, a high degree of accuracy is again necessary to
ensure that the lunar module gets into orbit close to the command
module (though again small adjustments can be made by bursts of
power), so that they can again be linked up into one space-craft. Once
they have been re-united and men, films and other souvenirs have been
transferred to the command module, the lunar module can itself be
discarded, and left to orbit or to hit the moon – this time probably at
speed!

The next stage is **a further and considerable boost to put what
remains of the space-ship out of moon orbit and on the return
path to earth** – once more really an elongated elliptical orbit with the
apogee this time near the earth. Although this has been described as a
considerable boost, the thrust required is nothing like so great as was
needed to start the craft on its journey to the moon because the neutral
point is now comparatively near, and once this has been passed the
earth's attraction will all the time be increasing, as will the speed of the
space-ship until it reaches something of the order of 10.46 km/s (more
than 37 000 km/h), the speed with which it started on its journey.

Now another reverse burst is needed to slow the craft down to approximately 7.5 km/s for a similar circular orbit, or partial orbit, and in the same direction too as that used after launch.

RE-ENTRY INTO THE ATMOSPHERE

On the Apollo missions, the craft re-entered the atmosphere after at least a partial orbit, and after discarding the larger part of what still remained of the space-craft leaving only the small command module, a mere 5 tonnes of the 3500 tonnes or more of the mass at launch. From re-entry to splashdown was one of the most difficult, and in some ways the crudest part of the whole procedure. Once again extreme accuracy was needed for the craft, by final use of the rocket power still available, **must enter the atmosphere, at some 400 000 feet, through a** 'window', as it is called, **only 8 kilometres wide, and at an angle of between 5.6° and 7.2° to the top of the atmosphere** – if it entered **too steeply it would have burned up, if too shallowly it would have bounced off again**. Not only is the angle of entry very critical, but the craft also had to be manoeuvred into such a position that it encountered **maximum drag (form drag rather than skin friction)** and so maximum retardation of about 6 g. Even so, the speed was so high, and the skin friction so great, that the heat generated was quite alarming, the surface of the craft was burnt and scarred, and the air ionised so that radio communication between the earth and the crew was temporarily interrupted. When denser air was reached, first a drogue parachute was released, followed by at least **three large parachutes**, and these reduced the velocity sufficiently for any surplus fuel to be jettisoned, and finally for a reasonably soft splash-down in the sea, again with reasonable accuracy of position.

In view of the crudeness of this method of approach and landing it is not surprising that a reusable vehicle, the space shuttle, was developed (Fig. 13E). This has the same problems – it gets hot and needs to slow down for landing. But using wings gives another way of controlling the trajectory and means that it can land on a runway. It is really a glider – albeit an unusual one!

FLIGHTS IN SPACE

We could go on to discuss orbits round the sun and the possibility of flights to and round the various planets. It is a fascinating subject, and becomes more and more so as the possibilities become practicabilities and then historical facts. But the principles of all such flights are the

Fig. 13E   The space shuttle
(By courtesy of NASA)

same as those we have mentioned, and the author must avoid the temptation of going any further into space. The reader who is as fascinated with the subject as is the author – and who, if he is interested in the mechanics of flight at all, is not? – must seek other books, though admittedly it is not easy to find books that are at the same time reasonably simple and reasonably sensible.

SUB–ORBITAL FLIGHT – THE AEROPLANE – MISSILE – SATELLITE

Have you ever realised that the 'lift' required to keep an aeroplane in the air depends upon the direction in which it is travelling? – e.g. whether it is going with the earth or against it. In an earlier paragraph we worked out the centripetal force on a body of mass 1 kg sitting on the earth's surface at the equator – sitting still, as it seems, but in fact behaving like a stone travelling at 1690 km/h on the end of a string of 6370 km radius. The answer didn't come to much – about 0.018 N – **but the principle is of extreme importance**.

The corresponding value for an aeroplane of mass 10 190 kg is about 400 N, still not much perhaps, but none the less **an appreciable and measurable quantity**. It means that if the real force of gravity on the aeroplane is 100 000 N, it would appear to weigh only 99 600 N – in fact, of course, we would call this the weight, it is the force we would have to exert to lift it.

But it is a solemn thought, though none the less a fact, that if this aeroplane were to fly against the direction of the earth's rotation, i.e. towards the west, at 1690 km/h, it would not require this centripetal force, and so would appear to weigh 100 000 N – and that is the lift the wings would have to provide. If, on the other hand, it flew towards the east at 1690 km/h, its real speed would be 3380 km/h, and the centrifugal force would be, no, not 800 N but 4 × 400, i.e. 1600 N (because the centripetal force depends on the square of the velocity), so the lift that the wings would have to provide would be 100 000 − 1600 = 98 400 N.

Similarly at a real speed of 6760 km/h (5070 km/h eastwards) the centripetal force would be 6400 N, and the necessary lift 93 600 N. At 12 800 km/h (11 200 km/h eastwards), the corresponding figures would be 25 600 N and 74 400 N; and at 25 600 km/h, 102 400 N and minus 2400 N! **At approximately 29 000 km/h the centripetal force is 100 000 N and the lift required nil.**

What does it all mean? Well, the reader who has followed the arguments in this chapter will surely know what it means – simply that the aeroplane travelling at 29 000 km/h near the earth's surface **is travelling at the circular velocity**, it doesn't need any lift from the wings, it will stay up of its own accord, it is a satellite. Nor when these velocities are reached does it make all that difference (only 1600 km/h each way) whether it travels east or west.

What it will need is colossal thrust to equal the colossal drag, which in any case will cause it to frizzle up.

But what if it flies higher – and higher – and higher? The drag for the same real speed will be less, less thrust will be needed, the circular velocity required for no lift conditions will be less, even the real force of gravity upon it will be less; **it won't even create a sonic boom at ground level**.

CAN YOU ANSWER THESE?

Now let us see what we know about this fascinating subject –

1. What is meant by escape velocity? What is its approximate value for the earth? Is it the same for the moon?
2. Is the escape velocity the same for a horizontal launch as for a vertical launch?

3. Distinguish between the perigee and the apogee in an elliptical orbit.
4. What is the particular significance of a satellite circling the earth at about 35 400 km from the centre of the earth?
5. What is the time of circular orbit of –
   (*a*) a satellite very near the earth's surface?
   (*b*) a satellite 1600 km from the earth's surface?
   (*c*) the moon?
6. Under what conditions is the path of a satellite parabolic? hyperbolic?

For numerical examples on missiles and satellites see Appendix 3.

AEROFOIL DATA

The aerofoil sections, of which particulars are given in the following pages, have been chosen from among the thousands that have been tested, as being **typical of the best that have been designed for particular purposes**.

Although the values have been taken from standard tests, they have been modified so as to bring them as far as possible into line with each other, and simplified so as to correspond with the symbols and methods used in this book. In thus modifying the figures **the aim has been to bring out the principles** even at the sacrifice of some degree of accuracy. For the purpose of this book nothing is lost by this simplification and, while it is right and proper that official results should be given to the accuracy with which they can be measured, the student should remember that they are, after all, taken from experimental figures and that there is a limit not only to the accuracy of such figures, but even more so to the various corrections that have to be applied to them.

Unfortunately it is not possible to obtain results for all the sections at the same Reynolds Number, but for each **the approximate Reynolds Number of the test has been given**, and where alternative results are available those at the highest Reynolds' Number have been chosen.

For reasons of security it is not possible to give test results for the most modern high-speed sections; but, even if they could be given, it is doubtful whether the tests would have been made at sufficiently high Mach Numbers and Reynolds Numbers to be reliable as a guide to full-scale performance. There is no difficulty in getting lift from bi-convex or double-wedge sections used at supersonic speeds, **the problem is to keep down the drag**.

For the benefit of those readers who would like to sketch out the shapes of the various aerofoil sections, the co-ordinates of the upper and lower surfaces are given; the measurements are expressed as percentages of the chord, negative values being below the chord line and positive values above it.

The tables give values of C_L and C_D at various angles of attack – from negative angles to above the stalling angle – but unfortunately values of C_D are not available for aerofoils with flaps down (except for No. 8, the model of the 'Lightning'). When a dash appears in the data columns it means that the figure is not available, or that for some reason it would be meaningless. Pitching moments are given about the leading edge, or about the quarter-chord, or about the aerodynamic centre; and an opportunity is given in the short questions that follow the data for the reader to work out one from the other. Similarly the position of the centre of pressure can be found from other data, the lift/drag ratio from C_L and C_D, and so on.

Most of the questions can be answered from the information given in Chapter 3, but the student may require a little guidance on the questions dealing with moment coefficients and centres of pressure, especially since the simplest method of reaching some of the answers is through the notation of differential calculus which has not been used in the text of the book.

For instance, to solve question (*c*) for RAF 15 (without slot):

On page 95 we arrived at the equation –

$$x/c = (C_{M.AC} - C_{M.LE})/C_L \tag{1}$$

or

$$C_{M.AC} = C_{M.LE} + (x/c) \cdot C_L \tag{2}$$

By differentiating (2) with respect to C_L we get the differential equation –

$$dC_{M.AC}/dC_L = dC_{M.LE}/dC_L + x/c \tag{3}$$

But, by definition, the moment coefficient about the aerodynamic centre does not change with the angle of attack (or with the lift coefficient), or expressed in mathematical terms –

$$dC_{M.AC}/dC_L = 0$$

Therefore

$$x/c = -dC_{M.LE}/dC_L \tag{4}$$

By drawing a graph of $C_{M.LE}$ against C_L we can determine the slope of the curve, i.e. $dC_{M.LE}/dC_L$ for any value of C_L, and so for any angle of attack. Thus we can get x/c from (4), and substitute in (2) to get $C_{M.AC}$. To solve question (*c*) for RAF 15 (with slot), we must first find $C_{M.LE}$ from the formula on page 95, i.e.

$$\text{CP position} = -C_{M.LE}/C_L \tag{5}$$

and then use the same method, as outlined above, to find $C_{M.AC}$.

The same method can be used to solve questions (*a*) and (*b*) on the Clark YH aerofoil; and for questions (*b*) and (*c*) on NACA 0009 we can start from –

$$x' = C_{M.C/4}/C_L \text{ as in (5) above.}$$

Finally it should be noted that formulae (1) and (2) are approximations based on the assumption that **the angle of attack is small**. The answers given in Appendix 4 have been arrived at by using these simplified formulae.

The more refined formula for (2) is –

$$C_{M.AC} = C_{M.LE} + x/c \, (C_L \cos \alpha + C_D \sin \alpha)$$

and the student is advised to work out one or two of the examples at the larger angles of attack with this formula, if only to satisfy himself that the use of the simplified formula is justified – at any rate for the angles of attack of normal flight.

When differentiating the full formula, it must be remembered that $\cos \alpha$, C_D and $\sin \alpha$ all vary with C_L and so the appropriate mathematical techniques must be used; these involve drawing graphs of $\cos \alpha$, C_D and $\sin \alpha$ against C_L, in order to determine $d \cos \alpha/dC_L$, dC_D/dC_L and $d \sin \alpha/dC_L$.

More extensive questions on aerofoils will be found in Appendix 3.

1. RAF 15
*1A. RAF 15 with slot

Chord line

Excellent general purpose aerofoil for use on biplanes.
Like Clark Y, used on many early types of aircraft.
Figures relate to aspect ratio of 6.
Reynolds Number of test $3\frac{1}{2}$ million; with slot 200 000.
Slot assumed to remain open at position giving maximum lift.

Distance from LE, % chord	Upper surface	Lower surface
0	1.50	1.50
1.25	3.14	0.76
2.5	3.94	0.50
5	5.00	0.18
7.5	5.37	0.02
10	6.09	0.02
15	6.67	0.18
20	6.96	0.53
30	6.94	1.02
40	6.63	1.02
50	6.13	0.71
60	5.52	0.33
70	4.79	0.06
80	3.91	0.04
90	2.81	0.21
95	2.17	0.32
100	0.94	0.94

Angle of attack	C_L	C_D	$C_{M.LE}$	$*C_L$	$*C_D$	$*CP$, fraction of chord
$-4°$	-0.14	0.014	-0.036	-0.27	0.120	–
$-2°$	$+0.02$	0.008	-0.052	-0.12	0.075	–
$0°$	0.14	0.008	-0.090	$+0.03$	0.056	–
$+2°$	0.32	0.012	-0.130	$+0.18$	0.050	0.70
$+4°$	0.46	0.020	-0.160	0.33	0.050	0.45
$6°$	0.60	0.030	-0.200	0.47	0.053	0.39
$8°$	0.76	0.044	-0.240	0.62	0.060	0.37
$10°$	0.90	0.060	-0.280	0.76	0.070	0.36
$12°$	1.04	0.070	-0.310	0.90	0.080	0.34
$14°$	1.16	0.096	-0.330	1.05	0.100	0.32
$15°$	1.22	0.110	-0.340	1.12	0.110	0.31
$16°$	1.16	0.140	-0.350	1.21	0.120	0.31
$18°$	1.02	0.210	-0.384	1.36	0.148	0.30
$20°$	0.94	0.260	-0.390	1.51	0.175	0.30
$24°$	–	–	–	1.70	0.234	0.30
$26°$	–	–	–	1.76	0.270	0.30
$28°$	–	–	–	1.76	0.304	0.30
$30°$	–	–	–	1.64	0.344	–
$34°$	–	–	–	1.30	0.430	–

(*a*) What is the max. value of L/D with slot? without slot?
(*b*) What is the stalling angle with slot? without slot?
(*c*) What is the value of $C_{M.AC}$ at $+4°$ with slot? without slot?
(*d*) What is the value of $C_L$max$/C_D$min with slot? without slot?
(*e*) What is the value of $C_L^{\frac{3}{2}}/C_D$ (without slot) at $4°$ and $8°$?

2. CLARK YH

Chord line

Excellent American general purpose aerofoil.
Modifications of Clark Y have been used on many types of
aircraft all over the world; Clark YH was one of the first of these
modifications.
Figures relate to aspect ratio of 6, and standard roughness.
Reynolds Number of test 7 million.

Distance from LE, % chord	Upper surface	Lower surface
0	3.50	3.50
1.25	5.45	1.93
2.5	6.50	1.47
5	7.90	0.93
7.5	8.85	0.63
10	9.60	0.42
15	10.68	0.15
20	11.36	0.03
30	11.70	0
40	11.40	0
50	10.51	0
60	9.15	0
70	7.42	0.06
80	5.62	0.38
90	3.84	1.02
95	2.93	1.40
100	2.05	1.85

Angle of attack	C_L	C_D	CP, *fraction of chord*	$C_{M.LE}$	L/D
−4°	−0.09	0.010	–	+0.030	−10
−2°	+0.05	0.009	0.74	−0.010	+5.2
0°	0.20	0.010	0.40	−0.046	19.3
2°	0.36	0.015	0.32	−0.072	23.2
4°	0.51	0.022	0.295	−0.116	23
6°	0.66	0.033	0.285	−0.150	20.6
8°	0.80	0.045	0.275	−0.184	17.7
10°	0.94	0.062	0.27	−0.220	15.2
12°	1.06	0.083	0.27	−0.244	13.3
14°	1.21	0.103	0.27	−0.276	11.8
16°	1.33	0.125	0.265	−0.320	11
18°	1.43	0.146	0.265	−0.352	9.9
19°	1.36	0.170	0.275	−0.356	8
20°	1.26	0.211	0.29	−0.354	7
25°	0.97	0.324	0.33	−0.354	2.9
30°	0.81	0.430	0.37	−0.352	1.9

(*a*) What is $C_{M.AC}$ at 0°, 4° and 8° for this aerofoil?
(*b*) Where is the aerodynamic centre of this aerofoil section?
(*c*) What is the stalling angle?
(*d*) What is the value of $C_L\max/C_D\min$?
(*e*) What is the value of $C_L^{\frac{3}{2}}/C_D$ at 4° and 8°?

3. NACA 0009
*3A. NACA 0009 with flap

A thin symmetrical section.
All figures relate to standard roughness.
Reynolds Number of test 6 million.
Position of aerodynamic centre 0.25 of chord from LE.
*With 20 per cent split flap set at 60°.

Distance from LE, % chord	*Upper and lower surfaces % chord*
0	0
1.25	1.42
2.5	1.96
5.0	2.67
7.5	3.15
10	3.51
15	4.01
20	4.31
30	4.50
40	4.35
50	3.98
60	3.50
70	2.75
80	1.97
90	1.09
95	0.61
100	0

Angle of attack	C_L	C_D	$C_{M.C/4}$	$^*C_L$	$^*C_{M.C/4}$
$-8°$	-0.88	0.022	0	$+0.45$	-0.200
$-6°$	-0.65	0.014	0	$+0.68$	-0.210
$-4°$	-0.45	0.011	0	$+0.90$	-0.216
$-2°$	-0.21	0.010	0	$+1.09$	-0.220
$0°$	0	0.009	0	$+1.29$	-0.216
$+2°$	$+0.21$	0.010	0	$+1.38$	-0.218
$4°$	$+0.43$	0.011	0	$+1.65$	-0.222
$6°$	$+0.64$	0.014	0	$+1.78$	-0.225
$8°$	$+0.85$	0.018	0	$+1.72$	-0.230
$10°$	$+0.90$	0.021	-0.002	$+1.58$	-0.275
$12°$	$+0.89$	0.028	-0.004	$-$	$-$
$14°$	$+0.87$	0.036	-0.012	$-$	$-$

(a) What is the value of $C_{M.AC}$ (without flap)?
   Is it the same at all angles, as it should be?
(b) What is the position of the CP (without flap) at $+4°$?
(c) What is the position of the CP (with flap) at $+4°$?
(d) What is the value of L/D (without flap) at $2°$, $6°$, $10°$?
(e) What is the stalling angle (i) without flap? (ii) with flap?

4. NACA 4412
*4A. NACA 4412 with flap

Medium thickness NACA 4-digit good all round section.
All figures relate to standard roughness.
Reynolds Number of test 6 million.
Position of aerodynamic centre 0.246 of chord from LE.
***With 20 per cent split flap set at 60°.**

Distance from LE, % chord	Upper surface	Lower surface
0	0	0
1.25	2.44	−1.43
2.5	3.39	−1.95
5.0	4.73	−2.49
7.5	5.76	2.74
10	6.59	−2.86
15	7.89	−2.88
20	8.80	−2.74
25	9.41	−2.50
30	9.76	−2.26
40	9.80	−1.80
50	9.19	−1.40
60	8.14	−1.00
70	6.69	−0.65
80	4.89	−0.39
90	2.71	−0.22
95	1.47	−0.16
100	0	0

Angle of attack	C_L	C_D	$C_{M.AC}$	$*C_L$	$*C_{M.C/4}$
−8°	−0.45	0.022	−0.097	+0.90	−0.287
−6°	−0.23	0.014	−0.092	+1.12	−0.297
−4°	−0.03	0.012	−0.092	+1.34	−0.302
−2°	+0.20	0.010	−0.092	+1.56	−0.305
0°	+0.38	0.010	−0.093	+1.75	−0.305
+2°	+0.60	0.010	−0.095	+1.95	−0.305
4°	+0.80	0.012	−0.098	+2.14	−0.305
6°	+1.00	0.014	−0.100	+2.43	−0.302
8°	+1.15	0.017	−0.100	+2.50	−0.300
10°	+1.27	0.022	−0.095	+2.65	−0.290
12°	+1.36	0.030	−0.092	+2.63	−0.275
14°	+1.35	0.042	−0.092	−	−
16°	+1.25	0.059	−0.095	−	−

(*a*) What is the value of $C_{M.C/4}$ (without flap) at 0° and 8°?
(*b*) Where is the CP (without flap) at these angles?
(*c*) What is the value of L/D (without flap) at these angles?
(*d*) What is the stalling angle (i) without flap? (ii) with 60° flap?
(*e*) What is the value of $C_L \max / C_D \min$ (without flap)?

5. NACA 23012

Chord
line

Medium thickness 5-digit section that has been much used.
Low drag; maximum camber well forward.
All figures relate to standard roughness.
Reynolds Number of test 6 million.
Position of aerodynamic centre 0.241 of chord from LE.

Distance from LE, % chord	Upper surface	Lower surface
0	0	0
1.25	2.67	−1.23
2.5	3.61	−1.71
5.0	4.91	−2.26
7.5	5.80	−2.61
10	6.43	−2.92
15	7.19	−3.50
20	7.50	−3.97
25	7.60	−4.28
30	7.55	−4.46
40	7.14	−4.48
50	6.41	−4.17
60	5.47	−3.67
70	4.36	−3.00
80	3.08	−2.16
90	1.68	−1.23
95	0.92	−0.70
100	0	0

Angle of attack	C_L	C_D	$C_{M.C/4}$	$C_{M.AC}$
$-8°$	-0.60	0.020	-0.018	-0.013
$-6°$	-0.43	0.014	-0.015	-0.013
$-4°$	-0.25	0.011	-0.013	-0.014
$-2°$	-0.08	0.010	-0.013	-0.016
$0°$	$+0.15$	0.010	-0.012	-0.016
$+2°$	$+0.36$	0.010	-0.010	-0.015
$4°$	$+0.55$	0.011	-0.008	-0.014
$6°$	$+0.75$	0.013	-0.010	-0.014
$8°$	$+0.96$	0.016	-0.013	-0.016
$10°$	$+1.14$	0.023	-0.014	-0.017
$12°$	$+1.23$	0.032	-0.012	-0.017
$14°$	$+0.82$	0.045	-0.013	$-$
$16°$	$+0.77$	0.065	-0.050	$-$

(*a*) What is the value of L/D for this aerofoil at $0°$, $4°$, $8°$?
(*b*) Where is the CP at these angles?
(*c*) Where is the maximum thickness?
(*d*) What is the stalling angle?
(*e*) What is the value of $C_L^{\frac{3}{2}}/C_D$ at $2°$, $4°$ and $6°$?

6. NACA 23018

Typical thick 5-digit section of the 230 series.
All figures relate to standard roughness.
Reynolds Number of test 6 million.
Position of aerodynamic centre 0.241 of chord from LE.

Distance from LE, % chord	Upper surface	Lower surface
0	0	0
1.25	4.09	−1.83
2.5	5.29	−2.71
5.0	6.92	−3.80
7.5	8.01	−4.60
10	8.83	−5.22
15	9.86	−6.18
20	10.36	−6.86
25	10.56	−7.27
30	10.55	−7.47
40	10.04	−7.37
50	9.05	−6.81
60	7.75	−5.94
70	6.18	−4.82
80	4.40	−3.48
90	2.39	−1.94
95	1.32	−1.09
100	0	0

Angle of attack	C_L	C_D	$C_{M.C/4}$	$C_{M.AC}$
−8°	−0.62	0.016	−0.018	−0.008
−6°	−0.47	0.014	−0.010	−0.007
−4°	−0.28	0.012	−0.008	−0.007
−2°	−0.09	0.011	−0.005	−0.007
0°	+0.12	0.010	−0.002	−0.007
+2°	+0.33	0.011	−0.001	−0.007
4°	+0.53	0.012	0	−0.007
6°	+0.72	0.014	+0.002	−0.007
8°	+0.90	0.016	+0.003	−0.007
10°	+1.01	0.020	+0.004	−0.008
12°	+1.06	0.028	+0.005	−0.008
14°	+0.75	0.040	+0.002	−
16°	+0.68	0.060	−0.020	−

(*a*) What is the maximum thickness? Where is it?
(*b*) What is the value of L/D at −4°, 0°, 4°, 8°, 12°?
(*c*) Where is the CP on this aerofoil at 2°, 4° and 6°?
(*d*) What is the stalling angle?
(*e*) What is the maximum lift coefficient?

7. NACA 65₁–212
*7A. NACA 65₁–212 with flap

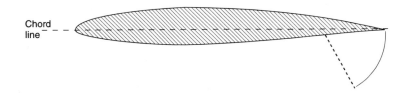

Typical of the NACA 6 series; medium thickness.
All figures relate to standard roughness.
Reynolds Number of test 6 million.
Position of aerodynamic centre 0.259 of chord from LE.
*With 20 per cent split flap set at 60°.

Distance from LE, % chord	Upper surface	Lower surface
0	0	0
0.5	0.970	−0.870
0.75	1.176	−1.036
1.25	1.491	−1.277
2.50	2.058	−1.686
5.00	2.919	−2.287
7.5	3.593	−2.745
10	4.162	−3.128
15	5.073	−3.727
20	5.770	−4.178
25	6.300	−4.510
30	6.687	−4.743
35	6.942	−4.882
40	7.068	−4.926
45	7.044	−4.854
50	6.860	−4.654
55	6.507	−4.317
60	6.014	−3.872
65	5.411	−3.351
70	4.715	−2.771
75	3.954	−2.164
80	3.140	−1.548
85	2.302	−0.956
90	1.463	−0.429
95	0.672	−0.040
100	0	0

Angle of attack	C_L	C_D	$C_{M.C/4}$	$*C_L$	$*C_{M.C/4}$
$-8°$	-0.68	0.020	-0.025	$+0.58$	-0.223
$-6°$	-0.50	0.015	-0.026	$+0.80$	-0.230
$-4°$	-0.33	0.013	-0.030	$+1.03$	-0.240
$-2°$	-0.10	0.010	-0.033	$+1.25$	-0.250
$0°$	$+0.12$	0.009	-0.035	$+1.45$	-0.260
$+2°$	$+0.35$	0.010	-0.037	$+1.63$	-0.265
$4°$	$+0.55$	0.011	-0.038	$+1.80$	-0.267
$6°$	$+0.80$	0.015	-0.039	$+1.87$	-0.264
$8°$	$+0.95$	0.023	-0.040	$+1.83$	-0.260
$10°$	$+1.07$	0.035	-0.040	$+1.70$	-0.255
$12°$	$+1.06$	0.050	-0.038	$+1.48$	-0.380
$14°$	$+1.01$	–	-0.035	–	–

(a) What is $C_{M.AC}$? Is it the same at all angles, as it should be?
(b) Where is the maximum thickness?
(c) What is the stalling angle (i) without flap? (ii) with 60° flap?
(d) What is the maximum value of L/D?
(e) What is the value of $C_L$max$/C_D$min (without flap)?

8. English Electric ASN/P1/3
*8A. English Electric ASN/P1/3 with flap

Chord
line

This is the symmetrical aerofoil section used on the BAC Mach
2+ 'Lightning'.
The wing is tapered and the ordinates relate to a section at 38.5
per cent of semi-span.
Reynolds Number of test $1\frac{1}{2}$ million (based on mean chord).
The values of coefficients refer to a complete model of the
aircraft, not to the wing section alone.
The 'Lightning' is a mid-wing monoplane with 60° sweepback on
leading edge (see Fig. 11B).
*Model with approx 25 per cent plain flaps set at 50°.

Distance from L.E. % chord	Upper and lower surface % chord
0	0
0.25	0.426
0.75	0.706
1.25	0.875
2.50	1.175
5.00	1.530
10	1.941
15	2.183
20	2.435
25	2.612
30	2.782
35	2.904
40	2.944
45	2.970
50	2.942
55	2.855
60	2.703
65	2.502
70	2.237
75	1.921
80	1.564
85	1.183
90	0.797
95	0.414
100	0.032

DATA FOR BAC 'LIGHTNING' MODEL – CLEAN AIRCRAFT

Angle of attack	C_L	C_D	C_M
0°	0	0.020	−0.017
2°	0.08	0.020	−0.013
4°	0.17	0.030	−0.008
6°	0.27	0.040	−0.006
8°	0.38	0.050	+0.005
10°	0.50	0.075	+0.010
12°	0.61	0.105	+0.016
14°	0.71	0.140	+0.026
16°	0.81	0.180	+0.040
18°	0.91	0.225	+0.055
20°	1.00	0.275	+0.070
22°	1.09	0.335	+0.088
24°	1.17	0.405	+0.108
26°	1.22	0.480	+0.124
28°	1.26	0.560	+0.132
30°	1.27	0.650	+0.140

DATA FOR BAC 'LIGHTNING' MODEL – FLAPS AT 50°

Angle of attack	C_L	C_D	C_M
0°	0.17	0.07	−0.072
2°	0.27	0.07	−0.068
4°	0.37	0.08	−0.062
6°	0.47	0.09	−0.060
8°	0.57	0.11	−0.053
10°	0.67	0.13	−0.040
12°	0.78	0.16	−0.032
14°	0.88	0.20	−0.025
16°	0.98	0.25	−0.015
18°	1.07	0.31	+0.005
20°	1.15	0.37	+0.013
22°	1.22	0.43	+0.020
24°	1.27	0.50	+0.037
26°	1.27	0.55	+0.045
28°	1.22	0.59	+0.074
30°	1.16	0.63	+0.074

The values of C_M are related to a point at 0.405 of mean chord. The position of the aerodynamic centre $= 0.405 + dC_M/dC_L$. The figures for the 'Lightning' were given by courtesy of the former British Aircraft Corporation, Preston.

(a) What is the thickness/chord ratio of this aerofoil?

(b) Where is the maximum thickness?

(c) What is the stalling angle of the aircraft model: (i) with flaps up? (ii) with flaps down?

(d) What are the values of L/D of the model at 4°, 12° and 20°: (i) with flaps up? (ii) with flaps down?

(e) What are the positions of the aerodynamic centre of the clean aircraft at 4°, 12° and 20°?

(*Note:* The answers involve the drawing of the curve of C_M against C_L and measuring the slopes of this curve at the specific angles.)

SCALE EFFECT AND REYNOLDS NUMBER

This Appendix is an amplification of the short note given about scale effect in Chapter 2 (p. 50). It may be of interest to those readers who would like to know more about a fascinating branch of the subject, but one that is not essential to the understanding of the remainder.

There are few of us who have not at some period of our lives played with models. To some it is just a passing fancy of childhood days, a cheap and safe way of playing at the real thing; to others it is a more serious hobby, one that persists through boyhood to manhood, sometimes even to old age, but always as a hobby, something quite apart from their daily work; to others it becomes a study, their job in life. It is as a study that we are interested in it here, and a very fascinating study it is. Much can be learnt from models, whether they be model locomotives, model ships, or model aeroplanes; but we must beware of certain traps, because otherwise we shall jump to wrong conclusions from our experiments with models.

The first difficulty in making a model is to make it accurate; that is part of the fascination. **The smaller we make the model, the more difficult it is to make it accurate.** For that very reason, curious creatures that we are, we delight in making small models, or in admiring the skill of those who are more patient and more clever than ourselves in making them. But clever though these small models may be, the attainment of real accuracy becomes a more practical proposition as the size of the model is increased, and in aeronautical work, for this reason among others, **we like our models to be large**.

Now, the majority of serious experiments on aeronautical models are done in wind tunnels. The air in a wind tunnel does not behave in quite the same way as the free air in which we fly; there are various reasons for this, but the most important one is that **the air is constrained by the wind tunnel walls.** The constraint affects the airflow, and the airflow affects the forces on the model and so may lead us to wrong conclusions. The obvious moral is to **make the tunnel large compared with the model.**

So our second difficulty about model work is that we need large wind tunnels. Large models, and much larger tunnels. Thus we find firms and even nations vying with each other to produce larger and larger tunnels. These large tunnels need enormous power and the cost is prodigious – but the race goes on.

But, however large our models, however large our wind tunnels, we shall still have to use models that are smaller than real aeroplanes, and it is very improbable that we shall be able to obtain the actual speeds of flight in these large wind tunnels.

And so we shall continue to be faced with the third and greatest problem, **that of scale effect**.

By means of photography, or by a scale drawing, we can reproduce a large body on a small piece of paper. A map is a good example. By trick photography we can make a small body appear large, or a large body small. In cinematography we can do even more; we can adjust both the size of a body and the speed at which it moves, and so make it appear to move slower or faster than it really does. But it is not so easy to play about with forces and accelerations. **A model ship tossing on a real sea does not look real, nor does a collision between two model cars, nor does the crash of a model aeroplane** – however clever the faking in a film, critical members of the audience can always spot the use of models designed to deceive them.

Why is this? Well, there are various ways of looking at it, and it is all really part of our problem of scale effect. Consider a 1/10th scale model. All the linear dimensions are 1/10th of those of the real thing, but the areas are 1/100th; and if it is constructed of the same materials, its mass is but 1/1000th of the real thing. So it is to scale in some respects, but not in others. This is one of the difficulties in trying to learn from flying models of aeroplanes; unless we are very careful in adjustment of the weights, we will get completely false results as regards behaviour in manoeuvre, spinning, crashes, and so on. It is all a question of adjustment, and another thing we must remember is that **we fly model aeroplanes in full-size air**, and sail model boats in full-size water; we do not adjust the air and the water to suit the models!

This is not a new problem. More than a hundred years ago a physicist, named Reynolds, was experimenting with the flow of fluids in pipes, and he made the important discovery that the flow changed from streamline to turbulent when the velocity reached a value that was inversely proportional to the diameter of the pipe. The larger the pipe, the less the velocity at which the flow became turbulent, e.g. if 10 m/s was the critical velocity in a pipe of 20 mm diameter, then 5 m/s would be the critical velocity in a pipe of 40 mm diameter. This illustrates the principle, but really the discovery was of much wider application than this. It did not only apply to the critical velocity, nor only to the diameter of the pipe, but to the flow past any body placed in the stream. **If the velocity multiplied by some linear dimension of the body, e.g. the diameter in the case of a sphere, came to the same value, then the flow pattern would be the same.**

We must emphasise this importance of the flow pattern. Since the time of Reynolds there has been ample confirmation of his experiments, and we know that if we get the same flow pattern over a model as over a full-scale body, then our laws of aerodynamics are true and we can scale up the forces with confidence; in other words, there will be no error due to scale effect. **For scale effect does not mean** – as we might justifiably suppose it to mean – **the effect of scale**. The laws of aerodynamics, the speed squared law, the law of dependence on area and density, all the laws summed up in the fundamental formulae $L = C_L . \frac{1}{2}\rho V^2 . S$ and $D = C_D . \frac{1}{2}\rho V^2 . S$, these are all the effects of scale; perhaps we should call them the primary effects of scale. Scale effect really means that **these laws are not strictly true unless certain conditions apply**, and the conditions are those founded on the experiments of Reynolds. If the conditions do not apply, then there are secondary effects of scale, errors are introduced, errors due to scale effect, and unfortunately these errors are apt to be erratic and difficult to predict.

To see how to avoid this scale effect, with its unpredictable errors, we must return to the principles discovered by Reynolds. The most simple of these is that we should keep the same value of **velocity multiplied by size** for the model

experiment as for full-scale flight. This may be referred to as the VL **law,** L
referring to some linear dimension of the body, usually taken as the chord in the case
of a wing section, the diameter in the case of a disc, sphere, or streamline body. This
law means that if we wish to experiment on a **1/10th scale model in order to
forecast what will happen on the full-size aeroplane at 200 knots, we should
test the model at 2000 knots** – thus keeping VL the same in both cases.

This conclusion is somewhat alarming, the more so if we remember that we must
also employ a large wind tunnel! In such a tunnel **we have never succeeded in
obtaining a speed of anywhere near 2000 knots,** indeed, it is doubtful whether
we could even obtain 200 knots. Furthermore, even if we could obtain this fantastic
speed, **we should have run into even more serious problems, because the
speed of sound would have been exceeded,** and for this reason the flow pattern
would have changed completely, so we would have defeated our own ends. But that
is not all; let us consider what the forces on the model would be. Because it is 1/10th
scale, the area is 1/100th, so the forces would be divided by 100 compared with full
scale; because we have 10 times the velocity, the forces should be multiplied by 100
(neglecting the effects of exceeding the speed of sound), so **the forces on the
model would be equal to those on the full-scale aeroplane.** Imagine this small
model wing supporting the same lift as the real aeroplane! – it just could not be
made strong enough to do it.

What then can we do? There does not seem much hope, unless it be to get as
large a VL as possible, and then trust that the error will be small, or perhaps we may
be able to estimate how much the error will be and so make allowances for it. This is
what we have tried to do, but it has not been altogether satisfactory. Let us consider
an example. Suppose we wish to estimate the total drag of a large airliner at 200
knots. We make a 1/10th-scale model – it will be of considerable size, and we must
have a much larger wind tunnel to test it in. We find its drag coefficient at, say, 50,
100, 150 and 200 knots. We cannot get any more speed in the tunnel; we are lucky to
get that. If there were no scale effect, the drag coefficient should be the same at the
various speeds, but we find that it is, in fact, increasing, and the values are, say,
0.050, 0.051, 0.052, and 0.053. That seems simple. We draw a graph and plot the
values of drag coefficient against VL; they lie on a straight line (they may, of course,
lie on a curve; a straight line has been chosen to show the difficulties of even the
simplest case). We can now extrapolate – a dangerous game! It is like forecasting
what is going to happen in the future, as a result of past history. To interpolate is
reasonably safe; that is only, as it were, to fill in gaps in past history, knowing what
came before and what came after. But extrapolation is a very different matter. We
can draw curves of wing loadings or of speeds of aeroplanes for the past forty years;
the curves show definite trends; they show where we are going; they show – if we
extrapolate – that in the year 2000 A.D. the wing loadings of aeroplanes will be more
than 2440 kgf/m² (23.9 kN/m²), and maximum speeds will be over 1000 knots. But
will they? That is the question. **Maybe, maybe not. (These words have
deliberately been left as they were written in the 6th edition of this book
published in 1950.)** In the intervening twenty or more years the maximum speed of
1000 knots predicted for A.D. 2000 has already been far exceeded, and the forecast
now for A.D. 2000 is. . . . No, **we have learnt the lesson of the dangers of
extrapolation,** as we have no intention of trying again!) But to return to the drag
coefficient – this is even worse than forecasting the future. At least we can feel sure
that the wing loadings and speeds in the future will be higher than they are to-day.
Not so with our drag coefficient. From 50 knots, to 200 knots, it has gone up steadily
by 0.001 for each 50 knots, so at 2000 knots it should be 0.089. Now 2000 **knots for**

the 1/10th-scale model is the same VL as 200 knots for the full-scale aeroplane, so if the drag coefficient is 0.089 for the model at 2000 knots it should be 0.089 for the full-scale aeroplane, and so we should be able to calculate the total drag of the aeroplane. **If**, that's the rub. And this is where we may find the result even worse than trying to forecast history; we may find that the actual drag coefficient of the full-scale aeroplane is **even less than 0.050**. So it was deceiving us by going up, and then it went down again. That is the sort of thing that sometimes happens – sometimes, not always! If it happened always, we should at least know what to expect. The moral is, of course, to go as high up the scale as possible, to test at the highest possible values of VL, but we have already seen how far they are off full-scale values.

Is there any other solution? Yes, there is at least a partial solution, but we would not be able to guess at it unless we went a little deeper into the experiments of Reynolds. The VL law is sound so long as we use the same fluid for our model experiments as for full-scale flight; this, of course, we do – in the ordinary wind tunnel. But Reynolds found that if different fluids were used, other properties of the fluid affected the type of flow; these properties, as one might almost expect, were the density and the viscosity of the fluid. To cut a long story short, he eventually summed it all up, as we intend to do, by saying that if the quantity

$$\frac{\text{Density} \times \text{Velocity} \times \text{Size}}{\text{Viscosity}}$$

is kept constant, the flow pattern will be similar, and there will be no error due to scale effect. This quantity, written in symbols

$$\frac{\rho VL}{\mu},$$

is called the **Reynolds Number** of the test; where ρ is the density of the fluid in kg/m³ (1.225 for air at sea-level); V is the velocity of the test in metres per sec; L is a dimension of the body in metres (for aerofoils and aeroplanes the chord is taken as the standard dimension); μ is the dynamic viscosity of the fluid which we defined when considering dimensions as force per unit area (N/m²), divided by the difference in velocity (m/s), and multiplied by the thickness (m); this comes to N-s/m² which, as we discovered in Appendix 2, is equivalent to kg/m/s when expressed in terms of mass, length and time. Its numerical value for air at normal temperature and pressure is 17.894×10^{-6} kg/m/s – it sounds a strange unit, but don't worry about that, it is really just the same, and has the same numerical value, when expressed as N-s/m². What is interesting and important, however, is that whereas the **Reynolds Number itself is non-dimensional, has no units**, and so has **the same numerical value whatever consistent system of units is used**, the dynamic viscosity (μ) has dimensions, the dimensions of $M/(L \cdot T)$, so its numerical value is different when expressed in a different system of units (in the old British system for instance it is 373.718×10^{-9} slugs/ft/sec).

Now, this idea gives us new hope, in that it suggests using a different fluid for the model test. Consider water, for instance. Water is about 815 times as dense as air at ground level, and only 64 times as viscous. So we immediately have a factor of about 12.8, and instead of experimenting on a 1/10th scale model at 2000 knots, we shall get the same Reynolds Number at 2000 knots/12.8, i.e. at about 156 knots. Unfortunately, this idea has only led us up the garden path. A little thought will make us realise that a large tunnel with a water flow of 156 knots is even more fantastic than 2000 knots for air, and a little calculation will show that the forces on the model, instead of equalling those of the full-size aeroplane, would be about five times as great.

What next? We are nearly beaten, but not quite. It may be surprising, but it is a fact that by compressing air we have little or no effect upon its viscosity. This offers a real ray of hope at last. Suppose we compress the air to, say, 25 atmospheres – and this can be done in the **compressed air tunnels** in Great Britain and in the USA – then we have a factor of 25 (as regards ρ), with no corresponding increase in the viscosity, μ. Thus a 1/10th-scale model can be tested at 2000 knots/25, i.e. at 80 knots. A practical solution at last – or very nearly so.

Why the qualification – very nearly so? Only because compressed air tunnels are considerable and costly engineering undertakings, and there seem to be very real limits to practicable sizes and speeds to be obtained in them. There are not as yet more than a dozen or so such tunnels in the world; in none of them could we test anything so large as a 1/10th-scale model of a large airliner, nor can we yet obtain speeds much higher than 100 knots when the compression is as high as 25 atmospheres. The modern tendency has been to reduce the compression ratio and raise the velocity. Within such limitations, however, the compressed air tunnel serves the problem, and it is particularly valuable for testing new devices when we are anxious to avoid the uncertainty of scale effect errors.

A small point to be kept in mind is that, although the density does not affect the viscosity of air, the temperature does, and quite considerably. Unlike liquids, which become less viscous with rise in temperature, air becomes more viscous, e.g. at 0°C the viscosity of air is 17.89×10^{-6} kg/m/s, whereas at 100°C it is 22.74×10^{-6} kg/m/s. Any increase in viscosity will mean so much less benefit from the increase in density; and therefore since the compression of the air will tend to raise its temperature, we must take precautions to cool it again.

If we work out the Reynolds Number of tests that can be done on models of the appropriate size in **atmospheric wind tunnels** (called 'atmospheric' to distinguish them from compressed air tunnels), we will find that the Reynolds Number varies **from about 100 000** for small slow-speed tunnels **to about 1 000 000 or 1 500 000** for large high-speed tunnels.

Similarly, the Reynolds Number of **full-scale flight** varies **from about 2 000 000** for small slow-speed aeroplanes **to about 20 000 000** or more for large high-speed aeroplanes.

The Reynolds Number that can be attained in **compressed air tunnels** varies **from about 7 000 000 to 12 000 000 or more** in the latest types.

These figures clearly illustrate the value of the compressed air tunnel, it being the only tunnel which gives a Reynolds Number even within the range of modern full-scale flight. In all atmospheric tunnels there is some risk of errors due to scale effect.

Let us see what sort of forces we may expect on the models used in a compressed air tunnel. Taking a 1/10th-scale model and a full-scale speed of 200 knots, a tunnel speed of 80 knots, and compression to 25 atmospheres, the forces will be

$$1/100\text{th (because of scale)} \times \frac{80^2}{200^2} \text{ (because of speed)} \times 25$$

(because of density), i.e. 1/25th of the full-scale forces.

This is a great improvement on the full-scale force for the atmospheric tunnel at 2000 knots and five times the full-scale force for the water tunnel at 156 knots. It is interesting, however, because even these forces, 1/25th of the full-scale forces, are so large as to cause serious distortions of the models unless they are made very strong.

The Reynolds Number has a marked effect on the behaviour of the boundary layer, and particularly on its thickness and separation from the surface over which it is flowing. For this reason it is usual to arrange to have a turbulent boundary layer over wind tunnel models, because a laminar layer would separate much earlier than the turbulent layer on the full-scale aircraft. Even with a turbulent boundary layer on the model, it is a thicker layer – owing to the lower Reynolds Number of the model test – and this thicker layer still separates earlier than on a full-scale aircraft, though not so early as a laminar one would. This is a very important point in connection with shock stalls when the rising pressure through the shock wave forces the boundary layer to separate – this depending critically on the strength of the shock and the thickness of the boundary layer.

As a result of this scale effect on the behaviour of the boundary layer, measurements of skin friction drag and form drag from wind tunnel tests at a low Reynolds Number need very careful correction if they are to be used to forecast the performance of a new type of aircraft – but estimates of wave drag are more predictable since the measurements are not so sensitive to scale effect.

Such then is the problem of scale effect, the meaning of Reynolds Number, and the reason for the large and expensive wind tunnels, and especially for the compressed air tunnel. It is a story that is very typical of progress in aeronautics, of new problems seemingly insoluble being solved by new applications of old ideas. It is a story that explains the varying reputation of the wind tunnel. At first, when the speeds in wind tunnels were as great as, or even greater than, the speeds of flight, wind tunnels gave good results, they won a good reputation and were blindly believed; then with progress in flight, practice outstripping theory, the speeds of flight far exceeding the speeds in wind tunnels, with scale effect at its worst and little understood, the wind tunnel was first mistrusted, then discredited, eventually debunked; then came the investigation into scale effect, the building of larger and better tunnels, then the compressed air tunnels, and once again a re-born faith in the wind tunnel, but this time with a full realisation of its limitations.

Now, of course, we are faced with new wind-tunnel problems in connection with flight at and above the speed of sound.

NUMERICAL QUESTIONS

The numerical examples that follow are given in order that the student may puzzle out some of the problems of flight for himself; they will also give him valuable practice in the use of SI units. It is recognised that, in order to solve some of the examples, assumptions must be made which can hardly be justified in practice, and these assumptions may have appreciable effect on the accuracy of the answers; but the benefit from solving these problems lies not so much in the numerical answers as in the considerations involved in obtaining them.

Unless otherwise specified, the following values should be used –

Density of water = 1000 kg/m³
Specific gravity of mercury = 13.6
Specific gravity of methylated spirit = 0.78
International nautical mile = 1852 m, or approx 6076 ft
1 knot = 0.514 m/s
1 ft = 0.3048 m
Radius of earth = 6370 km
Diameter of the moon = 3490 km
Distance of the moon from the earth = 385 000 km
Aerofoil data as given in Appendix 1
C_D **for flat plate at right angles = 1.2**
            cylinder = 0.6
            streamline shape = 0.06
            pitot tube = 1.00
Take the maximum length in the direction of motion for the length L in the Reynolds Number formula.
At standard sea-level conditions –
    Acceleration of gravity = 9.81 m/s²
    Atmospheric pressure = 101.3 kN/m², or 1013 mb, or 760 mmHg
    Density of air = 1.225 kg/m³ at 1013 mb and 288°K
    Speed of sound = 340 m/s = 661 knots = 1225 km/h
    Dynamic viscosity of air (μ) = 17.894 × 10⁻⁶ kg/ms
    For low altitudes one millibar change in pressure is equivalent to 30 feet change in altitude.
    International Standard Atmosphere as in Fig. 2.2.

CHAPTER 1. MECHANICS

1. A car is travelling along a road at 50 km/h. If it accelerates uniformly at 1.5 m/s² –
   (a) What speed will it reach in 12 s?
   (b) How long will it take to reach 150 km/h?
2. A train starts from rest with a uniform acceleration and attains a speed of 110 km/h in 2 min. Find –
   (a) the acceleration;
   (b) the distance travelled in the first minute;
   (c) the distance travelled in the two minutes.
3. If a motorcycle increases its speed by 5 km/h every second, find –
   (a) the acceleration in m/s²;
   (b) the time taken to cover 0.5 km from rest.
4. During its take-off run, a light aircraft accelerates at 1.5 m/s². If it starts from rest and takes 20 s to become airborne, what is its take-off speed and what length of ground run is required?
5. A boy on a bicycle is going downhill at 16 km/h. If his brakes fail and he accelerates at 0.3 m/s², what speed will he attain if the hill is 400 m long?
6. Assuming that the maximum deceleration of a car when full braking is applied is 0.8 g, find the length of run required to pull up from (a) 50 km/h, (b) 100 km/h.
7. A rifle bullet is fired vertically upwards with a muzzle velocity of 700 m/s. Assuming no air resistance, what height will it reach? and how long will it take to reach the ground again?
8. The landing speed of a certain aircraft is 90 knots. If the maximum possible deceleration with full braking is 2 m/s², what length of landing run will be required?
9. A 7000 kg aeroplane touches down at 100 knots and is brought to rest, the average resistance to motion due to brakes and aerodynamic drag being 6.867 kN. To reduce the landing run by 500 m a tail parachute is fitted. If the additional equipment increases the total mass of the aircraft by 200 kg and the landing speed by 5 knots, what additional average drag must the parachute supply?
10. An athlete runs 100 m in 11 seconds. Assuming that he accelerates uniformly for 25 m and then runs the remaining 75 m at constant velocity, what is his velocity at the 100 m mark?
11. An aircraft flying straight and level at a speed of 300 knots and at a height of 8000 m above ground level drops a bomb. Neglecting the effects of air resistance, with what speed will the bomb strike the ground? (*Remember that the final velocity will have to be found by compounding the vertical and horizontal velocities.*)
12. Two masses of 10 kg each are attached to the ends of a rope, and the rope is hung over a frictionless pulley. What is the tension in the rope?
13. One of the masses in Q12 is replaced by a 15 kg mass. What will be the tension in the rope when the system is released?
14. What force is necessary to accelerate a 133 kg shell from rest to a velocity of 600 m/s in a distance of 3.5 m?
15. What thrust is necessary to accelerate an aircraft of 5900 kg mass from rest to a speed of 90 knots in a distance of 750 m?
16. Calculate the thrust required to accelerate a rocket of 1 tonne mass from rest vertically upwards to a speed of 10 km/s in 10 s (*neglect air resistance*).
17. A train of 250 tonnes mass is moving at 100 km/h. What retarding force will be required to bring it to rest in 15 seconds?

18. A 76 kg man is standing on a weighing machine which is on the floor of a lift. What will the weighing machine record when –
    (a) the lift is ascending with velocity increasing at $0.6 \, \text{m/s}^2$?
    (b) the lift is ascending with velocity decreasing at $0.6 \, \text{m/s}^2$?
    (c) the lift is descending at a constant velocity of $1.2 \, \text{m/s}$?

19. An engine of 50 tonnes mass is coupled to a train of 400 tonnes mass. What pull in the coupling will be required to accelerate the train up a gradient of 1 in 100 from rest to 50 km/h in 2 min if the frictional resistance is 70 N per tonne?

20. An aircraft of 5000 kg mass is diving vertically downwards at a speed of 500 knots. The pilot operates the dive brakes at a height of 10 000 m and reduces the speed to 325 knots at 7000 m. If the average air resistance of the remainder of the aircraft during the deceleration is 15 kN, what average force must be exerted by the dive brakes? (*Assume that the engine is throttled back and is not producing any thrust.*)

21. A 1 tonne truck is pulled on a level track by a force of 245 N in excess of the frictional resistance. How far will it travel from rest in 30 seconds?

22. A horizontal jet of water from a nozzle 50 mm in diameter strikes a vertical wall. If the water is diverted at right angles and none splashes back, what force is exerted on the wall when the speed of the jet is 6 m/s?

23. A truck is standing on an incline of 1 in 80. If the frictional resistance is 50 N per tonne, how far will it travel in 15 s if released from rest?

24. If the air resistance of a 500 kg bomb is equal to $v^2/18 \, \text{N}$, where v is the velocity in m/s, what is the terminal velocity of the bomb?

25. An aircraft of 4000 kg mass has a take-off speed of 60 knots and a take-off run of 300 m in conditions of no wind. If the thrust delivered by the engine is 15 000 N, and the frictional resistance is 1000 N, what is the average aerodynamic resistance during the take-off?

26. What force is necessary to stop a 500 kg car in 10 m from a speed of 50 km/h?

27. A solid 1 kg shot is fired from a barrel of 100 kg mass with a muzzle velocity of 850 m/s. If the barrel is free to recoil against a resistance of 5 kN, how far will the barrel move back when the shot is fired?

28. A fighter aircraft of 5000 kg mass is fitted with four cannon each of which fires 600 rounds per minute with a muzzle velocity of 900 m/s. If each shot has a mass of 110 g, and if all the recoil is taken by the aircraft, find the loss in speed that the aircraft would experience in a 5 second burst of fire.

29. A 9000 kg aircraft is flying straight and level at 300 knots; what thrust is necessary to accelerate it to 450 knots in half a minute if the average air resistance of the aircraft between these speeds is 15 kN?

30. A rifle of 4 kg mass fires a 30 g bullet with a muzzle velocity of 750 m/s. Find the force a rifleman must exert on the butt of the rifle to limit the recoil to 40 mm.

31. An aircraft is fitted with brakes capable of exerting a force of 10 kN, and reversible pitch propellers capable of producing a backward thrust of 25 kN. If the aircraft has a mass of 10 000 kg and a landing speed of 110 knots, find the minimum length of runway required for the landing run. (*Neglect the effect of air resistance which will also help to decelerate the aircraft.*)

32. An aircraft carrier is steaming at 20 knots against a head wind of 30 knots. An aircraft of 9000 kg mass lands on the deck with an air speed of 100 knots; if the arrester gear must be sufficiently powerful to stop the aircraft in a distance of 25 m in these conditions, without any aid from the brakes or air resistance, find the retarding force that the gear must exert.

33. A 5000 kg aircraft touches down on the deck of an aircraft carrier with an air speed of 90 knots. If the carrier is heading into wind at 20 knots, and the wind speed

is 12 knots, what kinetic energy must be destroyed by the action of the arrester gear in bringing the aircraft to rest on the deck? If the average resistance exerted by the arrester gear is 55 kN, how far does the aircraft roll along the deck before coming to rest?

34. A propeller 3 m in diameter revolves at 2250 rpm. Find the angular velocity and the linear speed of the propeller tip.

35. The piston of an aircraft engine has a stroke of 150 mm and the engine runs at 3000 rpm. Find the angular velocity of the crankshaft and the average speed of the piston.

36. Find the acceleration of the propeller tip in Q34.

37. Find the acceleration of the crankpins in the engine in Q35.

38. A stone of mass 1 kg is whirled in a horizontal circle making 60 rpm at the end of a cord 1 m long; what is the pull in the string? If it is whirled in a vertical circle, what is the pull in the string (*a*) when the stone is at the top? (*b*) when the stone is at the bottom?

39. A mass of 50 kg travelling at 7.905 km/s maintains a circular path of radius 6370 km. What is its acceleration towards the centre?

40. A sphere of mass 500 kg is travelling on a circular path of 12 800 km radius with an acceleration of 2.45 m/s² towards the centre. How long does it take to complete one full circle?

41. At what speed (in km/h) is a bank angle of 45° required for an aeroplane to turn on a radius of 60 m?

42. An aeroplane has a mass of 1500 kg. It is turning on a horizontal circle of radius 100 m at an air speed of 80 knots. Calculate –
   (*a*) the centripetal force exerted by the air on the aircraft,
   (*b*) the correct angle of bank,
   (*c*) the total lift normal to the wings.

43. Find the work required to lift a mass of 5 tonnes to a height of 30 m. If this is done in 2 minutes, what power is being used?

44. Find the power required to propel a 3000 kg aircraft through the air at a speed of 175 knots if the air resistance is 3924 N.

45. Find the power required to propel the same aircraft at 350 knots when the air resistance is 14.715 kN.

46. A car of mass 750 kg can climb a gradient of 1 in 12 in top gear at 40 km/h. If the frictional resistance is 100 N per tonne, find the power developed by the engine in these conditions.

47. A projectile of mass 1 kg is fired from a gun with a muzzle velocity of 850 m/s. What is its kinetic energy? What will be its velocity when the kinetic energy has fallen to 90.3125 kJ?

48. The jet velocity of a certain gas turbine is 500 m/s when the engine is stationary on the ground; if the mass flow of jet gases is 15 kg/s, find the kinetic energy wasted to the atmosphere every minute.

49. A block of wood of mass 75 kg slides down a frictionless slope on to a rough level surface. The slope is 1 in 10 and 10 m long. If the frictional resistance on the level surface is 0.981 N/kg, how far will the block travel along the level surface?

50. Convert a pressure of 70 kN/m² into mm of mercury.

51. A rubber tube is connected to one limb of a U-tube containing mercury. A man blows down the rubber tube until there is a difference of level of 50 mm in the U-tube: what pressure is he exerting in kN/m²?

52. What pressure (in N/m²) is produced by a head of 160 mm of methylated spirit?

53. The air speed in a wind tunnel is measured by a pitot-static tube which is connected to a U-tube containing water. At a certain speed the difference of pressure between the pitot tube and the static tube is found to be 85 mm of water; express the difference of pressure in N/m^2.

CHAPTER 2. AIR AND AIRFLOW – SUBSONIC SPEEDS

54. At a certain height the barometric pressure is 830 mb and the temperature 227 K. Find the density of air at this height.

55. If one fifth of the air is oxygen, what will be the mass of oxygen in 1 m^3 of air at a temperature of $-33°C$ and a pressure of $40 \, kN/m^2$?

56. What is the total mass of air in a room 12 m long, 8 m wide and 4 m high in standard sea-level conditions?

57. What would be the total mass of air in the room mentioned in Q56 if the temperature rose from 15°C to 25°C and the pressure dropped from 760 mm to 735 mm of mercury? (*Assume that the room is not air-tight, and that therefore the air is free to enter or leave the room.*)

58. From Fig. 2.2 read the temperature, pressure and density of the air at sea-level. Taking these values and the corresponding values of temperature and pressure at (*a*) 10 000 ft and (*b*) 10 000 m, calculate the density at these two heights on the assumption that Boyle's Law and Charles' Law are true for air. Compare the calculated values with the corresponding values obtained from the relative density and density respectively given for the International Standard Atmosphere in Fig. 2.2.

59. During a gliding competition a barograph was installed in a glider to measure the altitude reached. On landing, the minimum pressure recorded by the barograph was 472 mb. Draw a graph of pressure against altitude from the values given in Fig. 2.2, and estimate the height reached by the competitor.

60. An aircraft is standing on an airfield 220 ft above sea-level on a day when the barometric pressure at ground level is 1004 mb. If the pilot sets the altimeter to read 220 ft on this day, what will it read if the barometric pressure drops to 992 mb?

61. An aircraft sets off from airfield A (126 ft above sea-level) where the ground pressure is 985 mb. If the pilot sets his alimeter (incorrectly) at 26 ft at A, what will it read when he lands at B?

62. The volume of the pressurised compartment of a jet aircraft is 336 m^3. If the pressurisation system has to maintain a temperature of 17°C and a cabin altitude of 10 000 ft when the aircraft is flying at 40 000 ft with a complete change of air every minute, calculate the mass of air per second which must be delivered to the pressurised compartment.

63. A light aircraft has a landing speed of 70 knots. A wind of 25 knots is blowing over the airfield. What is the ground speed of the aircraft when it touches down –
   (*a*) directly into wind?
   (*b*) at an angle of 30° to the wind?
   (*c*) at an angle of 60° to the wind?
   (*d*) directly down-wind?

64. A and B are two places 400 nautical miles apart. Find the total time taken by an aircraft flying at an air speed of 250 knots to fly from A to B and back to A –
   (*a*) if there is no wind,
   (*b*) if the wind is blowing at 30 knots from A towards B,

(c) if the wind is blowing at 30 knots at right angles to the line joining A and B.

65. A pilot must reach a destination 450 nautical miles away in one hour. He sets off at an air speed of 455 knots and after half an hour finds that he has covered only 212 nautical miles. Assuming constant wind velocity, at what air speed must he fly for the remaining half hour to reach his destination on time?

66. If the pilot in Q65 flew at an air speed of 465 knots for the first half hour, what air speed would be necessary for the remaining time to complete the journey in the hour?

67. An aircraft is taking part in a square search involving flying over the ground in the form of a square of 25 nautical miles side. If the aircraft cruises at 120 knots air speed, and there is a wind of 30 knots down one of the sides of the square, calculate the time of flight for each of the four sides.

68. A flat plate of area 0.25 m² is placed in a 60 knot airstream at right angles to the direction of the airflow. Calculate the air resistance of the plate in these conditions.

69. What would be the resistance of the flat plate of Q68 at 120 knots?

70. What will be the resistance of a sphere of radius 75 mm moving through air at 60 knots? ($C_D = 0.55$)

71. What would be the resistance of the same sphere moving at the same speed through water?

72. A 1/8th scale model of a streamlined body, when tested in a water tank at 5 m/s, had a resistance of 0.6 N. Neglecting any 'scale effect', what would be the resistance of the full-size body at 75 m/s in air?

73. A wind of 15 knots causes a pressure of 50 N/m² on a flat plate at right angles to it. What wind velocity would produce a total force of 1 kN on 3 m² of a similar plate?

74. Of two exactly similar parts of an aeroplane, one is situated in the slipstream from the propeller and the other is outside the slipstream. If the velocity of the slipstream is 1.4 times the velocity of the aeroplane and if the resistance of the part outside the slipstream is 100 N, what will be the resistance of the corresponding part within the slipstream?

75. The air resistance of the fuselage of an aircraft is 13.35 kN at ground level at an air speed of 175 knots. What will be the resistance of this fuselage at a true air speed of 220 knots at 20 000 ft, assuming that the density of air at this height is half the value at ground level?

76. If the undercarriage of an aircraft has a frontal area of 0.45 m², and a resistance of 475 N at a speed of 90 knots, what is the value of the drag coefficient?

77. A rough egg-shaped body with a circular cross-section 75 mm in diameter is tested in a wind tunnel at 100 knots and the air resistance is found to be 1.8 N. What is the value of the drag coefficient?

78. The drag of a loop aerial on an aircraft was found to be 400 N at a speed of 220 knots. In order to reduce this, a fairing of 0.1 m² cross-section and drag coefficient of 0.11 was fitted to the aerial. By how much did this reduce the drag at this speed?

79. A 1/5th scale model of an aeroplane is tested in a wind tunnel at a speed of 25 m/s, and the drag is found to be 56 N. What will be the drag of the full-size machine at 120 knots? (*Neglect any 'scale effect', and assume that the density of the air is the same in each case.*)

80. A 1/10th scale model of an aeroplane is tested in a wind tunnel, and the air

resistance is 65 N at a speed of 100 m/s. What would be the resistance of an 1/8th scale model at 130 m/s? (*Assume that the air density is the same for both tests.*)

81. A 1/10th scale model of a hull of a flying boat is tested in a water tank and has a resistance of 135 N at a speed of 12 m/s. What would be the resistance of the full-size hull in water at a speed of 45 knots?

82. A streamlined shape with a cross-sectional area of 0.01 m² is tested in a compressed air tunnel at a speed of 70 knots and a pressure of 25 atmospheres. If the resistance is 18 N, what is the value of the drag coefficient?

83. What would be the resistance of the same body in water at a speed of 3 m/s?

84. A 1/5th scale model has a resistance of 19.5 N when tested in a wind tunnel. What would be the resistance of a 1/2 scale model at half the speed in air five times as dense?

85. If the drag coefficient of the flaps used on an aircraft is 0.92, what would be the drag of these flaps at 100 knots if their area totalled 3 m²?

86. An aeroplane is to be modified, and in order to estimate the effect of the modification the drag of two models, one of the original and one of the proposed modified type, is measured in a wind tunnel.

The model of the original is 1/20th scale and, when tested in air of density 1.225 kg/m³ at 65 knots, the drag is 62 N.

The model of the modified type is 1/16th scale and, when tested in air of density 1.007 kg/m³ at 60 knots, the drag is 48 N.

By what percentage will the modification increase or decrease the drag coefficient of the aeroplane?

87. The resistance of a part of an aeroplane is 640 N when the aeroplane is flying at 150 knots near sea-level. What will be the resistance of this part at a height of 20 000 ft if the 'indicated' air speed is the same, i.e. 150 knots?

88. What would be the resistance of this part at 20 000 ft if the 'true' air speed were 150 knots?

89. If the static atmospheric pressure is 101.3 kN/m², and the air density is 1.225 kg/m³, what will be the pressure on the pitot side of the diaphragm in an air speed indicator when the forward speed of the aircraft is 100 knots?

90. An aircraft is flying at a true air speed of 270 knots at a height of 20 000 ft, where the air density is 0.653 kg/m³ and the pressure is 466 mb.

What are the pressures transmitted to the air speed indicator via (*a*) the static tube and (*b*) the pitot tube?

What will be the indicated air speed if the density assumed in the calibration of the instrument was 1.225 kg/m³?

91. If an aircraft stalls in straight and level flight at an indicated speed of 100 knots at sea-level, at what true air speed will it stall at –
    (*a*) 20 000 ft?   (*b*) 40 000 ft?

92. At what indicated air speed will it stall at –
    (*a*) 20 000 ft?
    (*b*) 40 000 ft?

93. An air speed indicator is being calibrated with a U–tube containing mercury. Calculate the speed that corresponds to a difference of pressure of 40 mm of mercury.

94. An aircraft flying at 10 000 ft runs into severe icing which blocks up the static tube, but leaves the pitot tube open. The aircraft descends and approaches to land at sea-level with the static tube still blocked by ice. If the pilot approaches at a true air speed of 60 knots, what speed will be indicated on the air speed indicator?

95. An aircraft is flying at sea-level at a true air speed of 150 knots. Calculate the static pressure at –
    (a) the stagnation point,
    (b) a point on the wing surface where the local flow velocity is double the free stream velocity.

96. A venturi tube is so designed that the ratio of the diameter at the throat to the diameter at the mouth is 0.6. The velocity of the airflow at the mouth is 80 knots and the static pressure there is 101.3 kN/m². Find the static pressure at the throat assuming that the air density is 1.225 kg/m³ at both mouth and throat.

97. The diameter at the throat of a venturi is one third of that at the mouth. What velocity of air at the mouth will be necessary to produce a decrease in pressure of 50 mm of mercury at the throat when the normal atmospheric pressure is 760 mm of mercury?

CHAPTER 3. AEROFOILS – SUBSONIC SPEEDS

98. The table shows the lift coefficient of a flat plate at angles of attack from 0° to 90°. A flat plate of 6 m span and 1 m chord is tested in an airstream of velocity 30 m/s. Plot a graph showing how the lift of such a plate varies as its angle to the airflow is increased from 0° to 90°.
    (a) What is the maximum lift obtained?
    (b) What would be the maximum lift of a RAF 15 aerofoil of the same area under similar conditions?

Angle of attack	0°	5°	10°	15°	20°	30°	40°	50°	60°	70°	80°	90°
Lift coefficient	0	0.36	0.68	0.80	0.78	0.80	0.76	0.68	0.56	0.38	0.20	0

99. If an aeroplane of mass 950 kg has a wing area of 20 m², what is the wing loading in N/m²?

100. A pressure plotting experiment is carried out in a wind tunnel on a model aerofoil of chord 350 mm, and large aspect ratio. Methylated spirit is used in the manometer. The table overleaf shows the distances of the holes a, b, c, d, etc., from the leading edge, and also the corresponding pressures recorded at these holes in millimetres of methylated spirit, the negative values representing pressures below the static pressure in the tunnel. The air speed was 45 m/s and the angle of attack 4°. Find the lift coefficient for the aerofoil at this angle of attack.

Note. Strictly speaking, from the data given, it is impossible to find the lift coefficient because we do not know the shape of the aerofoil section. The pressure forces act perpendicular to the surface of the aerofoil and therefore we do not know the true direction of the pressure forces relative to the airflow and so we cannot find the lift which, by definition, is that part of the total force which is at right angles to the airflow.

What we can do, however, is to find the normal force, i.e. the component of the total force which is at right angles to the chord of the aerofoil. From this we can calculate the normal force coefficient which, at this small angle of attack, will be of the same value as the lift coefficient to at least two places of decimals.

	Distance from leading edge (mm)	*Pressure,* mm *of methylated spirit*
Upper surface –		
Hole *a*	5	−228
Hole *b*	20	−203
Hole *c*	41	−195
Hole *d*	74	−153
Hole *e*	103	−112
Hole *f*	153	−76
Hole *g*	216	−64
Hole *h*	292	−25
Lower surface –		
Hole *k*	15	+97
Hole *l*	46	+99
Hole *m*	89	+86
Hole *n*	153	+56
Hole *o*	228	+25
Hole *p*	305	+8

The student is advised to work this question out because it will help him to understand several aspects of the subject. Proceed as follows –

Draw the chord line to some suitable scale, preferably on squared paper, marking off the position of each hole a', b', c', etc., as shown in the figure. (There is no need to incline the chord at 4°; exactly the same result will be obtained, rather more simply, if it is drawn horizontal.)

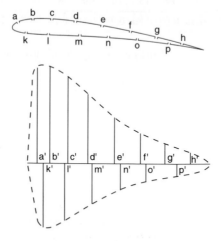

At each point set off a vertical line to some suitable scale to represent the pressure at the corresponding hole; the line should be upwards if the pressure is below atmospheric, and downwards if above atmospheric.

Draw a pressure distribution curve through the ends of these lines.

By one of the mathematical methods, or by counting the squares, or by using a planimeter, find the total area enclosed by this curve.

Divide the area by the length of the diagram, thus finding the average height; this represents the average pressure on the aerofoil – at right angles to the chord line – in millimetres of methylated spirit. (If the reader has worked on engines, he will recognise the similarity of this method to that of finding 'mean effective pressure' from an indicator diagram.)

Convert the average pressure into N/m^2 of wing area (S).

Then total force (normal to chord) = average pressure $\times S$.

This is called the normal force.

Just as lift = $C_L \cdot \frac{1}{2}\rho V^2 \cdot S$, so normal force = $C_Z \cdot \frac{1}{2}\rho V^2 \cdot S$, where C_Z is called the normal force coefficient.

Equating average pressure $\times S$ to $C_Z \cdot \frac{1}{2}\rho V^2 \cdot S$, the wing area (S) will cancel out; taking ρ as $1.225\,\mathrm{kg/m^3}$ and V as $45\,\mathrm{m/s}$ we can find C_Z which, as already explained, is very nearly the same as C_L.

101. Taking the values of the static pressures given in the table in Q100, find the speed of the airflow at each of the holes and construct velocity distribution diagrams for the upper and lower surfaces.

Note. The aerofoil acts like a venturi tube and Bernoulli's Principle (static pressure + dynamic pressure = constant) can be applied directly. Over the top surface the static pressure decreases, so the dynamic pressure and thus the speed of the airflow increases. On the under surface, the static pressure increases and the speed of the airflow decreases. The velocity distribution diagrams should be plotted with the aerofoil chord as abscissa (horizontal) and the velocity as ordinate (vertical), and the diagrams will be similar in shape to the pressure plotting diagrams. Assume the density of the air remains constant at $1.255\,\mathrm{kg/m^3}$.

102. When the angle of attack of a certain aerofoil is $12°$, the direction of the resultant lies between the perpendicular to the chord line and the perpendicular to the airflow, being inclined at $8°$ to the latter (see figure). If the total force is $700\,\mathrm{N}$, find its component in the direction OC.

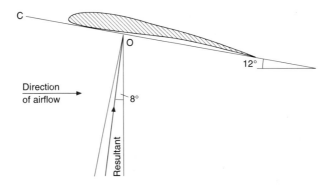

103. What are the lift and drag on the aerofoil in Q102?

104. From the tables given in Appendix 1 draw curves of lift coefficient against angle of attack for –

    (*a*) The 'general purpose' biplane aerofoil section, RAF 15.

    (*b*) The 'thin symmetrical' section, NACA 0009.

    (*c*) The 'glider and sailplane' section, NACA 23012.

    (*d*) The 'Lightning model' section, ASN/P1/3.

From your curves write down the maximum lift coefficients of these four aerofoil sections. (*These curves should be drawn to the same scale so that comparisons can be made between the various sections. The student will find it useful to keep these curves, and also those in the following questions, as they will be valuable in solving other problems.*)

105. Draw curves of drag coefficient against angle of attack for the four sections mentioned in Q104.

106. Draw curves of lift/drag ratio against angle of attack for the four sections mentioned in Q104.

107. Draw graphs to show how the centre of pressure moves as the angle of attack is altered on the following sections –

    (*a*) Clark YH.

    (*b*) NACA 0009.

    (*c*) NACA 4412.

    (*d*) NACA 23018.

Which of these four sections gives the least movement of the centre of pressure over the ordinary angles of flight?

108. A model aerofoil section (span 0.3 m, chord 50 mm) is tested in a wind tunnel at a velocity of 60 knots. The maximum lift obtained is 11 N. Find the value of the maximum lift coefficient.

109. From the curves drawn for Q159 find the lift of a RAF 15 wing of area 25 square metres at 75 knots, the angle of attack being 4°.

110. What would be the lift of the wing in Q109 if it had been of NACA 4412 section?

111. A sailplane of total mass 300 kg is fitted with a NACA 23012 section wing of area 20 square metres. From the graphs drawn for Q104 and 106, find the indicated air speed at which it should be flown so that the wings are operating at the maximum lift/drag ratio.

112. Draw curves of lift coefficient against angle of attack for the following –

    (*a*) NACA 4412.

    (*b*) NACA 4412 with 20% split flap set at 60°.

    (*c*) Lightning Model – clean aircraft.

    (*d*) Lightning Model with 25% flap set at 50°.

From these curves write down the values of the maximum lift coefficients, and the angles of attack at which they occur.

113. In a test on an aerofoil section the moment coefficient about the leading edge was found to be -0.005 at zero lift. At 6° angle of attack the moment coefficient about the leading edge was -0.240, and the lift and drag coefficients were 0.90 and 0.045 respectively. Find the position of the aerodynamic centre of this aerofoil section.

114. What is the position of the centre of pressure of the aerofoil section of Q113 at 6° angle of attack?

115. The aerodynamic centre of a certain aerofoil section is at 0.256 c. At an angle of attack of 8° the moment coefficient about the aerodynamic centre is -0.100 and the lift and drag coefficients are 1.05 and 0.035 respectively. What is the moment coefficient about the quarter-chord point at this angle of attack?

116. What is the position of the centre of pressure of the aerofoil section of Q115 at 8° angle of attack?

117. An elliptical planform wing has a span of 12 m and a chord of 2 m. What is the induced drag coefficient when the lift coefficient is 0.8?

118. What is the induced drag at 100 knots of a monoplane of mass 2400 kg having a wing span of 16 m?

119. A fighter aircraft has a mass of 7200 kg and a wing span of 12 m. If the wing

loading is $1.968 \, \text{kN/m}^2$, what is the induced drag when flying at speeds of 200 knots and 300 knots at sea-level?

120. A monoplane weighs 15 000 kgf, and has a span of 10 m. What is the induced drag at –

    (*a*) 120 knots,

    (*b*) 240 knots, at sea-level?

121. What would be the induced drag of the aircraft in Q120 at 40 000 ft at an indicated air speed of 240 knots?

122. An aircraft of mass 15 000 kg is a monoplane of 20 m span. Calculate the induced drag at sea-level at speeds ranging from 80 to 400 knots. Plot the results in the form of a graph of speed (abscissa) against induced drag (ordinate). (*The student will find it useful to keep this graph as it will be referred to in later questions.*)

123. Calculate the power (in kilowatts) required to overcome the induced drag of the aircraft at the speeds used in Q122, and plot the results in the form of a graph of speed (abscissa) against power (ordinate).

124. The sailplane in Q111 has an aspect ratio of 8. Find the percentage reduction in induced drag that would result from increasing the aspect ratio to 12.

125. The following table gives the values of all the drag except induced drag (i.e. of form drag + skin friction) of the aircraft in Q122 at various speeds at sea-level –

Speed	80	120	160	200	240	280	320	360	400 knots
Form drag + skin friction	2.4	5.4	9.6	15.0	21.6	29.4	38.4	48.6	60.0 kN

Plot these values on the graph drawn in Q122, and construct a graph showing the variation of the total drag with air speed.

What percentage of the total drag is induced drag at 120, 200, 280 and 360 knots?

126. At what speed is the induced drag exactly half the total drag? (*Note. This speed is the speed for minimum total drag, i.e. the speed for maximum range. Keep these graphs; they will be referred to in later questions.*)

CHAPTER 4. THRUST

127. If a propeller, on a stationary mounting, blows back 36 kg of air per second at a speed of 45 knots, what thrust does it produce?

128. An aircraft engine is being tested on the ground before take-off. If the propeller has a diameter of 3 m, and the velocity of the slipstream is 80 knots, what thrust is being produced? (*In calculating the mass flow of air past the propeller take the average velocity of the air well in front of and behind the propeller, i.e. 40 knots.*)

129. What thrust will be produced by the propeller in Q128 when the aircraft is flying at 160 knots at 5000 ft, if the velocity of the slipstream relative to the aircraft is now 180 knots?

130. A twin-engined aircraft, with propellers of 3 m diameter, is flying at 10 000 ft at an air speed of 350 knots. If the aircraft has a mass of 10 000 kg, and has a lift/drag ratio of 5 to 1 in these conditions, what thrust is being produced and what is the velocity of the slipstream relative to the aircraft?

131. The jet velocity of a gas turbine on the ground is 400 m/s; if the mass flow is 18 kg per second, what thrust does it produce?

132. An aircraft powered by two gas turbines is flying at 600 knots. If the jet velocity is 440 m/s, and the mass flow is 66 kg/s for each engine, what is the total thrust being developed?

133. What is the total power being developed by the gas turbines in Q132?

134. If a rocket can burn and eject 100 kg of fuel per second, what jet velocity is required to give a thrust of 50 kN?

135. A rocket, the total mass of which is 25 kg, contains 10 kg of fuel. If all the fuel is burnt in 2 seconds, and is ejected with a velocity of 500 m/s, what thrust is produced? What would be the initial acceleration if the rocket were fired off vertically upwards?

136. An aircraft using a rocket-assisted take-off system on an aircraft carrier has a mass of 4500 kg complete with rockets. It has a jet engine capable of producing 15 kN of thrust, and each rocket will produce 1566 N of thrust for 8 seconds. If the carrier steams at 20 knots into a head wind of 24 knots, how many rockets will have to be fired to get the aircraft airborne in a distance of 60 m along the deck if the take-off speed is 80 knots, and the average air and wheel resistance during take-off is 8.4 kN?

137. The pitch of a propeller is 2.5 m. If the slip is 15% when running at 1200 rpm, what is the speed of the aeroplane to which it is fitted?

138. The maximum speed of a light aeroplane is 80 knots when the engine is revolving at 3000 rpm; if the pitch of the propeller is 1.1 m, what is the percentage slip?

139. What is the torque of an engine which develops 1500 kW at 2400 rpm?

140. If the engine of Q139 is fitted with a propeller which has an efficiency of 83% at an advance per revolution of 3 m, what will be the thrust of the propeller?

141. When a certain aeroplane travels horizontally at an air speed of 300 knots, the engine develops 800 kW. If the propeller efficiency at this speed is 87%, find the thrust of the propeller.

142. The efficiency of a propeller is 78%. At what forward speed will it provide a thrust of 3 kN when driven by an engine of 210 kW power?

143. A propeller is revolving at 2600 rpm on an aircraft travelling at 280 knots. The thrust is 4 kN and the torque 2500 N-m. What is the efficiency of the propeller?

144. The diameter of a propeller is 3 m. The blade angle at a distance of 1 m from the axis is 23°. What is the geometric pitch of the propeller?

145. The geometric pitch of a propeller of 3 m diameter is to be 3.6 m, and should be constant throughout the blade. Find the blade angles at distances 0.8 m, 1 m, and 1.2 m respectively from the axis of the propeller.

CHAPTER 5. LEVEL FLIGHT

146. The mass of an aeroplane is 2000 kg. At a certain speed in straight and level flight the ratio of lift to drag of the complete aircraft is 7.5 to 1. If there is no force on the tail plane, what are the values of the lift, thrust and drag?

147. In a flying boat the line of thrust is 1.6 m above the line of drag. The mass of the boat is 25 000 kg. The lift/drag ratio of the complete aircraft is 5 to 1 in straight and level flight. If there is to be no force on the tail, how far must the centre of pressure of the wings be in front of the centre of gravity?

148. An aeroplane of 10 000 kg mass is designed with the line of thrust 0.9 m above the line of drag. In normal flight the drag is 18.2 kN and the centre of pressure on the main plane is 150 mm behind the centre of gravity. If the centre of pressure

on the tail plane is 10 m behind the centre of gravity, what is the load on the tail plane?

149. In a certain aeroplane the line of thrust is 100 mm below the line of drag. The mass of the aeroplane is 1500 kg, and the drag is 2.3 kN. If the aircraft is to be balanced in flight without any load on the tail, what must be the position of the centre of pressure relative to the centre of gravity?

150. In a certain aeroplane, which has a mass of 2000 kg, the centre of lift and the centre of gravity are in the same vertical straight line when in normal cruising flight. If the thrust is 4.5 kN, and is 180 mm below the centre of drag, what force must there be on the tail plane which is 6 m behind the centre of gravity?

151. What force would be required on the tail plane of the aeroplane in Q150 if the centre of lift had been 25 mm behind the centre of gravity?

152. A jet aircraft with a mass of 6000 kg has its line of thrust 150 mm below the line of drag. When travelling at high speed, the thrust is 18.0 kN and the centre of pressure is 0.5 m behind the centre of gravity. What is the load on the tail plane which is 8.0 m behind the centre of gravity?

153. An aircraft with a mass of 5500 kg is flying straight and level at its maximum speed. The thrust line is horizontal and is 0.3 m above the drag line which passes through the centre of gravity. If the drag is 12.0 kN, and the centre of pressure is 0.6 m behind the centre of gravity, find the load on the tail plane which is 5.5 m behind and on a level with the centre of gravity.

154. When the aircraft in Q153 is flying at its minimum speed, the thrust line is inclined at 25° to the horizontal. If the centre of pressure is now a horizontal distance of 0.1 m in front of the centre of gravity, what is the vertical load on the tail plane assuming that the drag (now 10 kN) still acts through the centre of gravity?

155. An aircraft with a mass of 14 000 kg is fitted with a wing of 60 m² area and of Clark YH section. Taking the values of the lift coefficient from the table on p. 449, calculate the indicated air speeds corresponding to the angles of attack from −2° to +25°; Draw the following graphs –

   (a) Indicated air speed (abscissa) v. Angle of attack (ordinate).
   (b) Indicated air speed (abscissa) v. Lift coefficient (ordinate).

At what speed must this aircraft fly to ensure that the wings are operating at the angle of attack giving the maximum lift/drag ratio? (*Note that this speed will be the aerodynamic range speed for the aircraft.*)

156. If the aircraft of Q155 uses 1500 kg of fuel during a flight, what should be the speed at the end of the flight to keep the wings operating at the angle of attack which gives maximum lift/drag ratio?

157. Draw a graph of the values of $C_L^{\frac{3}{2}}/C_D$ against angle of attack for the aerofoil section Clark YH. What angle of attack gives the maximum value of $C_L^{\frac{3}{2}}/C_D$? Refer to the graphs drawn in Q155 and determine the air speed that corresponds to this angle of attack for the aircraft in question. Compare this air speed with four-fifths of the range speed as found in Q155.

158. The aircraft in Q122 and Q125 is fitted with a 1000 kg load which is stowed internally. What is the new speed for maximum range? Compare the result with that obtained in Q126. (*Note. The maximum range speed is that which gives the minimum total drag of the aircraft. The curve of total drag is rather flat in this region, so it is not easy to estimate the minimum position accurately. A better method is to find the speed at which the induced drag is equal to all the other drag, i.e. form drag + skin friction, as this condition will also give the minimum total drag. All that is necessary in this question is to calculate the induced drag at the various speeds for this new load, and plot the results on*

the graph drawn for Q125. The intersection of the new induced drag curve and the old form drag + skin friction curve will give the new range speed.)

159. The aircraft of Q122 is a twin-engined aircraft, and on a certain flight one engine fails. The pilot continues flying with the propeller of the dead engine wind-milling. This increases the form drag + skin friction to the following values –

Speed	80	120	160	200	240	280	320	360	400	knots
Form drag + skin friction	3.0	6.7	12.0	18.7	27.0	36.8	48.1	60.8	75.2	kN

What is the speed for maximum range under these conditions? (*See note under Q158*)

160. The aircraft in Q158 is fitted with a 1000 kg torpedo instead of the previous 1000 kg load; the torpedo, however, has to be carried externally and this increases the form drag + skin friction to the following values –

Speed	80	120	160	200	240	280	320	360	400	knots
Form drag + skin friction	2.7	6.1	10.9	16.9	24.4	33.1	43.3	54.8	68.0	kN

What is the speed for maximum range under these conditions?

161. Taking the values of the total drag at sea-level found in Q125, determine the power needed to drive the aircraft through the air at speeds from 80 to 400 knots. Plot the results and find the speed at which minimum power is required, i.e. the speed for maximum endurance. Compare this speed with the range speed found in Q182.

162. From the graph drawn for Q161, read off the power required at sea-level to fly the aircraft at its maximum endurance speed, and calculate the power required for the same indicated air speed at 5000 ft, 10 000 ft and 15 000 ft above sea-level. (*Note. Remember that power depends on true air speed.*)

163. The following table gives the fuel consumption of a twin-engined training aircraft at various speeds at sea-level. Draw graphs of air speed against kg per hour, and air speed against air nautical miles per kg, and determine the speeds for maximum range and maximum endurance.

True air speed	80	90	100	110	120	130	140	150	160	knots
kg/hour		162	152	151	158	172	190	214	241	275	

164. Determine the speeds for maximum range over the ground for the aircraft of Q163 when flying –

   (*a*) against a head wind of 40 knots,
   (*b*) with a tail wind of 40 knots.

165. If the aircraft in Q125 is to be propelled by jet engines instead of piston/propeller engines, what will be –

   (*a*) the speed for maximum range?
   (*b*) the speed for maximum endurance?

166. The fuel consumption of a jet-propelled aircraft at 220 knots indicated air speed, at various altitudes, is as follows –

Sea-level	1240 kg/h
10 000 ft	1046 kg/h
20 000 ft	990 kg/h
30 000 ft	880 kg/h

Calculate the air nautical miles per kilogram for each altitude, and plot a graph showing the variation of air nautical miles per kilogram with altitude.

CHAPTER 6. GLIDING AND LANDING

167. When there is no wind, a certain aeroplane can glide (engine off) a horizontal distance of $1\frac{1}{2}$ nautical miles for every 1000 ft of height.
What gliding angle does this represent?

168. What will be the gliding angle, and what will be the horizontal distance travelled per 1000 ft of height, if the aeroplane in Q167 glides against a head wind of 20 knots? The air speed during the glide can be taken as 60 knots and the wind direction as horizontal.

169. An aeroplane glides with the engine off at an air speed of 80 knots, and is found to lose height at the rate of 1500 ft/min. What is the angle of glide? (*Assume conditions of no wind.*)

170. From the result of Q169, what is the value of the lift/drag for this aeroplane on this glide?

171. At angles of attack of 1°, 4° and 10° the values of the lift/drag ratio of a certain aeroplane when gliding with the engine off are 3.5, 8 and 4.5 respectively. What horizontal distance should a pilot be able to cover from a height of 10 000 ft if he glides at each of these angles of attack?

172. An aeroplane of 5000 kg mass is flying at 20 000 ft. It glides to 10 000 ft at an angle of attack giving a lift/drag ratio of 10 to 1. What is the horizontal distance covered?

173. A model of a sailplane is tested in a wind tunnel, and the following values of C_L and C_D are found at the angles of attack stated:

Angle of attack	0°	2°	4°	6°	8°	10°	12°
C_L	0.09	0.23	0.39	0.53	0.68	0.83	0.98
C_D	0.009	0.010	0.016	0.023	0.036	0.049	0.064

Neglecting any scale effect, what is the flattest possible gliding angle that should be obtainable with the full-size sailplane? and at what angle of attack should the sailplane be flown to cover the greatest horizontal distance? (*Assume conditions of no wind.*)

174. What angle of attack should be used with the sailplane of Q173 to give the minimum rate of descent? and what is the gliding angle under these conditions? (*Assume no wind.*)

175. A sailplane with an all-up weight of 2452 N has a lift/drag ratio of 24 to 1 when gliding for range at 40 knots. Calculate the angle of glide, and the sinking speed in ft/s.

176. When flying with a heavier pilot, the all-up weight of the sailplane of Q175 is 2600 N. At what speed must he fly to cover the same range (in still air)? and what will be the sinking speed?

177. When flying for endurance, the sailplane of Q175 has a lift/drag ratio of 21.5 and a speed of 34 knots. Calculate the angle of glide and the sinking speed, and compare these with the values found in Q175.

178. Find the minimum landing speed of an aeroplane of mass 500 kg and a wing area of 18.6 m². The maximum lift coefficient of the aerofoil section is 1.0.

179. When an aeroplane is fitted with an aerofoil having a maximum lift coefficient of 1.08, the minimum flying speed is 42 knots. If, by the use of slots, the maximum lift coefficient can be increased to 1.60, what will be the minimum flying speed when fitted with slots?

180. It is required that a light aeroplane of mass 400 kg should land at 40 knots. What wing area will be required if the following aerofoil sections are used –
  (*a*) Clark YH.
  (*b*) NACA 4412.
  (*c*) NACA 23012.
  (*d*) NACA 4412 with flap.

181. From curves for NACA 23018 aerofoil read the values of the lift coefficient at angles of attack of 2°, 6° and 12°. If an aeroplane of mass 1000 kg is fitted with this aerofoil, what air speed will be necessary for horizontal flight at each of these angles of attack, assuming that the effective wing area is 20 square metres?

182. The total weight of an aeroplane is 14 715 N, and its wing area is 40 m². The maximum lift coefficient of the aerofoil is 1.08. What will be the maximum speed of flight –
  (*a*) at sea-level?
  (*b*) at 10 000 ft?

183. The total loaded mass of an aeroplane is 1100 kg. When NACA 65$_1$-212 aerofoil section is used the minimum landing speed is 56 knots. With a view to decreasing this landing speed, the following alterations are considered –
  (*a*) the fitting of slots which will increase the maximum lift coefficient by 40 per cent,
  (*b*) the fitting of flaps which will increase the maximum lift coefficient by 80 per cent,
  (*c*) a 20 per cent increase in wing area.
It is estimated that the increase in total mass necessitated by these alterations would be (*a*) 25 kg, (*b*) 30 kg and (*c*) 50 kg. What will be the resulting reduction in landing speed which can be achieved by each of the three methods?

184. An aeroplane of mass 13 500 kg has a wing loading of 2.75 kN/m². At 8° angle of attack the lift coefficient is 0.61. What is the speed necessary to maintain horizontal flight at this angle of attack at sea-level?

185. An aircraft is to be fitted with a NACA 23018 aerofoil section, and flaps which increase the maximum lift coefficient by 60%. If the landing speed must not be more than 85 knots, what is the highest possible value of the wing loading?

186. An aircraft of 20 000 kg mass is to be fitted with a NACA 4412 aerofoil section with a 20 per cent split flap. If the area of the wing is to be 84 m², what will be the landing speed –
  (*a*) with flaps lowered to 60°?
  (*b*) without flap?

CHAPTER 7. PERFORMANCE

187. An aeroplane of mass 2700 kg has a wing area of 30 m² of NACA 65$_1$-212 aerofoil section. During the take-off the pilot allows the aeroplane to run along the ground with its tail up until it reaches a certain speed; he then raises the elevators and so increases the angle of attack of the main planes to 12° which just enables the aeroplane to leave the ground. What is the required speed?

188. What would have been the speed required for the aeroplane of Q187 if the pilot had allowed it to continue running along the ground until it took off at an angle of attack of 4°?

189. An experimental aircraft of total mass 45 000 kg has four normal engines, and four auxiliary lift engines to provide a short take-off when required. The average thrust produced by each of the normal engines during take-off is 12.5 kN, and the take-off run required in still air is 2000 m. If the average resistance to motion is 11.0 kN, what is the take-off speed?

190. When using the auxiliary lift engines, the aircraft of Q189 can take off in 1000 m at the same angle of attack in still air. What thrust must each lift engine produce?

191. An aeroplane of 3000 kg mass is climbing on a path inclined at 12° to the horizontal. Assuming the thrust to be parallel to the path of flight, what is its value if the drag of the aircraft is 5.0 kN?

192. An aeroplane of 12 000 kg mass climbs at an angle of 10° to the horizontal with a speed of 110 knots along its line of flight. If the drag at this speed is 36.0 kN, find –

(*a*) the power used in overcoming drag

(*b*) the power used in overcoming the force of gravity.

Hence find the total power required for the climb.

193. A jet aircraft with a wing loading of 2.4 kN/m², and a mass of 4500 kg, has a maximum thrust of 30 kN at sea-level. If the drag coefficient at a speed of 270 knots is 0.04, what will be –

(*a*) the maximum possible rate of climb?

(*b*) the greatest angle of climb?

at this speed.

194. An aircraft of 5000 kg mass is powered by an engine capable of producing 1500 kW power. Calculate the maximum angle of climb at an air speed of 130 knots if the efficiency of the propeller is 80 per cent, and the drag at this speed is 6.7 kN.

195. At sea-level the total drag of an aircraft of mass 5500 kg is 5.0 kN at a speed of 160 knots. Calculate the rate of climb and angle of climb at an indicated air speed of 160 knots at 10 000 ft, if the power available is 980 kW and the relative air density is 0.739.

196. For the aircraft of Q195 calculate the rate of climb and angle of climb at the same angle of attack, if the mass of the aircraft is increased to 7000 kg by the addition of internal load, and the power available at 10 000 ft remains unchanged.

197. A jet aircraft weighing 6000 kgf has a climbing speed of 250 knots. If the rate of climb is 9000 ft/min, and the drag of the aircraft in this condition is 8.2 kN, find the thrust being delivered by the engines.

198. The following table gives particulars relating to a certain aeroplane of 1050 kg mass –

Speed of level flight	45	50	55	60	65	70	75	80	85	90	95	100	knots
Power available from propeller	135	170	205	225	240	250	255	255	250	240	230	220	kW
Power required for level flight	250	115	93	90	100	120	150	180	215	255	300	350	kW

Estimate
(a) the minimum speed of level flight,
(b) the maximum speed of level flight,
(c) the best airspeed for climbing purposes,
(d) the power available for climbing at this airspeed and
(e) the maximum vertical rate of climb.

199. By throttling down the engine of the aeroplane of Q198 the power available is reduced by 30 per cent throughout the whole range of speed. Find the maximum and minimum speeds for level flight under these conditions.

200. An aeroplane of 1500 kg mass has a minimum landing speed of 50 knots, and the minimum power for level flight is 60 kW at 80 knots. If 250 kg of extra load is added, find the new minimum landing speed, and the new minimum power and speed for level flight, i.e. at the same angle of attack as before.

201. The following table gives particulars of a certain aeroplane of 6000 kg mass which is designed for either piston engine/propeller or jet propulsion –

Speed of level flight in knots	50	70	90	110	130	160	200	240	280	320	360	400	440
Power required for level flight in kW	2000	750	580	540	540	640	610	1340	2000	2870	4000	5500	7500
Power available from propellers in kW		900	1120	1300	1480	1640	1870	2130	2340	2550	2700	2800	2800	2650
Power available from jets in kW		660	950	1200	1500	1750	2170	2720	3260	3800	4340	4900	5450	6000

Estimate the maximum and minimum speeds for level flight when the aircraft is propelled by the piston engine/propeller combination.

202. Estimate the maximum and minimum speeds for level flight when the aircraft in Q201 is propelled by the jet engines, and compare the speeds with those found in Q201.

203. Estimate the best climbing speed and rate of climb at this speed for the aircraft in Q201 when propelled by the piston engine/propeller combination.

204. Estimate the best climbing speed and rate of climb at this speed when the aircraft in Q201 is propelled by the jet engines. Compare these answers with those obtained in Q203.

205. When the mass of the aircraft of Q201 is increased by carrying an extra 900 kg of fuel in overload tanks, the power required for level flight is increased to the following:

Speed of level flight in knots	60	70	90	110	130	160	200	240	280	320	360	400	440
Power required for level flight in kW	1600	1000	760	670	650	700	960	1385	2045	2910	4040	5535	7535

Estimate the maximum and minimum speeds for level flight in this condition when propelled by the piston engine/propeller combination.

206. Estimate the maximum and minimum speeds for level flight of the aircraft in Q205 when propelled by the jet engines.

207. The lift produced by the wings of an aircraft travelling at 250 knots at sea-level is 70 kN. At what speed must the aircraft travel at 45 000 ft to produce the same lift at the same angle of attack of the wings?

208. What will be the indicated air speed of the aircraft of Q207 at 45 000 ft?

209. If the aircraft of Q207 has a lift/drag ratio of 10 to 1 when travelling at an indicated air speed of 250 knots, calculate the power required to propel the aircraft at this indicated air speed –

    (*a*) at sea-level,

    (*b*) at 45 000 ft.

210. When flying at its endurance speed of 140 knots, an aircraft of 7250 kg mass has a drag of 13.4 kN. Calculate the power required to propel the aircraft at altitudes from sea-level to 45 000 ft (at 5000 ft intervals) and construct a graph of –

Altitude (abscissa) against Power Required (ordinate).

Estimate the altitude at which the power required is double that at sea-level.

211. The following table gives values of the maximum power available from the propeller fitted to the aircraft of Q210 when flying at the endurance speed at various altitudes –

Altitude (ft)	0	5000	10 000	15 000	20 000	25 000	30 000	35 000	40 000	45 000
Power available (kW)	2050	2095	2140	2180	2240	2085	1965	1840	1740	1630

Note that the rated altitude of the engine is 20 000 ft. Estimate –

    (*a*) the absolute ceiling of the aircraft,

    (*b*) the service ceiling (the altitude at which the maximum rate of climb is reduced to 100 ft/min).

212. If the engine fitted to the aircraft of Q210 uses fuel at the rate of 0.275 kg/kW hr, calculate the endurance of the aircraft –

    (*a*) at sea level, and (*b*) at 30 000 ft,

assuming that the aircraft has 1000 kg of fuel available for level flight, and that the propeller efficiency is 80 per cent at both altitudes.

213. An aircraft weighing 13 500 kgf has a range speed of 175 knots indicated. At this speed the lift/drag ratio is 8 to 1, and the propeller efficiency is 82 per cent. If the fuel consumption is 0.27 kg/kW hr, calculate the air nautical miles covered per kilogram of fuel at –

    (*a*) sea-level,

    (*b*) 20 000 ft,

    (*c*) 40 000 ft.

214. If the aircraft of Q213 is propelled by jet engines at the same indicated

speed, and if the fuel consumption is then 57 kg/kN of thrust hr, calculate the air nautical miles per kilogram of fuel at –
- (*a*) sea-level,
- (*b*) 20 000 ft,
- (*c*) 40 000 ft.

Compare the results with those obtained in Q213.

CHAPTER 8. MANOEUVRES

215. Find the correct angle of bank for an aeroplane of 1200 kg mass taking a corner of radius 60 m at 75 knots.

216. What will be the total lift in the wings of the aeroplane in Q215 while taking this corner at the correct angle of bank?

217. An aircraft with a mass of 1000 kg does a steady turn at 55 knots and an angle of bank of 45°. Calculate (*a*) the acceleration, (*b*) the force required to produce the acceleration, and (*c*) the wing loading on the aircraft during the turn if the wing area is 14.14 m².

218. If the aerofoil section used on the aeroplane of Q42 is NACA 23012, and if the total wing area is 25 metres, calculate –
- (*a*) the angle of attack required for normal horizontal flight at the same speed of 80 knots, and
- (*b*) the angle of attack required to produce the necessary lift when turning the corner at this speed.

219. An aeroplane of 1750 kg mass makes a horizontal turn at an angle of bank of 25°. If the speed in the turn is 85 knots, what is the radius of the turn?

220. Calculate the radius of turn and angle of bank of an aircraft doing a Rate 1 turn (180° per minute) at 550 knots. What is the acceleration in multiples of *g*? And if the mass of the aircraft is 5000 kg, what is the lift force on the wings during the turn?

221. Calculate the loading on an aircraft in a correctly banked turn at angles of bank of 60°, 75°, 83°, 84°. (*Note. The loading is expressed as a factor found by dividing the lift force on the wings by the force on the wings in straight and level flight, i.e. the weight.*)

222. Calculate the accelerations in multiples of *g* for each of the turns mentioned in Q221.

223. If the stalling speed of an aircraft is 60 knots in straight and level flight, what is the stalling speed in correctly banked turns at angles of bank of 45°, 60°, 75°, 83°, 84°?

224. Calculate the rate of turn of an aeroplane in degrees per minute if the acceleration during the turn is 4 *g* at speeds of 100 knots, 150 knots, 300 knots, 500 knots.

225. If the maximum load factor a certain aircraft can sustain without structural failure is 8, what is the maximum angle of bank it can use in a correctly banked turn?

226. If the stalling speed of the aircraft of Q225 is 80 knots in straight and level flight, what will the stalling speed be in the turn at the maximum permissible angle of bank?

227. What will be the radius of the turn of the aircraft of Q225 when it is flying on the stall at its maximum permissible angle of bank?

228. What angles of bank are required for Rate 1 turns (180° per minute) at 100 knots, 200 knots, 300 knots, 400 knots?

229. What are the loadings on the aircraft in the turns in Q228?

230. If an aircraft stalls at 110 knots in straight and level flight, what will be the load factor on the aircraft if it stalls in a turn at 246 knots?

231. What will be the radius of turn when the aircraft of Q230 stalls at 246 knots?

232. An aeroplane of 1500 kg mass performs a loop. If it is assumed that the top of the loop is in the form of a circle of radius 80 m, what must be the speed at the highest point in order that the loads on the aeroplane may be the same as those of normal horizontal flight? (*If the centripetal force on the aeroplane at the top of the loop is just equal to the weight, then there will be no lift; but if the centripetal force is double the weight, then the lift will be the same as in normal flight, and the pilot will be sitting on his seat with the usual force, but upwards!*)

233. If the terminal velocity of the aeroplane of Q232, in a vertical nose dive with engine off, is 365 knots, what is its drag when travelling at this speed?

234. A spherical ball of 1 kgf weight and diameter 75 mm is dropped from an aeroplane. What will be its terminal velocity in air of density 0.909 kg/m^3? (*Take C_D for the sphere as 0.8.*)

235. If, without any appreciable increase in weight, the ball is faired to a stream-line shape, with C_D of 0.05, what will be the terminal velocity in air of the same density?

236. An aircraft of 6000 kg mass completes a vertical loop at a constant speed of 425 knots, with a height range from top to bottom of the loop of 3000 m. What is the acceleration and, if the area of the wing is 40 square metres, what is the wing loading (*a*) at the top, (*b*) at the bottom of the loop? (*Assume that the loop is in the form of a perfect circle, which is very unlikely in practice.*)

237. An aircraft is in a vertical terminal velocity dive at 400 knots. If, in pulling out of the dive, it follows the arc of a circle, what will be the acceleration if the height lost is 5000 ft? What is the maximum loading during the pull-out?

238. What will be the acceleration and maximum loading of the aircraft in Q237 if the loss in height in the pull-out is only 2500 ft?

CHAPTER 9. STABILITY AND CONTROL

Note. Numerical questions on stability and control are too complex to be included in the scope of this book.

CHAPTER 10. A TRIAL FLIGHT

239. The landing speed of a light aircraft at sea-level is 50 knots. At what speed will it land on an airfield situated at an altitude of 5000 ft?

240. The normal stalling speed of an aircraft is 55 knots. At what ground speed would it stall if it were flying at low level (*a*) into wind, (*b*) down wind, if the wind speed was 20 knots?

241. A pilot is flying a low-level cross country exercise at a speed of 100 knots in a wind of 20 knots. He wishes to pass vertically over a point on the ground on completion of a Rate 1 turn of 90° from an into-wind to a cross-wind direction. How far from the point must the pilot start the turn?

242. A monoplane of span 30 m and aspect ratio 7 has a profile drag coefficient of 0.008, and a lift coefficient of 0.17, when flying at 400 knots at sea-level. Calculate the total drag of the monoplane under these conditions.

243. Find the power required from an engine to drive a propeller which is 80% efficient when it is producing 3.6 kN of thrust at 120 knots.

244. When a certain aeroplane is in horizontal flight, the thrust and drag lie along the same line which is 150 mm above the centre of gravity. The extreme positions of the centre of pressure for level flight are 25 mm in front of and 200 mm behind the centre of gravity. If the tail plane is 5.5 m behind the centre of gravity, what will be the load on the tail plane in each case? The mass of the aeroplane is 3500 kg, and the thrust required for both conditions is 7 kN.

245. The maximum value of the lift/drag ratio for a certain aeroplane is 5.5 to 1. Find the flattest possible gliding angle (in degrees) with the engine off.

246. The total loaded mass of a two-seater aeroplane is 1700 kg, and the corresponding minimum landing speed is 45 knots. What will be the minimum landing speed if it is flown as a single-seater, the reduction in mass being 80 kg?

247. The normal climbing speed of an aircraft of 2500 kg mass is 110 knots. At this speed 210 kW is required to overcome drag. With 15° of flap lowered, the climbing speed is 85 knots, and 220 kW is required to overcome the drag. If the maximum power available from the propeller is 375 kW, what is the angle of climb and rate of climb in each case?

248. If the stalling speed of an aeroplane in normal flight is 42 knots, what will be its stalling speed when executing a correctly banked turn at 45° angle of bank?

249. An aircraft completes a loop at a constant speed of 200 knots in 15 seconds. If the loop can be considered as a vertical circle, calculate the radius of the loop, and the maximum and minimum loadings during the manoeuvre.

CHAPTERS 11 AND 12. FLIGHT AT TRANSONIC AND SUPERSONIC SPEEDS

250. If the speed of sound is proportional to the square root of the absolute temperature, calculate the speed of sound at $-50°C$, $+50°C$ and $+100°C$.

251. An aircraft has a critical Mach Number of 0.85. If the pilot cannot control the aircraft at higher Mach Numbers than this, what is the maximum permissible speed of the aircraft (a) at sea-level? (b) at 40 000 ft?

252. What are the maximum indicated air speeds for the aircraft of Q251 (a) at sea-level? (b) at 40 000 ft?

253. If the temperature at sea-level on a certain day rises to 25°C, what will be the maximum speed of the aircraft of Q251 at sea-level on that day?

254. A certain aircraft has a critical Mach Number of 0.80, and a structural limitation which prevents it being flown at an indicated air speed greater than 438 knots. At what altitude are these two speeds equal? (*This question is best solved graphically.*)

255. The following table gives the values of the drag coefficient of a thin aerofoil at various Mach Numbers –

Mach Number	0.6	0.7	0.8	0.9	1.0	1.1	1.2	1.3	1.4
Drag Coefficient	0.01	0.011	0.019	0.05	0.068	0.063	0.052	0.044	0.037
Mach Number	1.5	1.6	1.7	1.8	1.9	2.0			
Drag Coefficient	0.032	0.028	0.026	0.022	0.021				

Calculate the drag of the aerofoil at sea-level at these Mach Numbers taking the area of the aerofoil as 40 square metres, and plot a graph of Drag (ordinate) *v.* Mach Number (abscissa).

256. Calculate the power necessary to propel the aerofoil of Q255 at speeds corresponding to the Mach Numbers given, and plot a graph of Power (ordinate) against Mach Number (abscissa). Compare the shape of this curve with that obtained for Q255.

257. Construct a graph similar to that in Fig. 11.9 for an aircraft which stalls at 115 knots at sea-level, and has a critical Mach Number of 0.85. Estimate the maximum height at which it can fly, assuming that it cannot fly at Mach Numbers above the critical Mach Number.

258. The diagram shows the variation of the specific excess power (P_s) in m/s for a supersonic fighter-type aircraft. Estimate –

Specific excess power (P_s) in m/s for supersonic aircraft

    (*a*) The speed range in Mach Nos. at 50 000 ft altitude.
    (*b*) The speed range in Mach Nos. at 60 000 ft altitude.
    (*c*) The highest altitude at which the aircraft can fly at $M = 2$.
(*Note. The student may like to consider how the pilot would get to 60 000 ft when flying at* $M = 1$ *at* 50 000 ft.)

259. The following table gives values of the drag coefficient for a supersonic aircraft together with the specific fuel consumption (kg/kN Thrust h) of the engines fitted to the aircraft for three cruising altitudes –

Mach Number	1.7	1.8	1.9	2.0	2.1	2.2
C_D	0.037	0.034	0.032	0.031	0.031	0.031
SFC at 50 000 ft	54	54	54	55	56	57
SFC at 55 000 ft	66	66	67	68	69	70
SFC at 60 000 ft	85	86	86	87	88	89

Determine the speed and altitude at which the aircraft should be flown for optimum range performance in still air.

260. Calculate the Mach Angle for flows at M 1, 2 and 3.

261. Calculate the rise in temperature of the surface of an aeroplane travelling at 650 knots at sea-level, using the $(V/100)^2$ formula given in Chapter 12, page 398.

262. Using the more general formula in which the temperature rise is given by $(M^2T)/5$, calculate the speed required at 40 000 ft to raise the temperature of the aircraft surface to 15°C.

CHAPTER 13. SPACE FLIGHT

263. The Law of Universal Gravitation can be expressed mathematically as –

$$F = 6.67 \times 10^{-11} m_1 m_2 / d^2$$

where F is the gravitational force in newtons between two masses of m_1 and m_2 kg separated by d metres. Calculate the force of attraction between (*a*) the earth and the moon, and (*b*) the earth and the planet Venus (*mass 0.81 that of the earth*).

264. Calculate the acceleration due to gravity at a distance of 12 000 km from the centre of the earth.

265. Calculate the acceleration of the moon towards the earth.

266. A communications satellite is positioned on a 'stationary orbit' over a geographical point on the equator. Calculate the radius of its orbit and its speed.

267. A satellite of 100 kg mass is circling the earth on a radius of 18 000 km from the centre of the earth. Calculate –

    (*a*) the force exerted on the satellite by the earth, and

    (*b*) the time taken for one complete orbit.

268. A satellite of 250 kg mass is required to orbit the moon at a distance of 1750 km from the centre of the moon; at what speed must it travel round its orbit? (*Take the mass of the moon as* 0.0123 *times the mass of the earth.*)

269. To what value must the speed of a 6000 kg space vehicle be reduced so that it will orbit 100 km above the surface of the moon?

270. Calculate the thrust to be produced by the retro-rocket, and the time for which it must operate, to put the space vehicle of Q269 into its desired orbit of the moon if its transit speed is 10.5 km/s and the deceleration is to be 50 m/s².

271. Calculate the thrust required to lift a 1500 kg space vehicle off the surface of the moon and accelerate it vertically upwards to a speed of 1.5 km/s in 30 s.

272. A satellite is circling the equator at a distance of 800 km from the surface of the earth, and in an east to west direction. If it crosses the Greenwich meridian at noon, when will it be vertically over the 180° line of longitude?

273. The radius of the planet Venus is about 6115 km, and its mass is about 0.81

times that of the earth. Calculate the acceleration due to gravity at the surface of Venus.

274. Find the escape velocity at a distance of 2000 km from the earth's surface.

275. The radius of Mars is about 3376 km, and its mass about 0.1 times that of the earth. What will be the velocity of a satellite circling Mars at a distance of 10 000 km from the centre of the planet?

276. Venus and Mars are approximately 42 000 000 and 56 000 000 km respectively from the earth. What are the radii of their zones of influence relative to the earth?

277. Using the data of the previous questions calculate the escape velocities at the surface of –

    (*a*) Venus,
    (*b*) Mars.

APPENDIX 1. AEROFOIL DATA

Five questions (*a*), (*b*), (*c*), (*d*), (*e*) are given at the end of the data for each aerofoil section.

APPENDIX 2. SCALE EFFECT

278. The viscosity of air varies with the temperature according to the formula –

$$\frac{\mu_1}{\mu_2} = \left(\frac{T_1}{T_2}\right)^{\frac{3}{4}}$$

where μ_1 and μ_2 are the viscosities at the absolute temperatures T_1 and T_2 respectively. Find the viscosity of air at $-25°C$, $-5°C$, $+5°C$, and $+25°C$.

279. An aircraft 12 m in length cruises at 150 knots at sea-level. Find its Reynolds Number under these conditions.

280. The average length of the chord of a wing of a certain aircraft is 3.05 m. Taking this as the length L, calculate the Reynolds Number when flying at sea-level at 200 knots.

281. A 1/10th scale model of the aircraft of Q280 is tested in a wind tunnel at 67 m/s, at a temperature of 15°C, the density of air in the tunnel being 1.225 kg/m³. What is the Reynolds Number of the test?

282. What will be the Reynolds Number if the test of Q281 is conducted in a compressed air tunnel in which the pressure is 1013 kN/m² and the temperature 25°C?

283. What must be the pressure in the compressed air tunnel of Q282 if the 1/10th scale model test is to have the same Reynolds Number as the full-scale aircraft in Q280, assuming that the maximum speed of the tunnel under these conditions is 75 m/s and the temperature 25°C?

284. By what percentage will the Reynolds Number be raised or lowered if the temperature in the compressed air tunnel in Q283 is reduced to 5°C, assuming that no more air is pumped into the tunnel during the cooling process?

285. Find the Reynolds Number of a test conducted in a decompressed air tunnel on an aerofoil of 0.46 m chord at a Mach Number of 0.85, a temperature of 15°C and a pressure of 20.26 kN/m².

ANSWERS TO NUMERICAL QUESTIONS

CHAPTER 1

1. (a) 114.8 km/h, (b) 18.5 s
2. (a) 0.25 m/s², (b) 458 m, (c) 1833 m
3. (a) 1.39 m/s², (b) 26.8 s
4. 58 knots, 300 m
5. 58 km/h
6. (a) 12.3 m, (b) 49.2 m
7. 24.97 km, 2 min 22.7 s
8. 536 m
9. 5.503 kN
10. 11.36 m/s
11. 425.4 m/s
12. 98.1 N
13. 117.7 N
14. 6.84 × 10⁶ N
15. 8.432 kN
16. 1009.8 kN
17. 462.96 kN
18. (a) 80.65 kgf, (b) 71.35 kgf, (c) 76 kgf
19. 113.5 kN
20. 16.62 kN
21. 110.25 m
22. 70.7 N
23. 8.17 m
24. 297.1 m/s
25. 7648 N
26. 4823 N

27. 722.5 mm
28. 7.7 knots
29. 38.15 kN
30. 1582 N
31. 457.5 m
32. 119.1 kN
33. 2226 kJ, 40.47 m
34. 236 rad/s, 353.6 m/s
35. 314 rad/s, 15 m/s
36. 83 350 m/s²
37. 7402 m/s²
38. 39.48 N, (a) 29.67 N, (b) 49.29 N
39. 9.81 m/s²
40. 3 h 58 min 50 s
41. 87.34 km/h
42. (a) 25.41 kN, (b) 59°55′, (c) 29.36 kN
43. 1471.5 kJ, 12.26 kW
44. 353.3 kW
45. 2650 kW
46. 7.65 kW
47. 361.25 kJ, 425 m/s
48. 112 500 kJ
49. 10 m
50. 525.2 mmHg
51. 6.665 kN/m²
52. 1223 N/m²
53. 833 N/m²

CHAPTER 2

54. $1.044\,\text{kg/m}^3$
55. $0.116\,\text{kg}$
56. $470.4\,\text{kg}$
57. $439.7\,\text{kg}$
58. (a) $0.906\,\text{kg/m}^3$ (0.905),
    (b) $0.414\,\text{kg/m}^3$ (0.414)
59. $19\,700\,\text{ft}$
60. $580\,\text{ft}$
61. $776\,\text{ft}$
62. $4.687\,\text{kg/s}$
63. (a) 45 kt, (b) 48.35 kt, (c) 57.5 kt,
    (d) 95 kt
64. (a) 3 h 12 min, (b) 3 h 15 min,
    (c) 3 h 13 min
65. 507 knots
66. 497 knots
67. 17 min, 13 min, 10 min, 13 min
68. $175\,\text{N}$
69. $700\,\text{N}$
70. $5.7\,\text{N}$
71. $4.632\,\text{kN}$
72. $10.6\,\text{N}$
73. 39 knots
74. $196\,\text{N}$

75. $10.549\,\text{kN}$
76. 0.804
77. 0.251
78. $314\,\text{N}$
79. $8.537\,\text{kN}$
80. $172\,\text{N}$
81. $50.24\,\text{kN}$
82. 0.091
83. $4.1\,\text{N}$
84. $152\,\text{N}$
85. $4.474\,\text{kN}$
86. Decrease by 29.3%
87. $640\,\text{N}$
88. $341\,\text{N}$
89. $102.9\,\text{kN/m}^2$
90. (a) $46.6\,\text{kN/m}^2$, (b) $52.9\,\text{kN/m}^2$,
    197 kt
91. (a) 137 kt, (b) 201 kt
92. (a) 100 kt, (b) 100 kt
93. 181 knots
94. 408 knots
95. (a) $104.9\,\text{kN/m}^2$, (b) $90.3\,\text{kN/m}^2$
96. $94.3\,\text{kN/m}^2$
97. 22.7 knots

CHAPTER 3

98. (a) $2.712\,\text{kN}$, (b) $4.035\,\text{kN}$
99. $466\,\text{N/m}^2$
100. 0.84
102. $48.8\,\text{N}$
103. $693\,\text{N}$, $97\,\text{N}$
104. (a) 1.22, (b) 0.90, (c) 1.23, (d) 1.27
105. (a) 0.0074, (b) 0.009, (c) 0.010,
     (d) 0.019
106. (a) 26.8 at 2.2°, (b) 47.3 at 7.5°,
     (c) 60.0 at 8°, (d) 7.6 at 8°
107. NACA 0009
108. 1.26
109. $10.486\,\text{kN}$
110. $18.236\,\text{kN}$
111. 31 knots

112. (a) 1.37 at 13°, (b) 2.66 at 10.5°,
     (c) 1.27 at 30°, (d) 1.28 at 25°
113. $0.261\,c$
114. $0.267\,c$
115. -0.106
116. $0.352\,c$
117. 0.034
118. $425\,\text{N}$
119. $1.700\,\text{kN}$, $756\,\text{N}$
120. (a) $29.516\,\text{kN}$, (b) $7.379\,\text{kN}$
121. $7.379\,\text{kN}$
124. $33\frac{1}{3}\%$
125. 57.7%, 15.1%, 4.4%, 1.7%
126. 130 knots

CHAPTER 4

127. 833 N
128. 7.336 kN
129. 6.719 kN
130. 19.62 kN, 367 knots
131. 7.2 kN
132. 17.336 kN
133. 5351 kW
134. 500 m/s
135. 2.5 kN, 90.19 m/s²
136. Four

137. 83 knots
138. 25.2%
139. 5996 N-m
140. 10.375 kN
141. 4.510 kN
142. 106 knots
143. 84.6%
144. 2.67 m
145. 35°36′, 29°48′, 25°31′

CHAPTER 5

146. 19.62 kN, 2.616 kN, 2.616 kN
147. 0.32 m
148. 3.157 kN downwards
149. 15.6 mm behind C.G.
150. 135 N upwards
151. 53.5 N upwards
152. 4.284 kN downwards
153. 7.341 kN downwards
154. 318 N upwards
155. 180 knots
156. 170 knots

157. $5\frac{1}{2}°$, 151 knots
158. 134 knots
159. 122 knots
160. 126 knots
161. 100 knots
162. 720 kW, 776 kW, 838 kW, 908 kW
163. 116 knots, 96 knots
164. (*a*) 130 knots, (*b*) 110 knots
165. (*a*) 173 knots, (*b*) 130 knots
166. 0.177, 0.245, 0.304,
     0.408 a.n.m. per kg

CHAPTER 6

167. 6°16′
168. 9°22′, 1847 m (0.9974 n.miles)
169. 10°40′
170. 5.3 to 1
171. 5.76, 13.17, 7.41 n.miles
172. 16.46 n.miles
173. 2°21′, 4°
174. 6°, 2°29′
175. 2°21′, 2.77 ft/s
176. 41.2 knots, 2.45 ft/s
177. 2°40′, 2.67 ft/s

178. 40.3 knots
179. 34.5 knots
180. (*a*) 10.58 m², (*b*) 11.12 m²,
     (*c*) 12.30 m², (*d*) 5.71 m²
181. 96 kt, 65 kt, 53 kt
182. (*a*) 46 kt, (*b*) 53 kt
183. (*a*) 8 kt, (*b*) 14 kt, (*c*) 4 kt
184. 167 knots
185. 1.986 kN/m²
186. (*a*) 74 kt, (*b*) 103 kt

CHAPTER 7

187. 71.7 knots
188. 99.5 knots
189. 114 knots
190. 55.2 kN

191. 11.12 kN
192. (*a*) 2037 kW, (*b*) 1157 kW,
     3194 kW
193. (*a*) 13 197 ft/min, (*b*) 28°51′

194. 13°15′
195. 1829 ft/min, 5°34′
196. 839 ft/min, 2°16′
197. 29.124 kN
198. 47.5 kt, 89 kt, 64 kt, 142 kW, 2714 ft/min
199. 79.5 kt, 49.5 kt
200. 54 kt, 75.6 kW, 86 kt
201. 312 kt, 59 kt
202. 398 kt, 64 kt
203. 175 kg, 4181 ft/min
204. 245 kt, 6438 ft/min

205. 309 kt, 67 kt
206. 397 kt, 72 kt
207. 568 knots
208. 250 knots
209. (*a*) 900 kW, (*b*) 2045 kW
210. 39 700 ft
211. (*a*) 37 000 ft, (*b*) 36 250 ft
212. (*a*) 3 h 1 min, (*b*) 1 h 51 min
213. (*a*) 0.36, (*b*) 0.36, (*c*) 0.36 a.n.m. per kg
214. (*a*) 0.19, (*b*) 0.25, (*c*) 0.37 a.n.m. per kg

CHAPTER 8

215. 68°26′
216. 32 kN
217. (*a*) 9.81 m/s^2, (*b*) 9.81 kN, (*c*) 981 N/m^2
218. (*a*) 4°, (*b*) 10°
219. 418 m
220. 5402 m, 56°30′, 1.51 *g*, 88.87 kN
221. 2.0, 3.86, 8.2, 9.53
222. 1.73, 3.73, 8.14, 9.51
223. 71 kt, 85 kt, 118 kt, 172 kt, 185 kt
224. 2621°/min, 174°/min, 874°/min, 524°/min
225. 82°49′
226. 226 knots

227. 174 m
228. 15°21′, 28°47′, 39°29′, 47°42′
229. 1.04, 1.14, 1.30, 1.49
230. 5
231. 333 m
232. 77 knots
233. 14.7 kN
234. 78 m/s
235. 312 m/s
236. 31.87 m/s^2, (*a*) 3.31 kN/m^2, (*b*) 6.25 kN/m^2
237. 27.8 m/s^2, 3.83
238. 55.6 m/s^2, 6.7

CHAPTER 10

239. 54 knots
240. (*a*) 35 knots, (*b*) 75 knots
241. 1191 m
242. 31.012 kN
243. 277.8 kW
244. 156.1 N up, 1.296 kN down

245. 10°18′
246. 43.9 knots
247. 6°50′, 1324 ft/min, 8°19′, 1244 ft/min
248. 49.9 knots
249. 245.5 m, 5.4, 3.4

CHAPTERS 11 AND 12

250. 582 kt, 700 kt, 752 kt
251. (*a*) 562 kt, (*b*) 488 kt
252. (*a*) 562 kt, (*b*) 242 kt
253. 572 knots
254. 10 000 ft
257. 73 500 ft
258. (*a*) M 0.66 to 1.0 and M 1.24 to

2.33, (*b*) M 1.54 to 2.11, (*c*) 63 000 ft
259. M 1.8 at 55 000 ft
260. 90°, 30°, 19°28′
261. 42°C
262. 734 knots

CHAPTER 13

263. (*a*) 1.984×10^{26} N,
     (*b*) 1.094×10^{18} N
264. 2.76 m/s^2
265. 0.0027 m/s^2
266. 41 850 km, 11 100 km/h
267. (*a*) 123 N, (*b*) 6 h 41 min
268. 1.673 km/s
269. 1.718 km/s
270. 300 kN for 2 min 55.64 s

271. 77.421 kN
272. 1247 hours G.M.T.
273. 8.62 m/s^2
274. 9.753 km/s
275. 1.995 km/s
276. Venus 19 900 000 km,
     Mars 13 450 000 km
277. (*a*) 10.27 km/s, (*b*) 4.856 km/s

APPENDIX 1

RAF 15
(*a*) (i) 11.3, (ii) 26.8
(*b*) (i) 27°, (ii) 15°
(*c*) (i) −0.085, (ii) −0.051
(*d*) (i) 36, (ii) 174
(*e*) (i) 15.60, (ii) 15.06

CLARK YH
(*a*) (i) 0.001, (ii) 0.004, (iii) 0.005
(*b*) $0.236 c$
(*c*) 18.2°
(*d*) 159
(*e*) (i) 16.55, (ii) 15.90

NACA 0009
(*a*) 0, Yes (from −8° to +8°)
(*b*) $0.25 c$
(*c*) $0.385 c$
(*d*) (i) 21.0, (ii) 45.7, (iii) 40.9
(*e*) (i) 9.6°, (ii) 6°

NACA 4412
(*a*) (i) −0.091, (ii) −0.095
(*b*) (i) $0.489 c$, (ii) $0.333 c$
(*c*) (i) 38.0, (ii) 67.6
(*d*) (i) 13°, (ii) 10°
(*e*) 137

NACA 23012
(*a*) (i) 15, (ii) 50, (iii) 60
(*b*) (i) $0.33 c$, (ii) $0.265 c$, (iii) $0.264 c$
(*c*) $0.30 c$
(*d*) 12°
(*e*) (i) 21.60, (ii) 37.08, (iii) 49.96

NACA 23018
(*a*) $0.18 c$ at $0.30 c$
(*b*) (i) −23.3, (ii) 12.0, (iii) 44.2,
     (iv) 56.3, (v) 40.8
(*c*) (i) $0.253 c$, (ii) $0.25 c$, (iii) $0.253 c$
(*d*) 12°
(*e*) 1.06

NACA 65_1-212
(*a*) −0.032 ± 0.002 between −8° and
     +8° angle of attack
(*b*) $0.40 c$
(*c*) (i) 10.5°, (ii) 6°
(*d*) 53.3
(*e*) 120

ASN/P1/3
(*a*) 0.0594
(*b*) $0.45 c$
(*c*) (i) 30°, (ii) 25°
(*d*) (i) 5.7, 5.8, 3.6, (ii) 4.6, 4.9, 3.1
(*e*) (i) $0.442 c$, (ii) $0.468 c$, (iii) $0.507 c$

APPENDIX 2

278. 15.99×10^{-6} N-s/m^2
     16.95×10^{-6} N-s/m^2

     17.43×10^{-6} N-s/m^2
     18.36×10^{-6} N-s/m^2

279. 63.39×10^6
280. 21.48×10^6
281. 1.399×10^6
282. 13.63×10^6

283. $1426\,kN/m^2$
284. Raised by 5.3%
285. 1.82×10^6

Tail end
Tailplane of the Antonov An–225 Mriya.

ANSWERS TO NON–NUMERICAL QUESTIONS

CHAPTER 1

1. The lift is decelerating in the downwards direction, so the acceleration is upwards.

2. (*a*) Pressure is a scalar quantity: it has no direction; pressure is measured by the force that it would produce on an area.

   (*b*) A moment is the product of a force and a distance (the moment arm). Momentum is the product of a mass and a velocity.

   (*c*) Energy is the rate of doing work: force × distance/time.

3. To pull a body up an inclined plane only requires that the force be equal and opposite to the component of the weight acting in the direction of the slope of the plane.

   The same work is done in each case; on the inclined plane, the force is smaller, but the distance moved is correspondingly larger.

4. Mass is a measure of the quantity of matter in a body. Weight is the force produced by the gravitational attraction between the body and a heavenly body (the earth, unless otherwise specified).

5. To accelerate the aircraft requires that the thrust exceeds the drag, but when a steady speed is reached, it will be maintained as long as the drag equals the thrust.

6. Yes, the aircraft is accelerating.

7. Yes, the centre of gravity of a ring is outside the material of the ring. The centre of gravity of a piece of bent wire will also not normally lie on the wire.

8. (*a*) Yes.

   (*b*) No, it will normally be accelerating.

9. (*a*) No.

   (*b*) Yes.

   (*c*) No.

   (*d*) No.

   (*e*) Yes.

10. Yes, otherwise they would be in equilibrium, and no movement would take place, so it would be stalemate.

11. (*a*) Yes.

    (*b*) Yes.

12. The flag will hang down limply, because there will be no relative motion between the balloon and the air.

CHAPTER 2

1. The pressure altimeter only tells you what height the external pressure would be given by in an International Standard Atmosphere.
2. Air density is a measure of the mass of air in a given volume.
3. The pressure.
4. The altimeter can be set so that it reads zero height when the pressure reaches the ground-level value at a specified location, usually the local airfield ground-level; the aircraft should then not hit the runway before the indicated height reaches zero. Alternatively, while en route, it can be set to read zero at the sea-level pressure, so that the pilot knows how high he is above features given on a map.
5. The temperature, density and pressure will all be lower.
6. A streamlined shape is one where the flow is able to follow the contours without separating.
7. See the definitions in the chapter.
8. This is the dynamic pressure.
9. This is the error on the air speed indicator reading caused by the static hole not being located at a position where the pressure is exactly equal to the free-stream static pressure.
10. The troposphere is the lowest part of the atmosphere, where the temperature varies almost linearly with height. The stratosphere is above the troposphere, and is the part where the temperature remains nearly constant with height.
11. Subsonic means that the speed of the object relative to the air is less than the local speed of sound. Supersonic means that it is greater.
12. The symbol q stands for dynamic pressure.

CHAPTER 3

1. As the angle of attack increases, a lift force develops due to the difference in pressure between upper and lower surfaces. This is mainly caused by a reduction in pressure on the upper surface. When the angle of attack increases too far the flow separates from the upper surface, this leads to a reduction in lift as the aerofoil stalls.
2. The centre of pressure is the point through which the aerodynamic force can be considered to act on the aerofoil.
3. Because the forces on an aerofoil are primarily dependent on the dynamic pressure and the wing area, if the lift or drag are divided by the product of these quantities, a force coefficient is obtained which depends mainly on angle of attack. This makes it easier to translate the results from, for example, an experiment to the full-scale aircraft. It must be remembered, though, that Reynolds number and Mach number will also have an effect (see Appendix 2 and Chapter 11).
4. The aerodynamic centre of an aerofoil section is the point about which the moment coefficient remains constant with changes in angle of attack.
5. The stalling angle of an aerofoil is the angle of attack at which the flow separates from the top surface, leading to a reduction in lift. The exact angle at which this happens is open to discussion – see the text! Apart from Reynolds number and Mach number effects (see question 3) this happens at a particular angle of attack independently of the air speed.
6. The aspect ratio of a wing is equal to the square of its span divided by its area. It is important because it influences the strength of the trailing vortices for a given lift, and hence the drag produced by these vortices.

CHAPTER 4

1. A ramjet has no turbine or compressor. Air is compressed purely by aerodynamic effects before it enters the combustion chamber. It is only of practical use at supersonic speeds where use can be made of shock wave compression.

2. The blade angle is the angle at which the propeller blade is set in relation to the plane of rotation. The angle reduces as the tip is approached because the speed at the tip is higher than it is near the spinner. The angle at which the blade is set must therefore be reduced as the tip is approached, to maintain the same local angle of attack at each section of the blade.

3. The advance per revolution is the actual forward distance travelled in one propeller revolution. The geometric pitch is equal to $2pr \tan \theta$ and is equivalent to the advance of a screw of 'blade' angle θ through a solid object. The experimental mean pitch is the advance per revolution at zero thrust.

4. Slip is the difference between the actual advance per revolution and the experimental mean pitch.

5. The angle of attack of a propeller blade is reduced by an increase in the aircraft air speed. To compensate for this it is desirable to use a higher blade angle (coarser pitch) as air speed increases.

6. Tip speed is important because it is at the tip that the local air speed is highest. Therefore the speed of sound may be reached at the tip long before the forward speed of the aircraft approaches this value. This may lead to the formation of shock waves over the tip with a marked loss of efficiency.

7. The solidity influences the ability of a propeller of given diameter and rotational speed to absorb power. It may be increased by increasing the number of blades or the chord of the blades.

8. Outside the atmosphere propulsion can be achieved by any system which ejects matter from the craft. It is necessary to employ a high velocity of projection so that adequate thrust can be obtained without ejecting too much mass. If the required ejection velocity is produced by combustion then the oxidant must be carried in the craft since no air is available. This is effectively rocket propulsion.

CHAPTER 5

1. The four forces are lift, weight, thrust, and drag.

2. For equilibrium the sum of the forces must be zero and they must produce no moment.

3. The weight of the aircraft will change as fuel is used or stores are dropped and this will also change the position of the centre of gravity. The lift will change to compensate for the weight change in level flight, or to provide additional lift for manoeuvres. The centre of lift will also change, particularly if the change in lift is produced by a change in angle of attack rather than speed. Changes in lift will be accompanied by changes in drag which in turn must be accompanied by an equal change in thrust to maintain equilibrium. Changes in the line of action of the latter forces are likely to be fairly small for conventional aircraft.

4. The lift of the aircraft depends on angle of attack and speed. The speed can be changed by changing the angle of attack to give the same overall lift as the speed is altered. Generally this will be accompanied by some change in drag so some adjustment will be needed to the throttle to compensate. The speed range can be

further extended by the use of high-lift devices to increase the maximum lift coefficient that can be obtained from the wing.

5. The relationship will be the same for indicated air speed but not for true air speed (see p. 171). At height a higher true air speed will be needed to give the same lift at a given angle of attack.

6. As the weight is increased a higher angle of attack will be needed at a given air speed.

7. (*a*) As far as the airframe is concerned the minimum drag will be approximately independent of altitude. The aircraft should then be flown at the best altitude for maximum engine efficiency, i.e. the height at which the throttle setting is fully open for the most economical fuel–air ratio. At greater heights the engine efficiency will be lower and the range consequently less.

(*b*) For maximum endurance with a piston engine we need to fly at minimum power as far as the airframe is concerned. Because the true air speed increases with height at the optimum angle of attack for minimum power, it is best to fly at low altitude for maximum endurance.

(*c*) Because the best angle of attack, and hence speed, for minimum power required by the airframe is different than that for minimum drag.

8. (*a*) Because the efficiency of a jet engine increases with speed a compromise between the best operating speed for the airframe and the engine is required. The aircraft must therefore be flown at a speed greater than that for minimum airframe drag.

(*b*) True air speed for a given angle of attack increases with height and temperature falls. Both of these improve engine efficiency, therefore it is best to cruise high for maximum range in a jet aircraft.

(*c*) Because the fuel flow in a jet aircraft is approximately proportional to thrust, the aircraft should be flown at the minimum drag speed for best endurance. There is some advantage in flying high from the point of view of improved engine efficiency.

CHAPTER 6

1. If the aircraft is operating at the minimum angle of glide then it is operating at its minimum drag for the weight. If the pilot pulls the nose up then the drag will increase and the glide angle steepen.

2. Lowering flaps during the glide will generally steepen the glide angle because the best lift to drag ratio is likely to be in the flaps-up configuration.

3. The load carried will not greatly alter the minimum glide angle, because the maximum lift to drag ratio, at which the minimum glide angle occurs, depends on angle of attack. To get the same lift at this angle of attack, however, the gliding speed must increase at heavier weights.

4. No! The flattest glide occurs at minimum drag. For maximum time in the air we need to operate at the minimum power condition so that the potential energy due to height is lost at the lowest possible rate.

5. (*a*) The true airspeed at stall will be higher the greater the altitude. However, the speed indicated on the air speed indicator will be approximately the same irrespective of height because the instrument works by sensing the dynamic pressure and no correction is made in its calibration for the change in density with height.

(*b*) The stalling angle will not change with height.

6. An engine-assisted approach allows the glide angle to be increased if necessary by a reduction in the throttle setting. If a 'go around' is required the engine can be set to full power quicker.

7. The lowering of flaps may change the position of the centre of pressure of the wing and produce a pitching moment (generally nose-down) which must be overcome by the tailplane.

8. A couple of things to get you started. After that you are on your own! Happy landings.

You need to fly at minimum glide angle, i.e. at minimum drag. Are the things you are carrying mounted outside the aeroplane and increasing the drag? Before you release them, is there a headwind? To reduce the effect of this you need a high gliding speed and so a lot of weight. If there is a tail wind you need a low speed to increase the effect of wind on range. Now it's up to you.

CHAPTER 7

1. When the aircraft is climbing its path is inclined to the horizontal. Because the lift is measured at right angles to the flight path, only the component of weight which is also normal to the flight path must be balanced by the lift. The other component must be supplied by the thrust.

If you still don't believe it, imagine an aircraft with an engine thrust greater than the weight. It can 'stand on its tail' and climb vertically with no lift at all.

2. For a jet aircraft maximum true air speed does not vary greatly over a wide altitude range. It is usually limited by compressibility effects (see Chapters 11 and 12). For a piston engine the relationship can be complicated by the question of supercharging, but will fall off at high altitude; engine power is limited and does not increase with forward speed like a jet engine. Because of the reducing density, the best true air speed as far as the airframe is concerned increases with altitude and the power available from the engine is insufficient to cope.

The minimum true air speed will increase with altitude, although the indicated minimum speed will be nearly constant.

3. Ceiling is the maximum height that can be reached. Service ceiling leaves something in hand for manoeuvres. It is usually specified as the height at which the rate of climb falls to a specified level.

4. No! It is the attitude of the aircraft that matters and this will require a change in true air speed at different heights.

5. An increase in weight will mean a reduction in range and endurance. There will also be an increase in minimum speed and a reduction in maximum rate of climb and ceiling.

CHAPTER 8

1. The degrees of freedom are –
   (*a*) Three translational degrees of freedom along the three aircraft axes.
   (*b*) Three rotational degrees of freedom along the same axes.
2. The radius of turn may be limited by –
   (*a*) Wing stalling because of increased lift required in the turn.
   (*b*) Engine power because the increased lift increases the drag and required power.

(*c*) The structural strength of the wing being exceeded because of increased lift.

3. This is a bit of a trick question – be careful!

Because the aircraft complete the circuit in the same time, the one at the greater radius is going faster. If you check the acceleration towards the centre for a constant circuit time, you will find it is proportional to the radius of the turn. The aircraft at the higher radius will therefore need to bank more than the one on the inside.

4. As the aircraft turns and climbs the upward motion reduces the angle of attack on the wings. Because the inner wing is turning on a smaller radius than the outer wing it suffers a greater reduction in angle of attack and a consequent loss of lift compared to the outer wing. This tends to increase the bank which must be held off by the ailerons. The reverse is clearly true in a gliding turn.

5. In a spin one wing is stalled with the aircraft describing a spiral path downwards. The variation of lift with angle of attack at the stall is such that the aircraft can become locked into this attitude and will not recover naturally.

CHAPTER 9

1. If the directional stability is very large and lateral stability small, then the aircraft will be prone to spiral instability.

2. If the lateral stability is large and the directional stability small, then spiral instability will not be a problem. However, another problem may be encountered in which there is a motion consisting of a combined rolling and yawing oscillation (known as Dutch roll).

3. In manually operated controls the aerodynamic forces acting on the control surface may make them very heavy to operate, especially at high speed. The use of suitable aerodynamic balancing can offset some of the hinge moment and make the control lighter to operate.

4. Aerodynamic balance is used to reduce the aerodynamic hinge moment on the control. Mass balancing is used to alter the inertial characteristic of the control to change its dynamic behaviour and prevent flutter of the control surface.

5. At high angles of attack, particularly, the down-going aileron produces a substantial increase in drag and hence yawing moment away from the intended turn.

If the angle of attack is too high it may even be that the downward deflection of the aileron is sufficient to reduce the lift rather than increasing it because of stall.

6. The yawing moment problem can be tackled by suitable design of the ailerons to ensure that the up-going aileron also produces a significant drag increment. A bigger rudder may also help.

The use of a spoiler instead of, or supplementing, the up-going aileron will produce the desired drag increment and can be used to drop that wing significantly with respect to the other, even with no aileron control on either wing.

7. The use of spoilers at low speed is explained above. They can also be a useful aid during the ground run on landing. At high speed they may be better than an aileron for roll control because they avoid imposing wing twist which is caused by the deflection of ailerons near the tips – a problem which is particularly severe on swept wings. Tip ailerons at transonic Mach numbers may also cause undesirable local changes in flow due to compressibility effects.

CHAPTER 11

1. The speed of sound in water is roughly four times the speed of sound in air.

2. The speed of sound reduces with height up to the stratosphere. It is proportional to the square root of the absolute air temperature, which reduces with height until the stratosphere is reached, where it remains approximately constant.

3. The shock forms first at the region where the local flow first becomes supersonic. This is usually on the upper surface of the wing.

4. The buffet boundary of a transonic aircraft is the Mach number at which an interaction between shock waves formed on the wing and the local boundary layer causes an unstable separation which causes a buffeting on the aircraft.

5. Mach number is air speed divided by the speed of sound. It is possible to talk of the Mach number at which an aircraft operates, or flight Mach number, as its true air speed divided by the speed of sound in the surrounding atmosphere. It is also possible to refer to a local Mach number relating to the flow over some particular part of the aircraft. This is the speed of the local flow divided by the local speed of sound. The local speed of sound will be different in different parts of the airflow because of temperature differences.

The critical Mach number can be defined as the flight Mach number at which local supersonic flow first appears somewhere on the aircraft.

A Mach meter is an instrument for measuring the flight Mach number of an aircraft.

6. A shock wave causes a rapid increase in pressure as the air passes through it. On a wing it is therefore likely to reduce the extent of the suction peak on the upper surface. It may further modify the pressure distribution by causing flow separation.

CHAPTER 12

1. A Mach line is the trace of a weak pressure wave produced by a body in a supersonic airstream. Its angle to the flow, the Mach angle, depends only on the Mach number. A Mach cone is a three-dimensional surface consisting of Mach lines. For a small body in a supersonic airstream this surface is conical.

2. Shock waves travel at a speed greater than the speed of sound. This means that a shock wave can form upstream of a body travelling at supersonic speed.

3. An expansion wave in supersonic flow is a region where the speed increases while the pressure, density and temperature decrease.

4. Sharp leading edges are used on supersonic wings to reduce the drag due to shock waves (the wave drag).

5. At cruising Mach numbers above about 2, the structure may be subjected to aerodynamic heating to a degree which makes the use of conventional aluminium alloys impossible. In this case alternative materials such as titanium must be used at critical parts of the structure where heating is severe. In the Concorde the problem is solved by employing the fuel as a heat sink to reduce the local structural temperatures.

6. In order to achieve low wave drag at supersonic speeds a slender wing (i.e. one which is long compared to its span) must be used. For efficient subsonic cruise a high aspect ratio wing is needed. This leads to the Concorde's highly swept 'ogee' planform. Because the high sweep leads to flow separation, a sharp leading edge is used so the wing has separated flow over the top surface at nearly all flight conditions. This separation, however, is in the form of two well-controlled vortices

above the wing which contribute to the lift and do not produce the buffet and increase in drag associated with the normal separation process leading to a conventional wing stall.

CHAPTER 13

1. Escape velocity is the velocity which is required by a body, with no assistance from propulsion, if it is to avoid being returned to the earth by gravitational attraction. The escape velocity required for a body starting on the earth's surface, and ignoring such important things as air resistance, can be calculated to be approximately 11.2 km/s. The escape velocity from the moon is less because the moon's mass, and hence gravitational attraction, is less than that of the earth.

2. Again, ignoring air resistance the escape velocity is the same for both horizontal and vertical launch, since the critical factor is the amount of kinetic energy possessed by the body. If we are clever, though, we can get part of the escape velocity on launch from the rotation of the earth if we launch near the equator and in the direction of rotation.

3. For an elliptical orbit round the earth the perigee is the point on the orbit nearest the earth and the apogee is the point furthest away.

4. This is the height at which the orbital period of the satellite is the same as the period of the earth's rotation. If the plane of the orbit is in the plane of the equator the satellite therefore remains stationary with respect to the earth's surface (called the geostationary orbit). This is clearly a help with communication satellites.

   5. (*a*) 1 hr 25 min.

   (*b*) 2 hr approx.

   (*c*) 28 days.

6. A satellite launched at the escape velocity has a parabolic path. Above this speed it is hyperbolic.

INDEX